Simulation of Stochastic Processes with Given Accuracy and Reliability

Series Editor
Nikolaos Limnios

Simulation of Stochastic Processes with Given Accuracy and Reliability

Yuriy Kozachenko
Oleksandr Pogorilyak
Iryna Rozora
Antonina Tegza

ISTE PRESS

ELSEVIER

First published 2016 in Great Britain and the United States by ISTE Press Ltd and Elsevier Ltd

ISTE Press Ltd
27-37 St George's Road
London SW19 4EU
UK

www.iste.co.uk

Elsevier Ltd
The Boulevard, Langford Lane
Kidlington, Oxford, OX5 1GB
UK

www.elsevier.com

British Library Cataloguing-in-Publication Data
A CIP record for this book is available from the British Library
Library of Congress Cataloging in Publication Data
A catalog record for this book is available from the Library of Congress
ISBN 978-1-78548-217-5

Contents

Introduction

The problem of simulation of stochastic process has been a matter of active research in recent decades. It has become an integral part of research, development and practical application across many fields of study. That is why one of the actual problems is to build a mathematical model of stochastic process and study its properties. Because of the powerful possibilities of computer techniques, the problems of numerical simulations have become especially important and allow us to predict the behavior of a random process.

There are various simulation methods of stochastic processes and fields. Some of them can be found in [OGO 96, ERM 82, CRE 93, KOZ 07a]. Note that in most publications dealing with simulation of stochastic processes, the question of accuracy and reliability is not studied.

In this book, the methods of simulation of stochastic processes and fields with given accuracy and reliability are considered. Namely, models are found that approximate stochastic processes and fields in different functional spaces. This means that at first we construct the model and then use some adequacy tests to verify it.

In most books and papers that are devoted to the simulation of stochastic processes, the modeling methods of exactly Gaussian processes and fields are studied. It is known that there is a need to simulate the processes that are equal to the sum of various random factors, in which effects of each other are independent. According to the central limit theorem, such processes are close to Gaussian ones. Hence, the problem of simulation of Gaussian stochastic processes and fields is a hot topic in simulation theory.

Let us mention that in this book only centered random processes and fields are considered, since simulation of determinate function can be made without any difficulties.

Note that all results in this book are applicable for Gaussian process.

Chapter 1 deals with the space of sub-Gaussian random variables and subclasses of this space containing strictly sub-Gaussian random variables. Different characteristics of these random variables are considered: sub-Gaussian standard, functional moments, etc. Special attention is devoted to inequalities estimating "tails" of the distribution of a random variable, or a sum of a random variable in the some functional spaces. These assertions are applied in investigation of accuracy and reliability of the model of Gaussian stochastic process.

In Chapter 2, general approaches for model construction of stochastic processes with given accuracy and reliability are studied. Special attention is paid to Karhunen–Loève and Fourier expansions of stochastic processes and their application to the simulation of stochastic processes.

Chapter 3 is devoted to the model construction of Gaussian processes, that is considered as input processes on some system of filter, with respect to output processes in a Banach space $C(T)$ with given accuracy and reliability. For this purpose, square-Gaussian random processes are considered; the concept of the space of square-Gaussian random variables is introduced and the estimates of distribution of a square-Gaussian process supremum are found. We also consider the particular case when the system output process is a derivative of the initial process.

Chapter 4 offers two approaches to construct the models of Gaussian stationary stochastic processes. The methods of model construction are generalized on the case of random fields. The proposed methods of modeling can be applied in different areas of science and technology, particularly in radio, physics and meteorology. The models can be interpreted as a set of signals with limited energy, harmonic signals and signals with limited durations.

In Chapter 5, the theorems on approximation of a model to the Gaussian random process in spaces $L_1([0,T])$ and $L_p([0,T])$, $p > 1$ with given accuracy and reliability are proved. The theorems are considered on estimates of the "tails" of norm distributions of random processes under different conditions in the space $L_p(T)$, where T is some parametric set, $p \geq 1$. These statements are applied to investigate the partition selection of the set $[0,\Lambda]$ such that the model approximates a Gaussian process with some accuracy and reliability in the space $L_p([0,T])$ when $p \geq 1$. A theorem on model approximation of random process with Gaussian with given accuracy and reliability in Orlicz space $L_U(\Omega)$ that is generated by the function U is also presented.

In Chapter 6, we introduce random Cox processes and describe two algorithms of their simulation with some given accuracy and reliability. The cases where an intensity of the random Cox processes are generated by log Gaussian or square

Gaussian homogeneous and inhomogeneous processes or fields are considered. We also describe two methods of simulation. The first one is more complicated to apply in practice because of technical difficulties. The second one is somewhat simpler and allows us to obtain the model of the Cox process as a model of Poisson random variables with parameters that depend on the intensity of the Cox process. The second model has less accuracy than the first model.

Chapter 7 deals with a model of a Gaussian stationary process with absolutely continuous spectrum that simulates the process with a given reliability and accuracy in $L_2(0, T)$. Under certain restrictions on the covariance function of the process, formulas for computing the parameters of the model are described.

Chapter 8 is devoted to simulation of Gaussian isotropic random fields on spheres. The models of Gaussian isotropic random fields on n-measurable spheres are constructed that approximate these fields with given accuracy and reliability in the space $L_p(S_n), \quad p \geq 2$.

1

The Distribution of the Estimates for the Norm of Sub-Gaussian Stochastic Processes

This chapter is devoted to the study of the conditions and rate of convergence of sub-Gaussian random series in some Banach spaces. The results of this chapter are used in other chapters to construct the models of Gaussian random processes that approximate them with specified reliability and accuracy in a certain functional space. Generally, the Gaussian stochastic processes are considered, which can be represented as a series of independent items. It should be noted that, as will be shown, these models will not always be Gaussian random processes. In Chapter 7, for example, the Gaussian models of stationary processes are sub-Gaussian processes. The accuracy of simulation is studied in the spaces $C(T)$, $L_p(T)$, $p > 0$, and Orlicz space $L_U(T)$, where T is a compact (usually segment) and U is some C-function. In addition, these models can be used to construct the models of sub-Gaussian processes that approximate them with a given reliability and accuracy in a case when the process can be performed as a sub-Gaussian series with independent items. Section 1.1 provides the necessary information from the theory of the sub-Gaussian random variables space. Sub-Gaussian random variables were introduced for the first time by Kahane [KAH 60]. Buldygin and Kozachenko in their publication [BUL 87] showed that the space of sub-Gaussian random variables is Banach space. The properties of this space are studied in the work of Buldygin and Kozachenko [BUL 00]. Section 1.2 deals with necessary properties of the theory for strictly sub-Gaussian random variables. In [BUL 00], this theory is described in more detail. Note that a Gaussian centered random variable is strictly sub-Gaussian. Therefore, all results of this section, as well as other results of this book, obtained for sub-Gaussian random variables and processes are also true for the centered Gaussian random variables and processes. In section 1.3, the rate of convergence of sub-Gaussian random series in the space $L_2(T)$ is found. Similar results are contained in [KOZ 99b] and [KOZ 07a]. Section 1.4 looks at the distribution estimate of the norm of sub-Gaussian random processes in space Lp (T). These estimates are also considered in [KOZ 07a]. For more general spaces, namely the spaces $Sub_\varphi(\Omega)$ such estimates can also be found [KOZ 09]. These estimates are used to find the rate of convergence of sub-Gaussian functional series in the norm of spaces $L_p(\Omega)$. Note that in the case where $p = 2$, the results of section 1.3 are better than section 1.4. In section 1.5, the estimates of distribution of the sub-Gaussian random processes norm in some Orlicz spaces are found; in section 1.6, these estimates are used to obtain the rate of convergence of sub-Gaussian random series in the norm of some Orlicz spaces. Similar estimates are contained in [KOZ 99b, KOZ 07a, KOZ 88, ZEL 88, RYA 90, RYA 91, TRI 91].

The results on the rate of convergence of sub-Gaussian random series in the Orlicz space that were received in section 1.6, are detailed in section 1.7 for the series with either uncorrelated or independent items. In sections 1.8 and 1.9, the rate of convergence for sub-Gaussian and strictly sub-Gaussian random series in the space $C(T)$ is obtained. Similar problems were discussed in [KOZ 99b] and [KOZ 07a].

Section 1.10 provides the distribution estimates for supremum of random processes in the space $L_p(\Omega)$.

1.1. The space of sub-Gaussian random variables and sub-Gaussian stochastic processes

This section deals with random variables that are subordinated, in some sense, to Gaussian random variables. These random variables are called sub-Gaussian (the rigorous definition is given below). Later, we will also study sub-Gaussian stochastic processes.

Let $\{\Omega, \mathcal{B}, P\}$ be a standard probability space.

DEFINITION 1.1.– *A random variable ξ is called sub-Gaussian, if there exists such number $a \geq 0$ that the inequality*

$$\mathbf{E}\exp\{\lambda\xi\} \leq \exp\{\frac{a^2\lambda^2}{2}\} \qquad [1.1]$$

holds true for all $\lambda \in \mathbb{R}$. The class of all sub-Gaussian random variables defined on a common probability space $\{\Omega, \mathcal{B}, P\}$ is denoted by $\mathrm{Sub}(\Omega)$.

Consider the following numerical characteristic of sub-Gaussian random variable ξ:

$$\tau(\xi) = \inf\left\{a \geq 0 \colon \mathbf{E}\exp\{\lambda\xi\} \leq \exp\{\frac{a^2\lambda^2}{2}\}, \lambda \in \mathbb{R}\right\}. \qquad [1.2]$$

We will call $\tau(\xi)$ *sub-Gaussian standard of random variable* ξ. We put $\tau(\xi) = \infty$ if the set of $a \geq 0$ satisfying [1.1] is empty. By definition, $\xi \in \mathrm{Sub}(\Omega)$ if and only if $\tau(\xi) < \infty$. The following lemma is clear.

LEMMA 1.1.– The relationships hold

$$\tau(\xi) = \sup_{\lambda\neq 0,\, \lambda\in\mathbb{R}} \left[\frac{2\ln \mathbf{E}\exp\{\lambda\xi\}}{\lambda^2}\right]^{\frac{1}{2}}. \qquad [1.3]$$

For all $\lambda \in \mathbb{R}$

$$\mathbf{E}\exp\{\lambda\xi\} \leq \exp\left\{\frac{\lambda^2\tau^2(\xi)}{2}\right\}. \qquad [1.4]$$

The sub-Gaussian assumption implies that the random variable has mean zero and imposes other restrictions on moments of the random variable.

LEMMA 1.2.– Suppose that $\xi \in \mathrm{Sub}(\Omega)$. Then

$$\mathbf{E}|\xi|^p < \infty$$

for any $p > 0$. Moreover, $\mathbf{E}\xi = 0$ and

$$\mathbf{E}\xi^2 \leq \tau^2(\xi).$$

PROOF.– Since as $p > 0$ and $x > 0$ the relationship $x^p \leq \exp\{x\}p^p\exp\{-p\}$ is satisfied. Hence, if instead of x we substitute $|\xi|$ and take the mathematical expectation then obtain that

$$\mathbf{E}|\xi|^p \leq p^p\exp\{-p\}\mathbf{E}\exp\{|\xi|\}.$$

Since

$$\mathbf{E}\exp\{|\xi|\} \leq \mathbf{E}\exp\{\xi\} + \mathbf{E}\exp\{-\xi\} \leq 2\exp\{\frac{\tau^2(\xi)}{2}\} < \infty,$$

then $\mathbf{E}|\xi|^p < \infty$. Further, by the Taylor formula, we obtain

$$\mathbf{E}\exp\{\lambda\xi\} = 1 + \lambda\mathbf{E}\xi + \frac{\lambda^2}{2}\mathbf{E}\xi^2 + o(\lambda^2),$$

$$\exp\{\frac{\lambda^2\tau^2(\xi)}{2}\} = 1 + \frac{\lambda^2}{2}\tau^2(\xi) + o(\lambda^2)$$

as $\lambda \to 0$. Then inequality [1.4] implies that $\mathbf{E}\xi = 0$ and $\tau^2(\xi) \geq \mathbf{E}\xi^2$. □

The following lemma gives an estimate for the moments of sub-Gaussian random variable.

LEMMA 1.3.– Let $\xi \in \mathrm{Sub}(\Omega)$, then

$$\mathbf{E}|\xi|^p \leq 2\left(\frac{p}{e}\right)^{p/2} (\tau(\xi))^p$$

for any $p > 0$.

PROOF.– Since for $p > 0$, $x > 0$ the inequality

$$x^p \leq \exp\{x\} p^p \exp\{-p\},$$

holds, then we can substitute $\lambda|\xi|$, $\lambda > 0$ for x and take the mathematical expectation of such a value. Hence,

$$\mathbf{E}|\xi|^p \leq \left(\frac{p}{\lambda e}\right)^p \mathbf{E}\exp\{\lambda|\xi|\}. \qquad [1.5]$$

Since

$$\mathbf{E}\exp\{\lambda|\xi|\} \leq \mathbf{E}\exp\{\lambda\xi\} + \mathbf{E}\exp\{-\lambda\xi\},$$

then it follows from [1.5] and [1.4] that for any $\lambda > 0$ the inequality

$$\mathbf{E}|\xi|^p \leq 2\left(\frac{p}{\lambda e}\right)^p \exp\{\frac{\lambda^2\tau^2(\xi)}{2}\}$$

is satisfied. The lemma will be completely proved if in the inequality above we substitute $\lambda = \frac{\sqrt{p}}{\tau(\xi)}$ under which the right-hand side of the equality is maximized. □

EXAMPLE 1.1.– Suppose that ξ is an $N(0, \sigma^2)$-distributed random variable, that is ξ has Gaussian distribution with mean zero and variance σ^2. Then

$$\mathbf{E}\exp\{\lambda\xi\} = \exp\left\{\frac{\sigma^2\lambda^2}{2}\right\},$$

meaning that ξ is sub-Gaussian and $\tau(\xi) = \sigma$.

Example 1.1 and lemma 1.1 show that a random variable is sub-Gaussian if and only if its moment generating function is majorized by the moment generating function of a zero-mean Gaussian random variable.This fact somewhat explains the term "sub-Gaussian". Note that a function $\mathbf{E}\exp\{\lambda\xi\}$ is called moment generating function of ξ.

EXAMPLE 1.2.– Suppose that ξ is a random variable that takes values 0 ± 1 with probabilities $\mathbf{P}\{\xi = \pm 1\} = p/2$, $\mathbf{P}\{\xi = 0\} = 1 - p$, $0 \le p \le 1$. Then for $p \ne 0$

$$\begin{aligned}
\mathbf{E}\exp\{\lambda\xi\} &= (1-p) + \frac{p}{2}\Big(\exp\{\lambda\} + \exp\{-\lambda\}\Big) \\
&= (1-p) + p\sum_{k=0}^{\infty}\frac{\lambda^{2k}}{(2k)!} \\
&= 1 + p\sum_{k=1}^{\infty}\frac{\lambda^{2k}}{(2k)!} = 1 + \sum_{k=1}^{\infty}\frac{\lambda^{2k}p^k}{2^k k!}\,\frac{2^k k!}{p^{k-1}(2k)!} \\
&\le \sum_{k=0}^{\infty}\left(\frac{\lambda^2 p}{2}\right)^k\left(\frac{1}{3p}\right)^{k-1}\frac{1}{k!}.
\end{aligned}$$

Therefore, $\mathbf{E}\exp\{\lambda\xi\} \le \exp\{\frac{\lambda^2 p}{2}\}$ as $p \ge \frac{1}{3}$, $\mathbf{E}\exp\{\lambda\xi\} \le \exp\{\frac{\lambda^2}{6}\}$ as $0 < p < \frac{1}{3}$, which means ξ is sub-Gaussian random variable. From lemma 1.2 it follows that for $p \ge \frac{1}{3}$ $\tau^2(\xi) = \mathbf{E}\xi^2 = p$, for $p < \frac{1}{3}$ $\tau^2(\xi) \le \frac{1}{3}$.

EXAMPLE 1.3.– Let ξ be uniformly distributed on $[-a, a]$ random variable, $a > 0$, then

$$\begin{aligned}
\mathbf{E}\exp\{\lambda\xi\} &= \frac{1}{2a}\int_{-a}^{a}\exp\{\lambda x\}\,dx = \frac{\sinh(\lambda a)}{\lambda a} = \sum_{k=0}^{\infty}\frac{(\lambda a)^{2k}}{(2k+1)!} \\
&\le 1 + \sum_{k=1}^{\infty}\frac{(\lambda a)^{2k}}{6^k k!} = \exp\{\frac{(\lambda a)^2}{6}\} = \exp\{\frac{\lambda^2}{2}\frac{a^2}{3}\}.
\end{aligned}$$

That is, ξ is a sub-Gaussian random variable, $\tau^2(\xi) = \mathbf{E}\xi^2 = \frac{a^2}{3}$.

LEMMA 1.4.– Let ξ be a bounded zero-mean random variable; that is, $\mathbf{E}\xi = 0$ and there exists $c > 0$ such that $|\xi| \le c$ almost surely. Then, $\xi \in \mathrm{Sub}(\Omega)$ and $\tau(\xi) \le c$.

PROOF.– Put $\psi(\lambda) = \ln \mathbf{E}\exp\{\lambda\xi\}$. Then

$$\psi'(\lambda) = \frac{\mathbf{E}\xi\exp\{\lambda\xi\}}{\mathbf{E}\exp\{\lambda\xi\}},$$

Since $\psi(\lambda) = \psi(0) + \psi'(0)\lambda + \frac{\psi''(\tilde{\lambda})}{2}\lambda^2$, then from last inequality and $\psi(0) = 0$, $\psi'(0) = 0$ follows that $\psi(\lambda) \le \frac{c^2\lambda^2}{2}$. □

Exponential upper bounds for "tails" of a distribution are of importance in various applications of sub-Gaussian random variables. An $N(0, \sigma^2)$-distributed random variable ξ satisfies the following inequality for $x > \sigma$:

$$\mathbf{P}\{\xi > x\} \le \exp\left\{-\frac{x^2}{2\sigma^2}\right\}.$$

A similar inequality also holds for a sub-Gaussian random variable. To avoid ambiguity, we put $\exp\{-u/0\} = 0$ for any $u > 0$.

LEMMA 1.5.– Suppose that $\xi \in \mathrm{Sub}(\Omega)$, $\tau(\xi) > 0$. Then, the following inequalities hold for $x > 0$:

$$\mathbf{P}\{\xi > x\} \leq V\left(\frac{x}{\tau(\xi)}\right) \quad \mathbf{P}\{\xi < -x\} \leq V\left(\frac{x}{\tau(\xi)}\right)$$

$$\mathbf{P}\{|\xi| > x\} \leq 2V\left(\frac{x}{\tau(\xi)}\right), . \qquad [1.6]$$

where $V(x) = \exp\{-\frac{x^2}{2}\}$.

PROOF.– By the Chebyshev–Markov inequality, we have

$$\mathbf{P}\{\xi > x\} \leq \frac{\mathbf{E}\exp\{\lambda\xi\}}{\exp\{\lambda x\}} \leq \exp\left\{\frac{\lambda^2\tau^2(\xi)}{2} - \lambda x\right\}$$

for any $\lambda > 0$ and $x > 0$. Minimizing the right-hand side in $\lambda > 0$ gives the first inequality in [1.6]. The proof of the second inequality is similar, and the third inequality follows from the former two:

$$\mathbf{P}\{|\xi| > x\} = \mathbf{P}\{\xi > x\} + \mathbf{P}\{\xi < -x\}.$$

□

THEOREM 1.1.– The space $\mathrm{Sub}(\Omega)$ is a Banach space with respect to the norm $\tau(\xi)$.

PROOF.– Prove now that $\mathrm{Sub}(\Omega)$ is a linear space and $\tau(\xi)$ is a norm. If $\xi = 0$ almost surely, then $\tau(\xi) = 0$ by the definition of sub-Gaussian standard. The converse is also true, since $\tau(\xi) = 0$ implies $\mathbf{E}\xi^2 = 0$ by lemma 1.2, almost surely giving $\xi = 0$.

By the definition of τ, we have

$$\tau(c\xi) = c\tau(\xi).$$

Let us prove triangle inequality $\tau(\xi + \eta) \leq \tau(\xi) + \tau(\eta)$. It is sufficient to prove this inequality in the case when $\xi \in \mathrm{Sub}(\Omega)$, $\eta \in \mathrm{Sub}(\Omega)$ and $\tau(\xi) \neq 0$, $\tau(\eta) \neq 0$. The Hölder inequality gives the following inequality for any $p > 1$, $q > 1$ such that $\frac{1}{p} + \frac{1}{q} = 1$ and any $\lambda \in \mathbb{R}$:

$$\mathbf{E}\exp\{\lambda(\xi + \eta)\} \leq \left[\mathbf{E}\exp\{p\lambda\xi\}\right]^{\frac{1}{p}} \cdot \left[\mathbf{E}\exp\{q\lambda\eta\}\right]^{\frac{1}{q}}$$

$$\leq \exp\left\{\frac{\lambda^2}{2}(p\tau^2(\xi) + q\tau^2(\eta))\right\}.$$

For a fixed $\lambda \in \mathbb{R}$, the minimum in $p > 1$ on the right-hand side of the last inequality attained at

$$p' = \frac{\tau(\xi) + \tau(\eta)}{\tau(\xi)}.$$

Substituting p' into the input inequality, we get

$$\mathbf{E} \exp\{\lambda(\xi + \eta)\} \leq \exp\left\{ \frac{\lambda^2}{2} (\tau(\xi) + \tau(\eta))^2 \right\},$$

that is $\tau(\xi + \eta) \leq \tau(\xi) + \tau(\eta)$. Show now that the space $\mathrm{Sub}(\Omega)$ is complete with respect to τ. Let $\{\xi_n, n \geq 1\} \in \mathrm{Sub}(\Omega)$ and

$$\lim_{n\to\infty} \sup_{m\geq n} \tau(\xi_n - \xi_m) = 0, \qquad [1.7]$$

it means that the sequence ξ_n is fundamental with respect to the norm τ. Since $\mathbf{E}(\xi_n - \xi_m)^2 \leq \tau^2(\xi_n - \xi_m) \to 0$, as $n, m \to \infty$, then the sequence ξ_n converges in mean square, hence and in probability. We denote limits as ξ_∞. For any $\lambda \in \mathbb{R}$ and $\varepsilon > 0$

$$\begin{aligned} \sup_{n\geq 1} \mathbf{E} \Big[\exp\{\lambda \xi_n\} \Big]^{1+\varepsilon} &= \sup_{n\geq 1} \mathbf{E} \exp\{\lambda(1+\varepsilon)\xi_n\} \\ &\leq \sup_{n\geq 1} \exp\left\{ \frac{\lambda^2(1+\varepsilon)^2\tau^2(\xi_n)}{2} \right\} < \infty. \end{aligned}$$

By the theorem on uniform integrability[LOE 60], we obtain

$$\mathbf{E} \exp\{\lambda \xi_\infty\} = \lim_{n\to\infty} \mathbf{E} \exp\{\lambda \xi_n\} \leq \exp\left\{ \frac{\lambda^2 \sup_{n\geq 1} \tau^2(\xi_n)}{2} \right\}.$$

It follows from [1.7] that $\sup_{n\geq 1} \tau^2(\xi_n) < \infty$. This means that ξ_∞ is a sub-Gaussian random variable and that

$$\tau(\xi_\infty) \leq \sup_{n\geq 1} \tau(\xi_n). \qquad [1.8]$$

The random variables $\xi_\infty - \xi_n$, $n \geq 1$, are also sub-Gaussian, since the space $\mathrm{Sub}(\Omega)$ is linear. In analogy with [1.8], we can show that

$$\tau(\xi_\infty - \xi_n) \leq \sup_{m\geq n} \tau(\xi_m - \xi_n).$$

By [1.7], we finally obtain

$$\lim_{n\to\infty} \tau(\xi_\infty - \xi_n) = 0.$$

□

From now the sub-Gaussian standard τ will also be called the sub-Gaussian norm.

1.1.1. *Exponential moments of sub-Gaussian random variables*

Let ξ be a Gaussian random variable with parameters 0 and $\sigma^2 > 0$. Then, it is easy to show that

$$\mathbf{E}\exp\left\{\frac{s\xi^2}{2\sigma^2}\right\} = (1-s)^{-\frac{1}{2}}. \qquad [1.9]$$

as $s < 1$. For sub-Gaussian random variables, this inequality is transformed to the following inequality.

LEMMA 1.6.– Assume that $\xi \in \mathrm{Sub}(\Omega)$ and $\tau(\xi) > 0$. Then, for all $0 \le s < 1$, we have

$$\mathbf{E}\exp\left\{\frac{s\xi^2}{2\tau^2(\xi)}\right\} \le (1-s)^{-\frac{1}{2}}. \qquad [1.10]$$

PROOF.– The result is obvious for $s = 0$. Let $F(x)$ be the cumulative distribution function of ξ. Inequality [1.4] can be rewritten as

$$\int_{-\infty}^{\infty} \exp\{\lambda x\}\, dF(x) \le \exp\left\{\frac{\lambda^2\tau^2(\xi)}{2}\right\}.$$

Let $s \in (0,1)$. The inequality above implies that

$$\int_{-\infty}^{\infty} \exp\{\lambda x\} \exp\left\{-\frac{\lambda^2\tau^2(\xi)}{2s}\right\} dF(x)$$
$$\le \exp\left\{\frac{\lambda^2\tau^2(\xi)}{2}\left(\frac{s-1}{s}\right)\right\}$$

for any $\lambda \in \mathbb{R}$. Integrating by λ in both sides of the last inequality gives

$$\int_{-\infty}^{\infty}\int_{-\infty}^{\infty} \exp\{\lambda x\} \exp\left\{-\frac{\lambda^2\tau^2(\xi)}{2s}\right\} d\lambda dF(x)$$
$$\le \int_{-\infty}^{\infty} \exp\left\{-\frac{\lambda^2\tau^2(\xi)}{2}\left(\frac{1-s}{s}\right)\right\} d\lambda. \qquad [1.11]$$

Now, we transform [1.11]. On the one hand, we have

$$\int_{-\infty}^{\infty} \exp\left\{-\frac{\lambda^2\tau^2(\xi)}{2}\left(\frac{1-s}{s}\right)\right\} d\lambda$$

$$= \frac{\tau(\xi)}{\sqrt{2\pi}}\left(\frac{1-s}{s}\right)^{\frac{1}{2}}\int_{-\infty}^{\infty}\exp\left\{-\frac{\lambda^2\tau^2(\xi)}{2}\left(\frac{1-s}{s}\right)\right\}d\lambda \qquad [1.12]$$

$$\times\frac{\sqrt{2\pi}}{\tau(\xi)}\left(\frac{s}{1-s}\right)^{\frac{1}{2}} = \frac{\sqrt{2\pi}}{\tau(\xi)}\left(\frac{s}{1-s}\right)^{\frac{1}{2}}.$$

On the other hand, we obtain

$$\int_{-\infty}^{\infty}\exp\{\lambda x\}\exp\left\{-\frac{\lambda^2\tau^2(\xi)}{2s}\right\}d\lambda$$

$$= \frac{\tau(\xi)}{\sqrt{2\pi s}}\int_{-\infty}^{\infty}\exp\{\lambda x\}\exp\left\{-\frac{\lambda^2\tau^2(\xi)}{2s}\right\}d\lambda\frac{\sqrt{2\pi s}}{\tau(\xi)}$$

$$= \frac{\sqrt{2\pi s}}{\tau(\xi)}\exp\left\{\frac{x^2 s}{2\tau^2(\xi)}\right\}.$$

Hence,

$$\int_{-\infty}^{\infty}\left(\int_{-\infty}^{\infty}\exp\{\lambda x\}\exp\left\{-\frac{\lambda^2\tau^2(\xi)}{2s}\right\}d\lambda\right)dF(x)$$

$$= \frac{\sqrt{2\pi s}}{\tau(\xi)}\int_{-\infty}^{\infty}\exp\left\{\frac{x^2 s}{2\tau^2(\xi)}\right\}dF(x) \qquad [1.13]$$

$$= \frac{\sqrt{2\pi s}}{\tau(\xi)}\mathbf{E}\exp\left\{\frac{s\xi^2}{2\tau^2(\xi)}\right\}.$$

If in [1.11] we substitute [1.12] and [1.13], then we obtain inequality [1.10]. □

The above exponential moment can be finite for $s \geq 1$. The random variable taking values ± 1 with probability $\frac{1}{2}$ is a simple example.

1.1.2. *The sum of independent sub-Gaussian random variables*

By theorem 1.1, the sub-Gaussian standard is a norm in the space $\mathrm{Sub}(\Omega)$. This means that for any $\xi_1, \ldots, \xi_n \in \mathrm{Sub}(\Omega)$, we have

$$\tau\left(\sum_{k=1}^{n}\xi_k\right) \leq \sum_{k=1}^{n}\tau(\xi_k).$$

This inequality can be made sharper when sub-Gaussian terms are independent.

LEMMA 1.7.– Assume that $\xi_1, \ldots, \xi_n$ are independent sub-Gaussian random variables. Then

$$\tau^2\left(\sum_{k=1}^{n}\xi_k\right) \leq \sum_{k=1}^{n}\tau^2(\xi_k). \qquad [1.14]$$

PROOF.– For any n and $\lambda \in \mathbb{R}$, we have

$$\mathbf{E}\exp\left\{\lambda \sum_{k=1}^{n} \xi_k\right\} = \prod_{k=1}^{n} \mathbf{E}\exp\{\lambda\xi_k\}$$
$$\leq \prod_{k=1}^{n} \exp\left\{\frac{\lambda^2\tau^2(\xi_k)}{2}\right\}$$
$$= \exp\left\{\frac{\lambda^2}{2}\sum_{k=1}^{n}\tau^2(\xi_k)\right\},$$

giving [1.14] by definition of τ. □

1.1.3. *Sub-Gaussian stochastic processes*

DEFINITION 1.2.– *A stochastic process $X = \{X(t),\, t \in T\}$, is called sub-Gaussian if for any $t \in T$ $X(t) \in \mathrm{Sub}(\Omega)$ and $\sup_{t\in T} \tau(X(t)) < \infty$.*

LEMMA 1.8.– Let ξ_i, $i = 1, \ldots, n$ be sub-Gaussian random variables, and $z = (\xi_1, \ldots, \xi_n)$ be a sub-Gaussian random vector with $\tau(\xi_i) = \tau_i$. Then, for all $t > 0$ the following inequality

$$E\exp\{t\|z\|\} = E\exp\left\{t\sum_{i=1}^{n}|\xi_i|\right\} \leq 2\exp\left\{\frac{t^2}{2}\left(\sum_{i=1}^{n}\tau_i\right)^2\right\} \qquad [1.15]$$

holds.

PROOF.– Let $t \geq 0$, $p_i > 1$, $i = 1, \ldots, n$, $\sum_{i=1}^{n}\frac{1}{p_i} = 1$. Then, it follows from the Hölder inequality that

$$E\exp\left\{t\left(\sum_{i=1}^{n}|\xi_i|\right)\right\} \leq \prod_{i=1}^{n}\left(E\exp\{p_i t|\xi_i|\}\right)^{\frac{1}{p_i}}$$
$$\leq 2\prod_{i=1}^{n}\exp\left\{p_i\tau_i^2\frac{t^2}{2}\right\} = 2\exp\left\{\frac{t^2}{2}\sum_{i=1}^{n}p_i\tau_i^2\right\}.$$

If in inequality above put $p_i = \tau_i^{-1}\sum_{j=1}^{n}\tau_j$, we get [1.15]. □

LEMMA 1.9.– Assume that the conditions of lemma 1.8 are satisfied. Then, for all $1 < \alpha \le 2$ and $s \in [0, 1)$ the inequality

$$E \exp \left\{ \frac{s}{\alpha} \frac{\sum_{i=1}^{n} |\xi_i|^\alpha}{\sum_{i=1}^{n} \tau_i^\alpha} \right\} \le \exp \left\{ \left(1 - \frac{\alpha}{2}\right) \frac{s}{\alpha} \right\} (1 - s)^{-\frac{1}{2}}.$$

holds.

PROOF.– Since η is a sub-Gaussian random variable with the norm $\tau(\eta) = \tau$, then it follows from lemma 1.6 that for $s \in [0, 1)$

$$E \exp \left\{ \frac{s\eta^2}{2\tau^2} \right\} \le (1 - s)^{-\frac{1}{2}}. \qquad [1.16]$$

In [LOE 60], the following inequality has been proved for $x > 0$ and $y > 0$

$$xy \le \frac{x^p}{p} + \frac{y^q}{q}, \quad \text{where} \quad \frac{1}{p} + \frac{1}{q} = 1, \ p > 1.$$

Let α be a number such that $1 < \alpha < 2$, then substituting in inequality above $p = \frac{2}{\alpha}$, $q = \frac{2}{2-\alpha}$ we obtain

$$xy \le \frac{\alpha}{2} x^{\frac{2}{\alpha}} + \frac{2 - \alpha}{2} y^{\frac{2}{2-\alpha}}. \qquad [1.17]$$

From [1.16] and [1.17] follows that for all $0 \le s < 1$

$$E \exp \left\{ \frac{s}{\alpha} \left(\frac{|\eta|}{\tau} \right)^\alpha \right\} \le E \exp \left\{ \frac{s\alpha\eta^2}{\alpha 2\tau^2} + \frac{(2 - \alpha)}{2} \frac{s}{\alpha} \right\}$$
$$= E \exp \left\{ \frac{s\eta^2}{2\tau^2} \right\} \cdot \exp \left\{ \frac{(2 - \alpha)s}{2\alpha} \right\} \le (1 - s)^{-\frac{1}{2}} \exp \left\{ \frac{(2 - \alpha)s}{2\alpha} \right\}. \qquad [1.18]$$

Let $r > 0$ and consider such values $p_i > 1$, $i = 1, \ldots, n$, that $\sum_{i=1}^{n} \frac{1}{p_i} = 1$.

Then

$$E \exp \left\{ \frac{1}{r} \sum_{i=1}^{n} |\xi_i|^\alpha \right\} \le \prod_{i=1}^{n} \left(E \exp \left\{ \frac{p_i |\xi_i|^\alpha}{r} \right\} \right)^{\frac{1}{p_i}} = I.$$

From [1.18], it follows that for $\dfrac{p_i\tau_i^{\alpha}\alpha}{r} < 1$

$$I = \prod_{i=1}^{n}\left(E\exp\left\{\frac{\alpha\tau_i^{\alpha}p_i}{r\alpha}\cdot\left(\frac{|\xi_i|}{\tau_i}\right)^{\alpha}\right\}\right)^{\frac{1}{p_i}}$$

$$\leq \prod_{i=1}^{n}\left(1-\frac{p_i\alpha\tau_i^{\alpha}}{r}\right)^{-\frac{1}{2p_i}}\exp\left\{\frac{(2-\alpha)\tau_i^{\alpha}}{2r}\right\},$$

and

$$\ln I \leq \sum_{i=1}^{n}\left(-\frac{1}{2p_i}\right)\cdot\ln\left(1-\frac{p_i\alpha\tau_i^{\alpha}}{r}\right)+\sum_{i=1}^{n}\frac{2-\alpha}{2r}\tau_i^{\alpha}$$

$$= \left(1-\frac{\alpha}{2}\right)\frac{1}{r}\sum_{i=1}^{n}\tau_i^{\alpha}+\sum_{i=1}^{n}\frac{1}{2p_i}\cdot\sum_{k=1}^{\infty}\frac{(p_i\alpha\tau_i^{\alpha})^k}{kr^k}.$$

If $p_i = \frac{1}{\tau_i^{\alpha}}\sum\limits_{j=1}^{n}\tau_j^{\alpha}$ and $r = \frac{\alpha}{s}\sum\limits_{i=1}^{n}\tau_i^{\alpha}$, $0 \leq s < 1$, then

$$\ln I \leq \frac{s}{\alpha}\left(1-\frac{\alpha}{2}\right)+\frac{1}{2}\sum_{k=1}^{\infty}\frac{s^k}{k} = \left(1-\frac{\alpha}{2}\right)\frac{s}{\alpha}-\frac{1}{2}\ln(1-s).$$

This implies the statement lemma for $1 < \alpha < 2$.

Let us study a case when $\alpha = 2$. Then, for $r > 0$, $p_i > 1$, $i = 1, \ldots, n$, $\sum\limits_{i=1}^{n}\frac{1}{p_i} = 1$ we have:

$$E\exp\left\{\frac{1}{r}\sum_{i=1}^{n}|\xi_i|^2\right\} \leq \prod_{i=1}^{n}\left(E\exp\left\{\frac{p_i|\xi_i|^2}{r}\right\}\right)^{\frac{1}{p_i}} = I\,.$$

From [1.16], it follows that for $\frac{2p_i\tau_i^2}{r} < 1$

$$I = \prod_{i=1}^{n}\left(E\exp\left\{\frac{2\tau_i^2p_i}{r2}\left(\frac{|\xi_i|}{\tau_i}\right)^2\right\}\right)^{\frac{1}{p_i}} \leq \prod_{i=1}^{n}\left(1-\frac{2\tau_i^2p_i}{r}\right)^{-\frac{1}{2p_i}}$$

and

$$\ln I \leq \sum_{i=1}^{n}\left(-\frac{1}{2p_i}\right)\ln\left(1-\frac{2\tau_i^2p_i}{r}\right) = \sum_{i=1}^{n}\frac{1}{2p_i}\sum_{k=1}^{\infty}\frac{(2\tau_i^2p_i)^k}{kr^k}.$$

If $p_i = \frac{1}{\tau_i^2} \sum\limits_{j=1}^{n} \tau_j^2$ and $r = \frac{2}{s} \sum\limits_{i=1}^{n} \tau_i^2$, then

$$\ln I \leq \frac{1}{2} \sum_{k=1}^{\infty} \frac{s^k}{k} = -\frac{1}{2} \ln(1-s).$$

Hence,

$$E \exp \left\{ \frac{s}{2} \frac{\sum\limits_{i=1}^{n} |\xi_i|^2}{\sum\limits_{i=1}^{n} \tau_i^2} \right\} \leq (1-s)^{-\frac{1}{2}}.$$

□

COROLLARY 1.1.– Let ξ_{ik}, $i = 1, \ldots, m_k$, $m_k \to \infty$ be sub-Gaussian random variables with $\tau_{ik} = \tau(\xi_{ik})$. If there exist the limits $\eta_1 = \lim\limits_{m_k \to \infty} \sum\limits_{i=1}^{m_k} |\xi_{ik}|$ (almost everywhere or with probability) and $\lim\limits_{m_k \to \infty} \sum\limits_{i=1}^{m_k} \tau_{ik} = \tau_1 > 0$, then for all $t \geq 0$

$$E \exp \{t\eta_1\} \leq 2 \exp \left\{ \frac{t^2 \tau_1^2}{2} \right\}.$$

If there exists $\eta_\alpha = \lim\limits_{m_k \to \infty} \sum\limits_{i=1}^{m_k} |\xi_{ik}|^\alpha$ (almost everywhere or with probability) and for $1 < \alpha \leq 2$, $\lim\limits_{m_k \to \infty} \sum\limits_{i=1}^{m_k} \tau_{ik}^\alpha = \tau_\alpha$, then for all $s \in [0, 1)$

$$E \exp \left\{ \frac{s\eta_\alpha}{\alpha\tau_\alpha} \right\} \leq (1-s)^{-\frac{1}{2}} \exp \left\{ \frac{(2-\alpha)s}{2\alpha} \right\}.$$

The assertion of this corollary follows from the Fatou lemma.

COROLLARY 1.2.– Assume that $X = \{X(t),\ t \in \mathbf{T}\}$ is a sub-Gaussian random process, where $(\mathbf{T}, \mathbf{A}, \mu)$ is a measurable space. Denote $\tau(t) = \tau(X(t))$. If for some $1 \leq \alpha \leq 2$ (with probability one or in mean square), there exist the integrals

$$\int_{\mathbf{T}} |X(t)|^\alpha d\mu(t),$$

and

$$\int_{\mathbf{T}} (\tau(t))^\alpha \, d\mu(t),$$

then for all $t \geq 0,\ (\alpha = 1)$

$$E\exp\left\{t\int\limits_{\mathbf{T}}|X(t)|d\mu(t)\right\} \leq 2\exp\left\{\frac{t^2}{2}\left(\int\limits_{\mathbf{T}}\tau(t)d\mu(t)\right)^2\right\}$$

or for all $s \in [0,1)$ $(1 < \alpha \leq 2)$

$$E\exp\left\{\frac{s}{\alpha}\frac{\int_{\mathbf{T}}|X(t)|^{\alpha}d\mu(t)}{\int_{\mathbf{T}}(\tau(t))^{\alpha}d\mu(t)}\right\} \leq (1-s)^{-\frac{1}{2}}\exp\left\{\frac{(2-\alpha)s}{2\alpha}\right\} \qquad [1.19]$$

The assertion of this corollary follows from corollary 1.1.

REMARK 1.1.– If for random variable $\theta > 0$ and all $t > 0$ the inequality holds

$$E\exp\{t\theta\} \leq 2\exp\left\{\frac{t^2}{2}b^2\right\},$$

then (see e.g. [BUL 00]) for all $x > 0$

$$P\{\theta > x\} \leq 2\exp\left\{-\frac{x^2}{2b^2}\right\}. \qquad [1.20]$$

Suppose that for random variable $\eta > 0$ and for all $s \in [0,1)$

$$E\exp\{\frac{s\eta}{\alpha}\} \leq (1-s)^{-\frac{1}{2}}\exp\left\{\frac{(2-\alpha)s}{2\alpha}\right\},$$

then from the Tchebyshev-Markov inequality it follows that for any $x > 0$

$$P\{\eta > x\} \leq E\exp\left\{\frac{s\eta}{\alpha}\right\}\cdot\exp\left\{-\frac{sx}{\alpha}\right\} \leq (1-s)^{-\frac{1}{2}}\exp\left\{s\left(\frac{1}{\alpha}-\frac{1}{2}-\frac{x}{\alpha}\right)\right\}.$$

Set $s = 1 - \left(2\left(\frac{x}{\alpha}+\frac{1}{2}-\frac{1}{\alpha}\right)\right)^{-1}$ (the minimum point of right-hand side of last inequality), then

$$P\{\eta > x\} \leq \sqrt{x\frac{2}{\alpha}+1-\frac{2}{\alpha}}\exp\left\{\frac{1}{\alpha}\right\}\cdot\exp\left\{-\frac{x}{\alpha}\right\}. \qquad [1.21]$$

1.2. The space of strictly sub-Gaussian random variables and strictly sub-Gaussian stochastic processes

Lemma 1.2 shows that for any sub-Gaussian random variable ξ, we have

$$\mathbf{E}\xi^2 \le \tau^2(\xi), \qquad \mathbf{E}\xi = 0,$$

where $\tau(\xi)$ is the sub-Gaussian standard. Now we consider a subclass of sub-Gaussian random variables where the above inequality becomes equality.

DEFINITION 1.3.– *A sub-Gaussian random variable ξ is called strictly sub-Gaussian if $\tau^2(\xi) = \mathbf{E}\xi^2$, that is, if the inequality*

$$\mathbf{E}\exp\{\lambda\xi\} \le \exp\left\{\frac{\lambda^2\sigma^2}{2}\right\}, \qquad [1.22]$$

$\sigma^2 = E\xi^2$, holds for all $\lambda \in \mathbb{R}$. The class of strictly sub-Gaussian random variables will be denoted by $\mathrm{SSub}(\Omega)$.

Each zero-mean Gaussian random variable is strictly sub-Gaussian (for example 1.1). A random variable from example 1.2 will be strictly sub-Gaussian as $p \ge \frac{1}{3}$. It can be shown [BUL 80b] that for $p < \frac{1}{3}$, $p \ne 0$ it is not strictly sub-Gaussian. Uniformly distributed on interval $[-a, a]$, a random variable is also strictly sub-Gaussian (for example 1.3). Sufficient conditions for a random variable to be strictly sub-Gaussian are given in [BUL 80b]. In [BUL 80b], it is also shown that the sum of strictly sub-Gaussian random variables need not be strictly sub-Gaussian. The next lemma points out an important situation where a sum of strictly sub-Gaussian random variables is strictly sub-Gaussian.

LEMMA 1.10.– Suppose that ξ is a strictly sub-Gaussian random variable and c is an arbitrary constant. Then, $c\xi$ is also a strictly sub-Gaussian random variable. Assume that ξ and η are independent strictly sub-Gaussian random variables. Then, the sum $\xi + \eta$ is also strictly sub-Gaussian.

The first statement of the lemma is obvious. The second statement follows from inequalities by lemma 1.7:

$$\begin{aligned}\mathbf{E}(\xi+\eta)^2 &\le \tau^2(\xi+\eta) \le \tau^2(\xi) + \tau^2(\eta)\\ &= \mathbf{E}\xi^2 + \mathbf{E}\eta^2 = \mathbf{E}(\xi+\eta)^2.\end{aligned}$$

DEFINITION 1.4.– *A family of random variables* Δ *from* $\text{Sub}(\Omega)$ *is called strictly sub-Gaussian if for any at most countable set of random variables* $\{\xi_i, i \in I\}$ *from* Δ *and for any* $\lambda_i \in \mathbb{R}$ *the relationship*

$$\tau^2\left(\sum_{i\in I}\lambda_i\xi_i\right) = \mathbf{E}\left(\sum_{i\in I}\lambda_i\xi_i\right)^2 \qquad [1.23]$$

holds true.

LEMMA 1.11.– Assume that Δ is a family of strictly sub-Gaussian random variables. Then, a linear closure of Δ in $L_2(\Omega)$ is strictly sub-Gaussian family.

PROOF.– Let $\xi_1, \xi_2, \dots, \xi_n$ be random variables from Δ, $\eta_i = \sum_{j=1}^n a_{ij}\xi_j$, $i = 1, \dots, m$ are the elements of linear span of Δ. Then

$$\begin{aligned}\tau^2\left(\sum_{i=1}^m \lambda_i\eta_i\right) &= \tau^2\left(\sum_{i=1}^m \lambda_i \sum_{j=1}^n a_{ij}\xi_j\right)\\ &= \tau^2\left(\sum_{j=1}^n\left(\sum_{i=1}^m \lambda_i a_{ij}\right)\xi_j\right) \qquad [1.24]\\ &= \mathbf{E}\left(\sum_{j=1}^n\left(\sum_{i=1}^m \lambda_i a_{ij}\right)\xi_j\right)^2 = \mathbf{E}\left(\sum_{i=1}^m \lambda_i\eta_i\right)^2.\end{aligned}$$

The assertion of the lemma for boundary elements of linear closure of Δ follows from [1.24] by limit transition (see theorem 1.1). □

DEFINITION 1.5.– *A strictly sub-Gaussian family of random variables, that is closed in* $L_2(\Omega)$*, is called a space of strictly sub-Gaussian random variables.*

The space of strictly sub-Gaussian random variables is denoted by $\text{SG}(\Omega)$.

EXAMPLE 1.4.– Suppose that $\Xi = \{\xi_k, k = 1, 2, \dots\}$ is a sequence of independent strictly sub-Gaussian random variables. A linear closure of Ξ in $L_2(\Omega)$ is a space of strictly sub-Gaussian random variables.

DEFINITION 1.6.– *A random vector* $\vec{\xi}^{\,T} = (\xi_1, \dots, \xi_n)$ *is called strictly sub-Gaussian if* ξ_k *are random variables from strictly sub-Gaussian family.*

LEMMA 1.12.– Let $\overrightarrow{\xi}^T = (\xi_1, \ldots, \xi_n)$ be a strictly sub-Gaussian vector with uncorrelated components, $\mathbf{E}\xi_k^2 = \sigma_k^2 > 0$. Then, for any s such that $0 \le s < 1$, and $N = 1, 2 \ldots$ the inequality

$$\mathbf{E}\exp\left\{\frac{s}{2R_N}\sum_{k=1}^{n}\xi_k^2\right\} \le \exp\left\{\frac{1}{2}\sum_{l=1}^{\infty}\frac{s^l}{lR_N^l}\sum_{k=1}^{n}\sigma_k^{2l}\right\} \qquad [1.25]$$

is satisfied, where $R_N = \left(\sum_{k=1}^{n}\sigma_k^{2N}\right)^{\frac{1}{N}}$.

PROOF.– It follows from the definition of a strictly sub-Gaussian vector and [1.24] that for all $\lambda_k \in \mathbb{R}$, $k = 1, 2, \ldots, n$, we have

$$\mathbf{E}\exp\left\{\sum_{k=1}^{n}\lambda_k\xi_k\right\} \le \exp\left\{\frac{1}{2}\mathbf{E}\left(\sum_{k=1}^{n}\lambda_k\xi_k\right)^2\right\}$$
$$= \exp\left\{\frac{1}{2}\sum_{k=1}^{n}\lambda_k^2\sigma_k^2\right\}. \qquad [1.26]$$

Suppose that s_k are such arbitrary numbers that $0 < s_k < 1$. Multiplying right-hand and left -hand sides of [1.26] by $\exp\{-\sum_{k=1}^{n}\frac{\lambda_k^2\sigma_k^2}{2s_k}\}$ and integrating both parts by λ_k, we obtain

$$\mathbf{E}\int_{-\infty}^{\infty}\cdots\int_{-\infty}^{\infty}\exp\left\{\sum_{k=1}^{n}(\lambda_k\xi_k - \frac{\lambda_k^2\sigma_k^2}{2s_k})\right\}d\lambda_1\ldots d\lambda_n$$
$$\le \int_{-\infty}^{\infty}\cdots\int_{-\infty}^{\infty}\exp\left\{\sum_{k=1}^{n}\frac{\lambda_k^2\sigma_k^2}{2}(1 - \frac{1}{s_k})\right\}d\lambda_1\ldots d\lambda_n. \qquad [1.27]$$

The left-hand side of [1.27] will be defined as A_n. It is easy to see that

$$A_n = \mathbf{E}\prod_{k=1}^{n}\int_{-\infty}^{\infty}\exp\left\{\lambda_k\xi_k - \frac{\lambda_k^2\sigma_k^2}{2s_k}\right\}d\lambda_k$$
$$= \prod_{k=1}^{n}\left(\frac{\sqrt{2\pi s_k}}{\sigma_k}\right)\cdot\mathbf{E}\exp\left\{\sum_{k=1}^{n}\frac{\xi_k^2 s_k}{2\sigma_k^2}\right\}. \qquad [1.28]$$

The right-hand side of [1.27] is defined as B_n. Then, it is clear that

$$B_n = \prod_{k=1}^{n}\int_{-\infty}^{\infty}\exp\left\{-\frac{\lambda_k^2\sigma_k^2(1 - s_k)}{2s_k}\right\}d\lambda_k \qquad [1.29]$$
$$= \prod_{k=1}^{n}\left(\frac{\sqrt{2\pi s_k}}{\sigma_k}\right)\cdot\prod_{k=1}^{n}(1 - s_k)^{-\frac{1}{2}}. \qquad [1.30]$$

[1.27]–[1.29] yield that for all $0 < s_k < 1$ the inequality

$$\mathbf{E}\exp\left\{\sum_{k=1}^{n}\frac{\xi_k^2 s_k}{2\sigma_k^2}\right\} \le \prod_{k=1}^{n}(1-s_k)^{-\frac{1}{2}} \qquad [1.31]$$

holds.

Denote now $s_k = \frac{s\sigma_k^2}{R_N}$, where s is a number such that $0 < s < 1$. From [1.31], it follows that

$$\mathbf{E}\exp\left\{\frac{s}{2R_N}\sum_{k=1}^{n}\xi_k^2\right\} \le \prod_{k=1}^{n}\left(1-\frac{\sigma_k^2 s}{R_N}\right)^{-\frac{1}{2}}.$$

The inequality above yields the following relationship

$$\ln\mathbf{E}\exp\left\{\frac{s}{2R_N}\sum_{k=1}^{n}\xi_k^2\right\} \le -\frac{1}{2}\sum_{k=1}^{n}\ln\left(1-\frac{\sigma_k^2 s}{R_N}\right)$$

$$= \frac{1}{2}\sum_{k=1}^{n}\sum_{l=1}^{\infty}\frac{\sigma_k^{2l}s^l}{lR_N^l} = \frac{1}{2}\sum_{l=1}^{\infty}\frac{s^l}{lR_N^l}\sum_{k=1}^{n}\sigma_k^{2l}.$$

The lemma is proved for $0 < s < 1$. In the case of $s = 0$, the inequality [1.25] is trivial. □

REMARK 1.2.– If in

$$\left(\sum_{i=1}^{n}x_i\right)^{\alpha} \le \sum_{i=1}^{n}x_i^{\alpha}, \quad x_i > 0, \quad 0 < \alpha < 1,$$

we put $x_i = \sigma_i^{2l}$, $\alpha = \frac{N}{l}$, $l \ge N$, then

$$\left(\sum_{k=1}^{n}\sigma_k^{2l}\right)^{\frac{N}{l}} \le \sum_{k=1}^{n}\sigma_k^{2N}. \qquad [1.32]$$

Hence, when $l \ge N$ $R_N^l \ge \sum_{k=1}^{n}\sigma_k^{2l}$. Therefore,

$$\sum_{l=N}^{\infty}\frac{s^l}{lR_N^l}\sum_{k=1}^{n}\sigma_k^{2l} \le \sum_{l=N}^{\infty}\frac{s^l}{l},$$

which means that the series

$$\sum_{l=N}^{\infty}\frac{s^l}{lR_N^l}\sum_{k=1}^{n}\sigma_k^{2l}$$

converges for any $N \geq 1$. Furthermore, [1.32] yields an inequality

$$R_N \geq R_I \quad \text{as} \quad I > N. \qquad [1.33]$$

LEMMA 1.13.– Assume that a random vector $\vec{\xi}^T = (\xi_1, \ldots, \xi_n)$ is strictly sub-Gaussian, $\operatorname{Cov} \vec{\xi} = \mathbf{E} \vec{\xi} \vec{\xi}^T = B$, A is symmetric positive definite matrix. Then for any $0 \leq s < 1$ and $N = 1, 2, \ldots$

$$\mathbf{E} \exp\left\{ \frac{s}{2Z_N} \vec{\xi}^T A \vec{\xi} \right\} \leq \exp\left\{ \frac{1}{2} \sum_{l=1}^{\infty} \frac{1}{l} \left(\frac{sZ_l}{Z_N} \right)^l \right\}, \qquad [1.34]$$

where

$$Z_l = \left(Sp(BA)^l \right)^{\frac{1}{l}}, \qquad [1.35]$$

Sp is a trace of matrix.

PROOF.– Assume that S is such a matrix that $S \cdot S = A$, $S = S^T$, O is an orthogonal matrix that transforms SBS to diagonal matrix $OSBSO^T = D = \operatorname{diag}(d_k^2)_{k=1}^n$. Let $\vec{\theta} = OS\vec{\xi}$. By lemma 1.11, $\vec{\theta}$ is also strictly sub-Gaussian vector.

$$\operatorname{Cov} \vec{\theta} = OS\operatorname{Cov} \vec{\xi} SO^T = OSBSO^T = D, \qquad [1.36]$$

$$\vec{\theta}^T \vec{\theta} = \sum_{k=1}^n \theta_i^2 = \vec{\xi}^T SO^T OS \vec{\xi} = \vec{\xi}^T A \vec{\xi}.$$

Therefore, from lemma 1.12 follows that for $0 \leq s < 1$ the relationships

$$\mathbf{E} \exp\left\{ \frac{s}{2\hat{R}_N} \vec{\xi}^T A \vec{\xi} \right\} = \mathbf{E} \exp\left\{ \frac{s}{2\hat{R}_N} \vec{\theta}^T \vec{\theta} \right\}$$
$$\leq \exp\left\{ \frac{1}{2} \sum_{l=1}^{\infty} \frac{s^l \hat{R}_l^l}{l \hat{R}_N^l} \right\}, \qquad [1.37]$$

holds true, where $\hat{R}_l = \left(\sum_{k=1}^n d_k^{2l} \right)^{\frac{1}{l}}$.

Since $\hat{R}_l^l = SpD^l$, then [1.36] yields that

$$SpD^l = Sp(OS(BA)^{l-1}BSO^T) = Sp(S(BA)^{l-1}BS)$$
$$= Sp((BA)^l) = Z_l^l.$$

Hence, from equality above and [1.37] follows the statement of the lemma. □

REMARK 1.3.– Since [1.33] holds for $\hat{R}_l$ and $Z_l = \hat{R}_l$, then

$$Z_N \geq Z_l \quad \text{at} \quad l > N, \qquad [1.38]$$

and the series $\sum_{l=1}^{\infty} \frac{s^l Z_l^l}{l Z_N^l}$ is convergent for arbitrary N. Moreover, $Z_1 = \mathbf{E}\overrightarrow{\xi}^T A \overrightarrow{\xi}$.

COROLLARY 1.3.– Suppose that the conditions of lemma 1.13 are satisfied. Then for all $0 \leq s < 1$ and $N = 1, 2, \ldots$, the inequality

$$\mathbf{E}\exp\left\{\frac{s}{2Z_N}\overrightarrow{\xi}^T A \overrightarrow{\xi}\right\} \leq \exp\left\{\frac{1}{2}\nu_N(s) + \omega_N(s)\right\} \qquad [1.39]$$

holds, where

$$\omega_N(s) = \frac{1}{2}\sum_{l=N}^{\infty}\frac{s^l}{l}, \quad \nu_1(s) = 0, \quad \nu_N(s) = \sum_{l=1}^{N-1}\frac{(sZ_l)^l}{lZ_N^l}, \qquad [1.40]$$

as $N > 1$.

COROLLARY 1.4.– Assume that the conditions of lemma 1.13 are satisfied. Then for any $0 \leq s \leq 1$, the inequality

$$\mathbf{E}\exp\left\{\frac{s \cdot \overleftarrow{\xi}^T A \overrightarrow{\xi}}{2 \cdot \mathbf{E}\overleftarrow{\xi}^T A \overrightarrow{\xi}}\right\} \leq \frac{1}{\sqrt{1-s}} \qquad [1.41]$$

holds.

PROOF.– The statement of corollary follows from lemma 1.13 (inequality [1.34]). Taking $N = 1$, by remark 1.3, we obtain

$$\sum_{l=1}^{\infty}\frac{1}{l}\left(\frac{sZ_l}{Z_1}\right)^l \leq \sum_{l=1}^{\infty}\frac{s^l}{l} = -\ln(1-s).$$

Moreover, we can readily show that $Z_1 = \mathbf{E}\overleftarrow{\xi}^T A \overrightarrow{\xi}$. □

COROLLARY 1.5.– Let $\eta_n = \overrightarrow{\xi}_n^T A_n \overrightarrow{\xi}_n$, where $\overrightarrow{\xi}_n$ are strictly sub-Gaussian random vectors, $B_n = \operatorname{Cov}\overrightarrow{\xi}_n$, A_n are symmetric positive definite matrices, $Z_{l,n} = \left(Sp(B_n A_n)^l\right)^{\frac{1}{l}}$. If η is a random variable such that $\eta_n \to \eta$ in probability as $n \to \infty$ and the condition

$$\lim_{n\to\infty} Z_{l,n} = Z_l > 0 \quad \text{at} \quad l = 1, 2, \ldots, N \qquad [1.42]$$

is satisfied, then the inequality

$$\mathbf{E}\exp\left\{\frac{s\eta}{2Z_N}\right\} \leq \exp\left\{\frac{1}{2}\nu_N(s) + \omega_N(s)\right\} \qquad [1.43]$$

holds for any $0 < s < 1$, where $\omega_N(s)$, $\nu_N(s)$ are defined in [2.25],

$$\nu_1(s) = 0, \quad \nu_N(s) = \sum_{l=1}^{N-1} \frac{(sZ_l)^l}{lZ_N^l}, \quad \text{when} \quad N > 1.$$

PROOF.– Inequality [1.39] and the Fatou lemma yield the statement of corollary. □

LEMMA 1.14.– Assume that either the conditions of lemma 1.13, $\eta = \overrightarrow{\xi}^T A \overrightarrow{\xi}$ are satisfied, where Z_l is defined in [1.35], or the conditions of corollary 1.5 are satisfied, where η and Z_l are defined in corollary 1.5. Then for any $N = 1, 2, \ldots$, $x > 0$, $0 \leq s < 1$

$$\mathbf{P}\{\eta > x\} \leq W_N(s, x), \qquad [1.44]$$

where

$$W_N(s, x) = \exp\left\{-\frac{sx}{2Z_N}\right\} \exp\left\{\frac{1}{2}\nu_N(s) + \omega_N(s)\right\},$$

$\omega_N(s)$ and $\nu_N(s)$ are defined in [2.25].

PROOF.– The Chebyshev inequality and either [1.39] or [1.43] imply that the inequality

$$\mathbf{P}\{\eta > x\} = \mathbf{P}\left\{\frac{\eta s}{2Z_N} > \frac{sx}{2Z_N}\right\}$$
$$\leq \mathbf{E}\exp\left\{\frac{s\eta}{2Z_N}\right\} \exp\left\{-\frac{sx}{2Z_N}\right\} \leq W_N(s, x)$$

holds for $x > 0$, $0 \leq s < 1$. □

REMARK 1.4.– From [1.38] follows that the greater N, the more accurate inequality [1.44] is for a large enough x. Furthermore, [1.44] implies

$$\mathbf{P}\{\eta > x\} \leq \inf_{0 \leq s < 1} W_N(s, x). \qquad [1.45]$$

EXAMPLE 1.5.– By putting $N = 1, \omega_1(s) = -\frac{1}{2}\ln(1-s)$, the inequality [1.44] will be expressed as

$$\mathbf{P}\{\eta > x\} \le \exp\left\{-\frac{sx}{2Z_1}\right\}(1-s)^{-\frac{1}{2}}. \qquad [1.46]$$

By minimizing right-hand side of [1.46] with respect to s, we obtain that the inequality

$$\mathbf{P}\{\eta > x\} \le e^{\frac{1}{2}}\left(\frac{x}{Z_1}\right)^{\frac{1}{2}} \exp\left\{-\frac{x}{2Z_1}\right\} \qquad [1.47]$$

holds true for $x > Z_1$.

EXAMPLE 1.6.– Taking $N = 2$, $\omega_2(s) = -\frac{1}{2}\ln(1-s) - \frac{s}{2}$ inequality [1.44] has been expressed as

$$\mathbf{P}\{\eta > x\} \le \exp\left\{-\frac{s(x-Z_1)}{2Z_2}\right\}\exp\{-\frac{s}{2}\}(1-s)^{-\frac{1}{2}}. \qquad [1.48]$$

By minimizing the right-hand side of [1.48] with respect to s,we obtain that the inequality

$$\mathbf{P}\{\eta > x\} \le (\frac{x-Z_1}{Z_2}+1)^{\frac{1}{2}} \exp\left\{-\frac{x-Z_1}{2Z_2}\right\} \qquad [1.49]$$

holds as $x > Z_1$.

1.2.1. *Strictly sub-Gaussian stochastic processes*

DEFINITION 1.7.– *A stochastic process $X = \{X(t),\, t \in T\}$, is called strictly sub-Gaussian, if the family of random variables $\{X(t), t \in T\}$ is strictly sub-Gaussian.*

EXAMPLE 1.7.– Suppose that $X(t) = \sum_{k=1}^{\infty} \xi_k f_k(t)$, where $\xi = \{\xi_k, k = 1, 2, \ldots\}$ is a family of strictly sub-Gaussian random variables and the series

$$\sum_{l=1}^{\infty}\sum_{k=1}^{\infty} \mathbf{E}\xi_k\xi_l f_k(t) f_l(t),$$

is convergent for any $t \in T$, then from lemma 1.11 it follows that $X(t)$ is a strictly sub-Gaussian stochastic process. Lemmas 1.10 and 1.11 imply that in the case of independent strictly sub-Gaussian random variables ξ_k and

$$\sum_{k=1}^{\infty} \mathbf{E}\xi_k^2 f_k^2(t) < \infty$$

for all $t \in T$, $X(t)$ is a strictly sub-Gaussian stochastic process.

DEFINITION 1.8.– *Stochastic processes* $X_i = \{X_i(t),\, t \in T, \quad i = 1, 2, \ldots, M\}$, *are called jointly strictly sub-Gaussian, if the family of random variables* $\{X_i(t),\, t \in T,\, i = 1, 2, \ldots, M\}$ *is strictly sub-Gaussian.*

REMARK 1.5.– A zero-mean Gaussian stochastic process is sub-Gaussian.

LEMMA 1.15.– Let $X = \{X(t), t \in T\}$ be a strictly sub-Gaussian stochastic process and $(T, \mathcal{L}, \mu)$ be a measurable space. Assume that Lebesgue integral $\int_T (\mathbf{E}X^2(t))d\mu(t)$ is finite. Then, there exists almost sure the integral $\int_T X^2(t)d\mu(t)$ and the inequality

$$\mathbf{E}\exp\left\{\frac{s}{2}\frac{\int_T X^2(t)d\mu(t)}{\int_T (\mathbf{E}X^2(t))d\mu(t)}\right\} \le (1-s)^{-1/2}$$

holds for all $0 \le s \le 1$.

PROOF.– From the Fatou lemma follows the existence of integral $\int_T X^2(t)d\mu(t)$ with probability 1. Since $\int_T X^2(t)d\mu(t)$ can be represented as the limit

$$\int_T X^2(t)d\mu(t) = \lim_{n\to\infty} \sum_{k=1}^{n} c_k^2 \mu(A_k),$$

where $A_k \in \mathcal{L}, \quad A_k \cap A_j = \emptyset$, $k \ne j$ and $\bigcup_{k=1}^{n} A_k = T$, c_k are the value of $X(t)$ in some points from A_k and

$$\int_T \mathbf{E}X^2(t)d\mu(t) = \lim_{n\to\infty} \sum_{k=1}^{n} \mathbf{E}c_k^2 \mu(A_k),$$

then the statement of lemma follows from corollaries 1.4 and 1.5 and the Fatou lemma. □

COROLLARY 1.6.– Suppose that for $X = \{X(t), t \in T\}$ the conditions of lemma 1.15 are satisfied. Then, the inequality

$$\mathbf{P}\{\int_T X^2(t)d\mu(t) > \varepsilon\} \le e^{\frac{1}{2}} \left(\frac{\varepsilon}{\int_T (\mathbf{E}X^2(t)d\mu(t))}\right)^{\frac{1}{2}} \exp\left\{\frac{-\varepsilon}{2\int_T (\mathbf{E}X^2(t)d\mu(t))}\right\} \qquad [1.50]$$

holds as

$$\varepsilon > \int_T (\mathbf{E}X^2(t))d\mu(t).$$

PROOF.– The proof is similar to the proof of example 1.6. □

1.3. The estimates of convergence rates of strictly sub-Gaussian random series in $L_2(T)$

In this section, the results of previous section are used for finding norm distribution estimates in $L_2(T)$ for residuals of strictly sub-Gaussian stochastic series.

Assume that $(T, \mathfrak{A}, \mu)$ is some measurable space. Consider a stochastic series in the form

$$S(t) = \sum_{k=1}^{\infty} \xi_k f_k(t), \qquad t \in T, \tag{1.51}$$

where $\xi = \{\xi_k, k = 1, 2, \ldots\}$ is a family of strictly sub-Gaussian random variables and $f = \{f_k(t), k = 1, 2, \ldots\}$ is a family of function from $L_2(T)$. Suppose that the following condition holds: for all $t \in T$

$$\sum_{k=1}^{\infty}\sum_{l=1}^{\infty} \mathbf{E}\xi_k\xi_l f_l(t) f_k(t) < \infty, \tag{1.52}$$

which means that the series in [3.1] is mean square convergent for all $t \in T$ and $S(t)$ is strictly sub-Gaussian process.

Denote for $1 \le n \le m \le \infty$

$$S_n^m(t) = \sum_{k=n}^{m} \xi_k f_k(t).$$

DEFINITION 1.9.– *The series [3.1] mean square converges in space $L_2(T)$, if*

$$\mathbf{E}\int_T \left(S(t) - S_1^n(t)\right)^2 d\mu(t) = \mathbf{E}\|S(t) - S_1^n(t)\|^2_{L_2(T)} \to 0$$

as $n \to \infty$.

We can readily show the next assertion.

LEMMA 1.16.– The series [3.1] is mean square convergent in $L_2(T)$ if and only if either

$$\mathbf{E}\|S_n^m(t)\|_{L_2(T)}^2 = \sum_{l=n}^{m}\sum_{k=n}^{m} \mathbf{E}\xi_k\xi_l \int_T f_l(t)f_k(t)\,d\mu(t) \to 0$$

as $n, m \to \infty$, or

$$\sum_{k=1}^{\infty}\sum_{l=1}^{\infty} \mathbf{E}\xi_k\xi_l \int_T f_l(t)f_k(t)\,d\mu(t) < \infty \qquad [1.53]$$

The following theorem gives a convergence rate estimate of the series [3.1] in $L_2(T)$.

THEOREM 1.2.– Let $A_{n,m} = \|a_{kl}\|_{k,l=n}^m$,

$$a_{kl} = \int_T f_l(t)f_k(t)\,d\mu(t), \quad B_{n,m} = \|\mathbf{E}\xi_k\xi_l\|_{k,l=n}^m,$$

$$b_{kl} = \mathbf{E}\xi_k\xi_l.$$

If the condition [1.53] is satisfied and for $l = 1, 2, \ldots, N$ there exists a limit

$$\lim_{m\to\infty}\Big(Sp\big(B_{nm}A_{nm}\big)^l\Big)^{\frac{1}{l}} = J_{nl},$$

then the inequality

$$\mathbf{P}\Big\{\big\|S_n^\infty(t)\big\|_{L_2(T)} > x\Big\} = \mathbf{P}\Big\{\int_T \big|S_n^\infty(t)\big|^2\,d\mu(t) > x^2\Big\} \le V_N(s, x^2) \qquad [1.54]$$

holds true for any $x > 0$, $0 \le s < 1$, $N = 1, 2, \ldots$, $n = 1, 2, \ldots$, where

$$S_n^\infty(t) = \sum_{k=n}^{\infty} \xi_k f_k(t),$$

$$V_N(s, x) = \exp\{-\frac{sx}{2J_{nN}}\}\exp\Big\{\frac{1}{2}\tilde{\nu}_N(s) + \omega_N(s)\Big\},$$

$\omega_N(s)$ is defined in [2.25], $\tilde{\nu}_1(s) = 0$,

$$\tilde{\nu}_N(s) = \sum_{l=1}^{N-1} \frac{(sJ_{nl})^l}{lJ_{nN}^l}, \qquad N > 1.$$

PROOF.– The assertion of the theorem follows from lemma 1.14. A random variable $\left\|S_n^\infty(t)\right\|_{L_2(T)}$ is mean square limit as $m \to \infty$, that is why in probability also, of random variables

$$\left\|S_n^m(t)\right\|_{L_2(T)} = \sum_{l=n}^{m} \sum_{k=n}^{m} \xi_l \xi_k a_{lk}.$$

Really, from [1.53] it follows that

$$\mathbf{E}\left|\left\|S_n^\infty(t)\right\|_{L_2(T)} - \left\|S_n^m(t)\right\|_{L_2(T)}\right|^2$$
$$\leq \mathbf{E}\left\|S_n^\infty(t) - S_n^m(t)\right\|^2_{L_2(T)} \to 0$$

as $m \to \infty$. □

COROLLARY 1.7.– Suppose that either random variables in [3.1] are uncorrelated, $\mathbf{E}\xi_k^2 = \sigma_k^2$, or the system of function $f_k(t)$ is orthogonal

$$\int_T f_k(t) f_l(t)\, d\mu(t) = \delta_k^l a_k^2,$$

where δ_k^l is Kronecker symbol. If the series

$$\sum_{k=1}^{\infty} \sigma_k^2 a_k^2 < \infty \qquad [1.55]$$

converges, where

$$a_k^2 = \int_T f_k^2(t)\, d\mu(t),$$

then the series [3.1] is mean square convergent in $L_2(T)$ and for all $x > A_n^{\frac{1}{2}}$, $n = 1, 2, \ldots$, the inequality

$$\mathbf{P}\left\{\left\|S_n^\infty(t)\right\|_{L_2(T)} > x\right\} \leq \exp\{\frac{1}{2}\} \frac{x}{A_n^{\frac{1}{2}}} \exp\{-\frac{x^2}{2A_n}\} \qquad [1.56]$$

holds, where $A_n = \sum_{k=n}^\infty \sigma_k^2 a_k^2$.

PROOF.– Since under condition of the corollary

$$\sum_{l=1}^{\infty} \sum_{k=1}^{\infty} \mathbf{E}\xi_l \xi_k \int_T f_k(t) f_l(t)\, d\mu(t) = \sum_{k=1}^{\infty} \sigma_k^2 a_k^2 < \infty$$

then the series [3.1] converges in mean square. By theorem 1.2 as $N = 1$, the estimate [1.54] is satisfied, where

$$J_{n1} = \lim_{m\to\infty} \sum_{k=n}^{m} \sigma_k^2 a_k^2 = \sum_{k=n}^{\infty} \sigma_k^2 a_k^2 = A_n.$$

Minimizing the expression [1.54] in $0 \le s < 1$ (see example 1.5), we obtain inequality [1.56]. □

COROLLARY 1.8.– Assume that random variables in [3.1] are uncorrelated and the system of function $f_k(t)$ is orthogonal, $\mathbf{E}\xi_k^2 = \sigma_k^2$, $\int_T |f_k(t)|^2\, d\mu(t) = a_k^2$. If

$$\sum_{k=1}^{\infty} \sigma_k^2 a_k^2 < \infty, \qquad [1.57]$$

then the series [3.1] is mean square convergent in $L_2(T)$ and for all $x > 0$, $0 \le s < 1$, $N = 1, 2, \ldots$, $n = 1, 2, \ldots$, the inequality

$$\mathbf{P}\Big\{ \big\| S_n^{\infty}(t) \big\|_{L_2(T)} > x \Big\} \le V_N(s, x^2) \qquad [1.58]$$

holds, where the function $V_N(s, x)$ is from [1.54], and

$$J_{nl} = \hat{J}_{nl} = \left(\sum_{k=n}^{\infty} \sigma_k^{2l} a_k^{2l} \right)^{\frac{1}{l}}. \qquad [1.59]$$

PROOF.– The assertion of corollary 1.8 follows from theorem 1.2. In this case, equity [1.59] holds and condition [1.57] provides the convergence of the series [1.59] for all $l \ge 1$. □

EXAMPLE 1.8.– If in corollary 1.8 we put $N = 2$ and minimize the right-hand side of [1.58] in $0 \le s < 1$, then we obtain (see, example 1.6) that the inequality

$$\mathbf{P}\Big\{ \big\| S_n^{\infty}(t) \big\|_{L_2(T)} > x \Big\} \le \left(\frac{x^2 - \hat{J}_{n1}}{\hat{J}_{n2}} + 1 \right)^{\frac{1}{2}} \exp\left\{ -\frac{x^2 - \hat{J}_{n1}}{2\hat{J}_{n2}} \right\} \qquad [1.60]$$

holds true for all $x^2 > \hat{J}_{n1}$.

1.4. The distribution estimates of the norm of sub-Gaussian stochastic processes in $L_p(T)$

In this section, the estimates for distribution of the norm in $L_p(T)$ of sub-Gaussian random processes are found. Obtained results are applied to strictly sub-Gaussian stochastic processes and series.

Let $(T, \mathfrak{A}, \mu)$, $\mu(T) < \infty$, be some measurable space. $L_p(T)$ is a space integrated in the power of p measurable functions

$$f = \{f(t), t \in T\}, \qquad \|f\|_{L_p} = \left(\int_T |f(t)|^p \, d\mu(t)\right)^{\frac{1}{p}}.$$

Assume that $X = \{X(t), t \in T\}$ is a sub-Gaussian stochastic process,

$$\sup_{t \in T} \tau(X(t)) = \tau < \infty.$$

Since

$$\mathbf{E} \int_T |X(t)|^p \, d\mu(t) = \int_T \mathbf{E}|X(t)|^p \, d\mu(t),$$

and from lemma 1.2 it follows that there exists such constant c_p that $\sup_{t \in T} \mathbf{E}|X(t)|^p < c_p$, then

$$\mathbf{E} \int_T |X(t)|^p \, d\mu(t) \leq c_p \mu(T) < \infty.$$

Therefore, $\int_T |X(t)|^p \, d\mu(t) < \infty$ with probability 1, it means that $X \in L_p(T)$ almost surely.

THEOREM 1.3.– The inequality

$$\mathbf{P}\{\|X\|_{L_p} > x\} \leq 2 \exp\left\{-\frac{x^2}{2\tau^2 [\mu(T)]^{\frac{2}{p}}}\right\} \qquad [1.61]$$

holds true for any $p \geq 0$ and

$$x \geq p^{1/2} [\mu(T)]^{\frac{1}{p}} \tau.$$

PROOF.– By the Lyapunov moment inequality for $s \geq p > 0$, we have

$$\mathbf{E}\|X\|_p^s = \mathbf{E}\Big(\int_T |X(t)|^p \, d\mu(t)\Big)^{s/p} = \mathbf{E}\Big[\int_T |X(t)|^p \, d\Big(\frac{\mu(t)}{\mu(T)}\Big)\Big]^{s/p} (\mu(T))^{s/p}$$

$$\leq \mathbf{E}\Big[\int_T |X(t)|^s \, d\Big(\frac{\mu(t)}{\mu(T)}\Big)\Big](\mu(T))^{s/p} = \int_T \mathbf{E}|X(t)|^s \, d\mu(t)(\mu(T))^{(s/p)-1}.$$

Then, it follows from the Chebyshev inequality that

$$\mathbf{P}\{\|X\|_{L_p} > x\} \leq \frac{\mathbf{E}\|X\|_p^s}{x^s}$$

and lemma 1.3 implies that

$$\mathbf{P}\{\|X\|_{L_p} > x\} \leq 2s^{s/2}a^s,$$

where $a = \frac{\tau \cdot (\mu(T))^{1/p}}{x\sqrt{e}}$. Let $s = a^{-2}e^{-1}$ (it is a point that minimizes the right-hand side of inequality above). Then, for $s = \frac{1}{a^2 e} \geq p$, that is for $x \geq p^{1/2}(\mu(T))^{1/p}\tau$, the inequality

$$\mathbf{P}\{\|X\|_{L_p} > x\} \leq 2\exp\Big\{-\frac{1}{2a^2 e}\Big\}$$

is satisfied. So, the theorem is completely proved. □

Consider now the random series

$$S(t) = \sum_{k=1}^{\infty} \xi_k f_k(t), \qquad [1.62]$$

where $\xi = \{\xi_k, k = 1, 2, \ldots\}$ is a family of strictly sub-Gaussian random variables, $f = \{f_k(t), t \in T, k = 1, 2, \ldots\}$ is a family of measurable functions. The following theorem gives convergence rate estimates of the series [1.62] in $L_p(T)$.

THEOREM 1.4.– Assume that the condition

$$\sup_{t\in T} \sum_{k=1}^{\infty} \sum_{l=1}^{\infty} \mathbf{E}\xi_k\xi_l f_l(t) f_k(t) < \infty \qquad [1.63]$$

holds true. Then, for arbitrary $p > 0$ and

$$x \geq p^{1/2}[\mu(T)]^{\frac{1}{p}}\sigma_n \leq p \leq 2$$

where

$$\sigma_n^2 = \sup_{t\in T} \sum_{k=n}^{\infty} \sum_{l=n}^{\infty} \mathbf{E}\xi_k\xi_l f_l(t) f_k(t),$$

the following inequality

$$\mathbf{P}\{\|S_n^\infty(t)\|_{L_p} > x\} \leq 2\exp\left\{-\frac{x^2}{2\sigma_n^2[\mu(T)]^{\frac{2}{p}}}\right\} \qquad [1.64]$$

holds.

$$\text{where} \quad S_n^\infty(t) = \sum_{k=n}^{\infty} \xi_k f_k(t).$$

PROOF.– Since stochastic process $S_n^\infty = \{S_n^\infty(t), t \in T\}$ is sub-Gaussian with $\tau^2(S_n^\infty(t)) = \sigma_n^2$, then theorem 1.4 is a corollary of theorem 1.3. □

COROLLARY 1.9.– Suppose that random variables in ξ_k [1.62] are uncorrelated. Then, the assertion of theorem 1.4 holds, where

$$\sigma_n^2 = \sup_{t\in T} \sum_{k=n}^{\infty} \mathbf{E}\xi_k^2 f_k^2(t).$$

COROLLARY 1.10.– Let the series in [1.62] be stationary. That is

$$S(t) = \sum_{k=1}^{\infty} (\xi_k \cos\lambda_k t + \eta_k \sin\lambda_k t),$$

where $\mathbf{E}\xi_k\xi_l = \mathbf{E}\eta_k\eta_l = 0$, when $k \neq l$, for all k and l, $\mathbf{E}\xi_k\eta_l = 0$, $\mathbf{E}\xi_k^2 = \mathbf{E}\eta_k^2 = b_k^2$. Then, the assertion of theorem 1.4 holds, where $\sigma_n^2 = \sum_{k=n}^{\infty} b_k^2$.

REMARK 1.6.– It is clear that the estimates in [1.64] are the best for stationary random series. In other cases they can be improved, which will be done in the following sections.

1.5. The distribution estimates of the norm of sub-Gaussian stochastic processes in some Orlicz spaces

In this section, the estimates for norm distribution of sub-Gaussian stochastic processes in Orlicz space are found such that generated functions increase not faster than the function $U(x) = \exp\{x^2\} - 1$. The results are applied to strictly sub-Gaussian processes and series.

Let $(T, \mathfrak{U}, \mu)$, $\mu(T) < \infty$, be some measurable space, $L_U(T)$ be Orlicz space with generated C-function $U = \{U(x), x \in \mathbb{R}\}$. Remember that a continuous even convex function $U(\cdot)$ is called C- function if it is monotone increasing, $U(0) = 0$, $U(x) > 0$, as $x \neq 0$. For example, $U(x) = \exp\{|x|^\alpha\} - 1$, $\alpha \geq 1$.

The Orlicz space, generated by the function $U(x)$, is defined as the family of functions $\{f(t), t \in T\}$ where for each function $f(t)$ there exists a constant r such that

$$\int_T U\left(\frac{f(t)}{r}\right) d\mu(t) < \infty.$$

The space $L_U(T)$ is Banach with respect to the norm

$$\|f\|_{L_U} = \inf\left\{r > 0 \colon \int_T U\left(\frac{f(t)}{r}\right) d\mu(t) \leq 1\right\}. \qquad [1.65]$$

A norm $\|f\|_{L_U}$ is called the Luxemburg norm.

Let $X = \{X(t), t \in T\}$ be a sub-Gaussian stochastic process, $\sup_{t \in T} \tau(X(t)) = \tau < \infty$.

THEOREM 1.5.– Assume that the Orlicz C-function $U = \{U(x), x \in \mathbb{R}\}$ is such that

$$G_U(t) = \exp\{\left(U^{(-1)}(t-1)\right)^2\}, \quad t \geq 1,$$

is convex as $t \geq 1$, ($U^{(-1)}(t)$ is an inverse function to $U(t)$). Then, for all x such that

$$x \geq \hat{\mu}(T)\tau(2 + (U^{(-1)}(1))^{-2})^{\frac{1}{2}}, \qquad [1.66]$$

where $\hat{\mu}(T) = \max(\mu(T), 1)$ the inequality

$$\mathbf{P}\{\|X(t)\|_{L_U} > x\}$$
$$\leq \exp\left\{\frac{1}{2}\right\} \frac{xU^{(-1)}(1)}{\hat{\mu}(T)\tau} \exp\left\{-\frac{x^2(U^{(-1)}(1))^2}{2(\hat{\mu}(T))^2\tau^2}\right\} \qquad [1.67]$$

holds.

PROOF.– Note that from the definition of the function $G_U(t)$ follows that for all $z \geq 0$ the equality

$$G_U(U(z) + 1) = \exp\{z^2\} \qquad [1.68]$$

holds true. Equality [1.68], the definition of the norm $\|\cdot\|_{L_U}$ (see [1.65]), the Chebyshev inequality and the Jensen one for any $p \geq 1$, $x \geq 0$, imply the relationships:

$$\begin{aligned}\mathbf{P}\{\|X(t)\|_{L_U} > x\} &= \mathbf{P}\left\{\int_T U\left(\frac{X(t)}{x}\right) d\mu(t) \geq 1\right\}\\ &= \mathbf{P}\left\{\int_T U\left(\frac{X(t)}{x}\right)\mu(T)\frac{d\mu(t)}{\mu(T)} \geq 1\right\}\\ &= \mathbf{P}\left\{\int_T \left(U\left(\frac{X(t)}{x}\right)\mu(T)+1\right)\frac{d\mu(t)}{\mu(T)} \geq 2\right\}\\ &\leq \frac{\mathbf{E}\left(G_U\left(\int_T\left(U\left(\frac{X(t)}{x}\right)\mu(T)+1\right)\frac{d\mu(t)}{\mu(T)}\right)\right)^p}{(G_U(2))^p}\\ &\leq \frac{\mathbf{E}\left(G_U\left(\int_T U\left(\frac{X(t)}{x}\hat{\mu}(T)\right)+1\right)\frac{d\mu(t)}{\mu(T)}\right)^p}{(G_U(2))^p}\\ &\leq \frac{\int_T \mathbf{E}\left(G_U\left(U\left(\frac{X(t)\hat{\mu}(T)}{x}\right)+1\right)\right)^p\frac{d\mu(t)}{\mu(T)}}{(G_U(2))^p}\\ &= \frac{\int_T \mathbf{E}\exp\left\{\frac{p(X(t))^2(\hat{\mu}(T))^2}{x^2}\right\}\frac{d\mu(t)}{\mu(T)}}{\exp\{p(U^{(-1)}(1))^2\}}. \end{aligned} \qquad [1.69]$$

If now in [1.69] we set $p = sx^2(\hat{\mu}(T)\tau\sqrt{2})^{-2}$, where $0 \leq s < 1$, then for such x ($p \geq 1$) that

$$x^2 > \frac{(\hat{\mu}(T))^2 2\tau^2}{s}, \qquad [1.70]$$

from lemma 1.6 we have

$$\mathbf{E}\exp\left\{\frac{p(X(t))^2(\hat{\mu}(T))^2}{x^2}\right\} = \mathbf{E}\exp\left\{\frac{s(X(t))^2}{2\tau^2}\right\} \leq \frac{1}{\sqrt{1-s}}. \qquad [1.71]$$

Hence, [1.69] and [1.71] yield

$$\mathbf{P}\{\|X(t)\|_{L_U} > x\} \leq \frac{1}{\sqrt{1-s}}\exp\left\{-\frac{sx^2(U^{(-1)}(1))^2}{2(\hat{\mu}(T))^2\tau^2}\right\}. \qquad [1.72]$$

If we minimize the right-hand side of the inequality above with respect to s, $0 \le s < 1$, it means that

$$s = 1 - \frac{(\hat{\mu}(T))^2\tau^2}{x^2(U^{(-1)}(1))^2}, \qquad [1.73]$$

then for x from [1.70] and since $s > 0$

$$x^2 > \frac{(\hat{\mu}(T))^2\tau^2}{(U^{(-1)}(1))^2} \qquad [1.74]$$

inequality [1.67] is obtained.

To complete the proof of the theorem, it is enough to remark that for s in equality [1.73], inequality [1.70] holds true if and only if [1.66] is satisfied. It is clear that for such x inequality [1.74] holds true. □

REMARK 1.7.– From [1.67] it follows also that the trajectories of the process X almost surely belong to the Orlicz space $L_U(\Omega)$.

REMARK 1.8.– It is easy to obtain from theorem 1.5 the estimates of distribution $\|X(t)\|_{L_p}$, because $L_p(T)$ is the Orlicz space generated by C-function $U(x) = |x|^p$, $p \ge 1$. But in this case, we should consider a C-function, that is equivalent to $U(x) = |x|^p$ and the assumptions of the function $G_U(x)$ hold true. However, the estimates of section 1.4 are more precise.

EXAMPLE 1.9.– The conditions of theorem 1.5 are satisfied for C-function

$$U_\alpha(x) = \exp\{|x|^\alpha\} - 1, \qquad 1 \le \alpha \le 2.$$

In this case

$$U_\alpha^{(-1)}(t) = (\ln(t+1))^{\frac{1}{\alpha}}, \qquad U_\alpha^{(-1)}(1) = (\ln 2)^{\frac{1}{\alpha}},$$

$$G_{U_\alpha}(t) = \exp\{(\ln t)^{\frac{2}{\alpha}}\}.$$

EXAMPLE 1.10.– The conditions of theorem 1.5 hold for C-function

$$U_\alpha(x) = \begin{cases} \left(\frac{e\alpha}{2}\right)^{\frac{2}{\alpha}} x^2, & |x| \le \left(\frac{2}{\alpha}\right)^{\frac{1}{\alpha}}, \\ \exp\{|x|^\alpha\}, & |x| > \left(\frac{2}{\alpha}\right)^{\frac{1}{\alpha}}, \end{cases}$$

where $0 < \alpha < 1$. In such a case, $U^{(-1)}(1) = \left(\frac{2}{e\alpha}\right)^{\frac{1}{\alpha}}$.

Consider a random series [1.62]

$$S(t) = \sum_{k=1}^{\infty} \xi_k f_k(t),$$

where $\xi = \{\xi_k, k = 1, 2, \ldots\}$ is a family of strictly sub-Gaussian random variables. The following theorem gives convergence rate estimates of the series [1.62] in the norm of space $L_U(\Omega)$.

THEOREM 1.6.– Let a C-function $U(x)$ satisfy all conditions of theorem 1.5. Assume that the condition [1.63] also holds true. Then for

$$x \geq \hat{\mu}(T)\sigma_n(2 + (U^{(-1)}(1))^{-2})^{\frac{1}{2}}, \qquad [1.75]$$

where

$$\sigma_n^2 = \sup_{t\in T} \sum_{k=n}^{\infty} \sum_{l=n}^{\infty} \mathbf{E}\xi_k \xi_l f_k(t) f_l(t),$$

the inequality

$$\mathbf{P}\{\|S_n^{\infty}(t)\|_{L_U} > x\} \leq \exp\Big\{\frac{1}{2}\Big\} \frac{xU^{(-1)}(1)}{\hat{\mu}(T)\sigma_n} \exp\Big\{-\frac{x^2(U^{(-1)}(1))^2}{2(\hat{\mu}(T))^2\sigma_n^2}\Big\} \qquad [1.76]$$

holds.

REMARK 1.9.– If ξ_k are uncorrelated, then

$$\sigma_n^2 = \sup_{t\in T} \sum_{k=n}^{\infty} \mathbf{E}\xi_k^2 f_k^2(t).$$

And if $S(t)$ is a stationary process, then $\sigma_n^2 = \sum_{k=n}^{\infty} b_k^2$. Here, remark 1.6 can be applied.

1.6. Convergence rate estimates of strictly sub-Gaussian random series in Orlicz spaces

In this section, the estimates for the norm distribution in Orlicz space of residuals of strictly sub-Gaussian random series from classes $D_U(c)$ are found. Examples of such series will be considered later. In contrast to previous sections, here the Orlicz spaces are considered that are generated by C-functions, which grow faster than the function $U_2(x) = \exp\{x^2\} - 1$. Note that in some cases, even when generated C-functions grow not faster than $U_2(x)$, and also for $L_p(T)$ these estimates can be better than estimates discussed in previous section.

Let $(T, \mathfrak{A}, \mu)$ be some measurable space, $L_U(T)$, is the Orlicz space that generated C-function $U = \{U(x), x \in \mathbb{R}\}$.

DEFINITION 1.10.– *Assume that* $f = \{f_k(t), t \in T, k = 1, 2, \ldots\}$ *is the family from the space* $L_U(T)$. *This family belongs to the class* $D_U(c)$, *if there exists such numerical sequence* $c = \{c_k, k = 1, 2, \ldots\}$, $0 \le c_k \le c_{k+1}$, *that for any sequence* $r = \{r_k, k = 1, 2, \ldots\}$ *the inequality*

$$\left\| \sum_{k=1}^{n} r_k f_k(t) \right\|_{L_U} \le c_n \left\| \sum_{k=1}^{n} r_k f_k(t) \right\|_{L_2} \qquad [1.77]$$

holds true.

REMARK 1.10.– In definition 1.10, the sequence $c = \{c_k, k = 1, 2, \ldots\}$ is the same for any sequence r. It means that c depends only on f and U.

Consider random series (process)

$$S(t) = \sum_{k=1}^{\infty} \xi_k f_k(t), \qquad [1.78]$$

where $\xi = \{\xi_k, k = 1, 2, \ldots\}$ is a family of strictly sub-Gaussian random variables. Suppose that $f = \{f_k(t), t \in T, k = 1, 2, \ldots\}$ is the family of functions from the space $L_U(T)$ that belong to the class $D_U(c)$. Assume that the condition [1.52] is also satisfied. Hence, the series [1.78] converges in mean square. Denote for $1 \le m \le n \le \infty$

$$S_m^n(t) = \sum_{j=m}^{n} \xi_j f_j(t).$$

Assume that $a = \{a_k, k = 1, 2, \ldots\}$ is some sequence such that $0 \le a_k \le a_{k+1}$, $a_k \to \infty$ at $k \to \infty$. For $1 \le n \le m$, we denote

$$S_m^n(a, t) = \sum_{j=m}^{n} a_j \xi_j f_j(t).$$

It is easy to check that the equity

$$\int_T \left(S_m^k(a,t) \right)^2 d\mu(t) = \sum_{j=m}^{k} \sum_{i=m}^{k} a_j a_i \xi_i \xi_j \int_T f_i(t) f_j(t)\, d\mu(t)$$
$$= \vec{\xi}_{mk}^T A_{mk}(f) \vec{\xi}_{mk} \qquad [1.79]$$

holds true, where $m \leq k$, $\vec{\xi}_{mk}^T = (\xi_m, \xi_{m+1}, \ldots, \xi_k)$,

$$A_{mk}(f) = \|a_{ij}(f)\|_{i,j=m}^k, \quad a_{ij}(f) = a_i a_j \int_T f_i(t) f_j(t)\, d\mu(t).$$

Let

$$B_{mk} = \mathrm{Cov}\vec{\xi}_{mk} = \left\| \mathbf{E}\xi_i \xi_j \right\|_{i,j=m}^k.$$

Define

$$J_l(m,k,a) = \left(Sp(B_{mk} A_{mk}(f))^l \right)^{\frac{1}{l}}.$$

LEMMA 1.17.– Suppose that the sequence $a = \{a_k,\ k = 1,2,\ldots\}$ is such that $a_k \leq a_{k+1}$ for any $0 \leq s < 1$, $N = 1,2,\ldots$ and $m \leq n$. Then, the inequality

$$\mathbf{E}\exp\left\{ \frac{\left(s\left\| S_m^n(t) \right\|_{L_U} \right)^2}{2(B_N(m,n,a))^2} \right\} \leq \exp\left\{ \frac{A_N(m,n,a,s)}{2B_N(m,n,a)} + \omega_N(s) \right\} \qquad [1.80]$$

is satisfied, where

$$A_N(m,n,a,s) = \sum_{k=m}^{n} b_{kn} (J_N(m,k,a))^{\frac{1}{2}} \sum_{l=1}^{N-1} \frac{(sJ_l(m,k,a))^l}{lJ_N^l(m,k,a)},$$

as $N > 1$ and $A_1(m,n,a,s) = 0$,

$$B_N(m,n,a) = \sum_{k=m}^{n} b_{kn} (J_N(m,k,a))^{\frac{1}{2}},$$

$b_{kn} = c_k d_{kn}$, $d_{kn} = (a_k^{-1} - a_{k+1}^{-1})$, $k = m, m+1, \ldots, n-1$, $d_{nn} = a_n^{-1}$, $\omega_N(s)$ is defined in [2.25].

PROOF.– The equity (Abel transform) is carried out

$$S_m^n(t) = \sum_{k=m}^{n-1}(a_k^{-1} - a_{k+1}^{-1})S_m^k(a,t) + a_n^{-1}S_m^n(a,t)$$
$$= \sum_{k=m}^{n} d_{kn}S_m^k(a,t). \quad [1.81]$$

From [1.81] and definition 1.10 follows inequality

$$\|S_m^n(t)\|_{L_U} \le \sum_{k=m}^{n} d_{kn}\|S_m^k(a,t)\|_{L_U}$$
$$\le \sum_{k=m}^{n} d_{kn}c_k\|S_m^k(a,t)\|_{L_2} \quad [1.82]$$
$$= \sum_{k=m}^{n} b_{kn}\|S_m^k(a,t)\|_{L_2}.$$

Let $\delta_k > 0$, $k = m, m+1, \dots, n$ be such numbers that $\sum_{k=m}^{n} \delta_k = 1$. Suppose $W_{m,n}$ is an arbitrary number. A convexity of the function $y = x^2$ and the Hölder inequality imply the relationships:

$$I_m^n = \mathbf{E}\exp\left\{\left(\frac{\sum_{k=m}^{n} b_{kn}\|S_m^k(a,t)\|_{L_2}}{W_{m,n}}\right)^2\right\}$$
$$= \mathbf{E}\exp\left\{\left(\sum_{k=m}^{n} \frac{\delta_k b_{kn}\|S_m^k(a,t)\|_{L_2}}{\delta_k W_{m,n}}\right)^2\right\} \quad [1.83]$$
$$\le \mathbf{E}\exp\left\{\sum_{k=m}^{n} \delta_k \left(\frac{b_{kn}\|S_m^k(a,t)\|_{L_2}}{\delta_k W_{m,n}}\right)^2\right\}$$
$$\le \prod_{k=m}^{n}\left(\mathbf{E}\exp\left\{\frac{b_{kn}^2\|S_m^k(a,t)\|_{L_2}^2}{\delta_k^2 W_{m,n}^2}\right\}\right)^{\delta_k}.$$

Denote

$$\delta_k = \frac{\sqrt{2}b_{kn}((J_N(m,k,a))^{\frac{1}{2}}}{\sqrt{s}W_{m,n}},$$

$$W_{m,n} = \sqrt{\frac{2}{s}}\sum_{k=m}^{n} b_{kn}((J_N(m,k,a))^{\frac{1}{2}}.$$

Then, [1.79] and corollary 1.3 as $N > 1$ yield inequality:

$$\mathbf{E}\exp\left\{\frac{b_{kn}^2\|S_m^k(a,t)\|_{L_2}^2}{\delta_k^2 W_{m,n}^2}\right\} = \mathbf{E}\exp\left\{\frac{s\|S_m^k(a,t)\|_{L_2}^2}{2J_N(m,k,a)}\right\} \quad [1.84]$$

$$\leq \exp\left\{\frac{1}{2}\sum_{l=1}^{N-1}\frac{(sJ_l(m,k,a))^l}{lJ_N^l(m,k,a)} + \omega_N(s)\right\}.$$

It follows from [1.82]–[1.84] that

$$\mathbf{E}\exp\left\{\frac{s\|S_m^n(t)\|_{L_U}^2}{2\left(\sum_{k=m}^n b_{kn}(J_N(m,k,a))^{\frac{1}{2}}\right)^2}\right\}$$

$$\leq I_m^n \leq \exp\left\{\frac{1}{2}\sum_{k=m}^{n}\sum_{l=1}^{N-1}\frac{(sJ_l(m,k,a))^l}{lJ_N^l(m,k,a)}\delta_k + \omega_N(s)\right\}.$$

If now in above inequality we substitute the value δ_k, then inequality [1.80] is obtained.The proof will be the same when $N = 1$. □

LEMMA 1.18.– Suppose that the assumptions of lemma 1.17 are satisfied, then for any $x > 0$, $0 \leq s < 1$

$$\mathbf{P}\{\|S_m^n(t)\|_{L_U} > x\} \quad [1.85]$$

$$\leq \exp\left\{-\frac{sx^2}{2(B_N(m,n,a))^2}\right\}\exp\left\{\frac{A_N(m,n,a,s)}{2B_N(m,n,a)} + \omega_N(s)\right\}.$$

PROOF.– From the Chebyshev–Markov inequality follows

$$\mathbf{P}\{\|S_m^n(t)\|_{L_U} > x\}$$

$$= \mathbf{P}\left\{\frac{s\|S_m^n(t)\|_{L_U}^2}{2(B_N(m,k,a))^2} > \frac{sx^2}{2(B_N(m,k,a))^2}\right\}$$

$$\leq \exp\left\{-\frac{sx^2}{2(B_N(m,k,a))^2}\right\}\cdot\mathbf{E}\exp\left\{\frac{s\|S_m^n(t)\|_{L_U}^2}{2(B_N(m,n,a))^2}\right\}.$$

The above inequality and [1.80] yield inequality [1.85]. □

THEOREM 1.7.– Suggest that the assumptions of lemma 1.17 are satisfied. Assume that for some sequence $a = \{a_k, k = 1, 2, \ldots\}$, such that $a_k < a_{k+1}$, $a_k \to \infty$ as $k \to \infty$, for some integer $N \geq 1$, all $0 \leq s < 1$ the conditions

$$B_N(m,n,a) \to 0 \quad \text{as} \quad m,n \to \infty, \quad [1.86]$$

$$A_N(m,n,a,s) \to 0 \quad \text{as} \quad m,n \to \infty \qquad [1.87]$$

hold, where $B_N(m,n,a)$, $A_N(m,n,a,s)$ are defined in [1.80]. Then, a stochastic process

$$S(t) = S_1^\infty(t) = \sum_{k=1}^{\infty} \xi_k f_k(t)$$

almost surely belongs to the space $L_U(\Omega)$. If in this case, there exists a limit

$$B_N(m,a) = \lim_{n\to\infty} B_N(m,n,a) < \infty,$$

$$A_N(m,a,s) = \limsup_{n\to\infty} A_N(m,n,a,s) < \infty,$$

then for any $x > 0$, $0 \le s < 1$ and $m = 1, 2, \ldots$ the inequality

$$\mathbf{P}\{\|S_m^\infty(t)\|_{L_U} > x\} \qquad [1.88]$$

$$\le \exp\left\{-\frac{sx^2}{2(B_N(m,a))^2}\right\} \cdot \exp\left\{\frac{A_N(m,a,s)}{2B_N(m,a)} + \omega_N(s)\right\}$$

holds, where $\omega_N(s)$ is defined in [2.25].

PROOF.– It follows from [1.85] that for all $x > 0$, $0 \le s < 1$,

$$\mathbf{P}\{\|S_m^n(t)\|_{L_U} > x\}$$

$$\le \exp\left\{-\frac{sx^2 - A_N(m,n,a,s)B_N(m,n,a)}{2(B_N(m,n,a))^2}\right\} \cdot \exp\{\omega_N(s)\}.$$

Hence, [1.86] and [1.87] imply that for any $x > 0$

$$\mathbf{P}\{\|S_m^n(t)\|_{L_U} > x\} \to 0 \quad \text{as} \quad m,n \to \infty,$$

it means that

$$\|S_m^n(t)\|_{L_U} \to 0 \quad \text{as} \quad m,n \to \infty \quad \text{in probability.}$$

Then, there exists a sequence $m_k < n_k$, $k = 1, 2, \ldots$, such that

$$\|S_{m_k}^{n_k}(t)\|_{L_U} \to 0 \quad \text{as} \quad m_k \to \infty \quad \text{almost surely.}$$

That is $S(t)$ almost surely belongs to the space $L_U(\Omega)$. Since $S_m^\infty(t) = S(t) - S_1^{m-1}(t)$ almost surely also belongs to $L_U(\Omega)$, then it is easy to show that

$$\|S_m^\infty(t) - S_m^n(t)\|_{L_U} \to 0 \quad \text{as} \quad n \to \infty \quad \text{in probability.} \qquad [1.89]$$

Therefore, for arbitrary $x > 0$

$$\mathbf{P}\{\|S_m^n(t)\|_{L_U} > x\} \to \mathbf{P}\{\|S_m^\infty(t)\|_{L_U} > x\} \qquad \text{as} \quad n \to \infty.$$

From [1.89], inequality [1.85] and the conditions of the theorem follows inequality [1.88]. □

REMARK 1.11.– As $N = 1$ $A_1(m, n, a, s) = 0$, that is why to obtain theorem 1.7 it is sufficient to satisfy the condition

$$B_1(m, n, a) \to 0 \quad \text{at} \quad m, n \to \infty.$$

As $N > 1$ (see remark 1.3) from [1.87] follows [1.86].

REMARK 1.12.– It is easy to show (see remark 1.4) that under large enough x inequality [1.88] the better the larger N. But to apply this inequality under large N is too complicated, because of cumbrous calculation. It can be minimized with respect to s right-hand side of [1.88] as either $N = 1$ or $N = 2$ and obtain quite precise simple inequalities.

COROLLARY 1.11.– Suppose that the conditions of lemma 1.17 hold. If for some sequence $a = \{a_k, k = 1, 2, \ldots\}$, such that $a_k < a_{k+1}$, $a_k \to \infty$ as $k \to \infty$, for any $m = 1, 2, \ldots$ there exists a limit

$$B_1(m, a) = \lim_{n\to\infty} \sum_{k=m}^{n} b_{kn}(J_1(m, k, a))^{\frac{1}{2}} \qquad [1.90]$$

and the condition [1.86] as $N = 1$ is satisfied, then stochastic process $S(t)$ almost surely belongs to Orlicz space $L_U(\Omega)$. In this case for arbitrary $x \geq B_1(m, a)$, the inequality

$$\mathbf{P}\{\|S_m^\infty(t)\|_{L_U} > x\}$$
$$\leq \exp\Big\{\frac{1}{2}\Big\}\frac{x}{B_1(m, a)} \exp\Big\{-\frac{x^2}{2(B_1(m, a))^2}\Big\} \qquad [1.91]$$

holds.

PROOF.– The condition [1.87] of theorem 1.7 holds because of $A_1(m, n, a, s) = 0$. That is why the assertion of the corollary about $S(t)$ almost surely belonging to the space $L_U(\Omega)$ is carried out. Inequality [1.88] in this case has a view (see example 1.5)

$$\mathbf{P}\{\|S_m^\infty(t)\|_{L_U} > x\} \leq \frac{1}{\sqrt{1-s}} \exp\Big\{-\frac{sx^2}{2(B_1(m, a))^2}\Big\}. \qquad [1.92]$$

Minimizing [1.92] with respect to $0 \leq s < 1$, i.e. setting $s = 1 - \frac{(B_1(m,a))^2}{x^2}$, we obtain inequality [1.91] as $x \geq B_1(m, a)$. □

COROLLARY 1.12.– Let the assumptions of lemma 1.17 hold true. If for some sequence $a = \{a_k, k = 1, 2, \ldots\}$, such that $a_k < a_{k+1}$, $a_k \to \infty$ as $k \to \infty$, and for any $m = 1, 2, \ldots$ there exist the limits

$$B_2(m, a) = \lim_{n\to\infty} \sum_{k=m}^{n} b_{kn}(J_2(m, k, a))^{\frac{1}{2}} < \infty,$$

$$C_2(m, a) = \limsup_{n\to\infty} \sum_{k=m}^{n} b_{kn} \frac{J_1(m, k, a)}{(J_2(m, k, a))^{\frac{1}{2}}} < \infty \qquad [1.93]$$

and

$$\sum_{k=m}^{n} b_{kn} \frac{J_1(m, k, a)}{(J_2(m, k, a))^{\frac{1}{2}}} \to 0 \quad \text{at} \quad m, n \to \infty \qquad [1.94]$$

holds, then stochastic process $S(t)$ almost surely belongs to Orlicz space $L_U(\Omega)$. And for any $x \geq (B_2(m, a) \cdot C_2(m, a))^{\frac{1}{2}}$, the inequality

$$\mathbf{P}\{\|S_m^\infty(t)\|_{L_U} > x\}$$
$$\leq \frac{(x^2 - C_2(m, a)B_2(m, a) + (B_2(m, a))^2)^{\frac{1}{2}}}{B_2(m, a)} \qquad [1.95]$$
$$\times \exp\left\{-\frac{x^2}{2(B_2(m, a))^2} + \frac{C_2(m, a)}{2B_2(m, a)}\right\}$$

holds true.

PROOF.– Corollary 1.12 follows from theorem 1.7 if we consider $N = 2$. Really, we can easily show that

$$B_2(m, n, a) = \sum_{k=m}^{n} b_{kn}(J_2(m, k, a))^{\frac{1}{2}},$$

$$A_2(m, n, a) = s \sum_{k=m}^{n} b_{kn} \frac{J_1(m, k, a)}{(J_2(m, k, a))^{\frac{1}{2}}}.$$

That is why [1.94] yields [1.86] and [1.87] as $N = 2$ (see remark 1.3). Therefore, $S(t)$ almost surely belongs to $L_U(\Omega)$. In this case, inequality [1.88] for arbitrary $x > 0$, $0 \leq s < 1$ and $m = 1, 2, \ldots$ has the following representation:

$$\mathbf{P}\{\|S_m^\infty(t)\|_{L_U} > x\}$$
$$\leq \frac{1}{\sqrt{1-s}} \exp\left\{-\frac{s}{2}\right\} \exp\left\{-\frac{sx^2}{2(B_2(m, a))^2}\right\} \exp\left\{\frac{sC_2(m, a)}{2B_2(m, a)}\right\}.$$

Minimizing right-hand side of last inequality with respect to $0 \le s < 1$, i.e. setting

$$s = 1 - \left(\frac{x^2 - C_2(m,a)B_2(m,a) + B_2^2(m,a)}{B_2^2(m,a)} \right)^{-1},$$

as $x^2 \ge C_2(m,a)B_2(m,a)$, inequality [1.95] is obtained. □

1.7. Strictly sub-Gaussian random series with uncorrelated or orthogonal items

In this section, the results of the previous section are applied to the series with uncorrelated or orthogonal items.

Consider stochastic series [1.78]

$$S(t) = \sum_{k=1}^{\infty} \xi_k f_k(t),$$

where $f = \{f_k(t), t \in T, \ k = 1, 2, \ldots\}$ is a family of functions from the space $L_U(T)$ that belongs to $D_U(c)$, and $\xi = \{\xi_k, k = 1, 2, \ldots\}$ is a family of strictly sub-Gaussian random variables.

This section deals with convergence rate of the series [1.78] if either random variables ξ_k are uncorrelated or the functions $f_k(t)$ are orthogonal. In this case, the estimates of previous section are essentially simplified.

THEOREM 1.8.– Consider random series (process) [1.78]. Suppose that the assumptions of lemma 1.17 are satisfied, random variables $\xi = \{\xi_k, k = 1, 2, \ldots\}$ are uncorrelated or the functions $f = \{f_k(t), t \in T, k = 1, 2, \ldots\}$ are orthogonal

$$\left(\int_T f_k(t) f_l(t)\, d\mu(t) = 0, \quad k \ne l \right);$$

$$\mathbf{E}\xi_k^2 = \sigma_k^2 > 0, \qquad \int_T |f_k(t)|^2\, d\mu(t) = b_k^2 > 0.$$

If there exists such sequence $a = \{a_k, k = 1, 2, \ldots\}$ that $0 \le a_k \le a_{k+1}$, $a_k \to \infty$ as $k \to \infty$, and

$$\sum_{k=1}^{\infty} c_k (a_k^{-1} - a_{k+1}^{-1}) \left(\sum_{j=1}^{k} \sigma_j^2 b_j^2 a_j^2 \right)^{\frac{1}{2}} < \infty, \qquad [1.96]$$

then stochastic process $S(t)$ almost surely belongs to the space $L_U(\Omega)$ and for any $x > \tilde{B}_1(m,a)$ the inequality

$$\mathbf{P}\{\|S_m^\infty(t)\|_{L_U} > x\} \tag{1.97}$$
$$\leq \exp\Big\{\frac{1}{2}\Big\}\frac{x}{\tilde{B}_1(m,a)}\exp\Big\{-\frac{x^2}{2(\tilde{B}_1(m,a))^2}\Big\}$$

holds, where

$$\tilde{B}_1(m,a) = \sum_{k=m}^{\infty} c_k(a_k^{-1} - a_{k+1}^{-1})\Big(\sum_{j=m}^{k}\sigma_j^2 b_j^2 a_j^2\Big)^{\frac{1}{2}}.$$

PROOF.– The assertion of theorem follows from corollary 1.11. Really, it is easy to check that under conditions of the theorem

$$J_1(m,k,a) = \sum_{j=m}^{k}\sigma_j^2 b_j^2 a_j^2.$$

That is why the equality

$$B_1(m,n,a) \tag{1.98}$$
$$= \sum_{k=m}^{n-1} c_k(a_k^{-1} - a_{k+1}^{-1})\Big(\sum_{j=m}^{k}\sigma_j^2 b_j^2 a_j^2\Big)^{\frac{1}{2}} + \frac{c_n}{a_n}\Big(\sum_{j=m}^{n}\sigma_j^2 b_j^2 a_j^2\Big)^{\frac{1}{2}}$$

holds. The condition [1.96] implies that

$$\sum_{k=m}^{n-1} c_k(a_k^{-1} - a_{k+1}^{-1})\Big(\sum_{j=m}^{k}\sigma_j^2 b_j^2 a_j^2\Big)^{\frac{1}{2}}$$
$$\leq \sum_{k=m}^{n-1} c_k(a_k^{-1} - a_{k+1}^{-1})\Big(\sum_{j=1}^{k}\sigma_j^2 b_j^2 a_j^2\Big)^{\frac{1}{2}} \to 0 \tag{1.99}$$

as $m, n \to \infty$. Similarly, the next relationships follow:

$$\frac{c_n}{a_n}\Big(\sum_{j=m}^{n}\sigma_j^2 b_j^2 a_j^2\Big)^{\frac{1}{2}} \leq \frac{c_n}{a_n}\Big(\sum_{j=1}^{n}\sigma_j^2 b_j^2 a_j^2\Big)^{\frac{1}{2}}$$
$$= \sum_{k=n}^{\infty}(a_k^{-1} - a_{k+1}^{-1})c_n\Big(\sum_{j=1}^{n}\sigma_j^2 b_j^2 a_j^2\Big)^{\frac{1}{2}} \tag{1.100}$$
$$\leq \sum_{k=n}^{\infty}(a_k^{-1} - a_{k+1}^{-1})c_k\Big(\sum_{j=1}^{n}\sigma_j^2 b_j^2 a_j^2\Big)^{\frac{1}{2}} \to 0$$

as $n \to \infty$. Hence, [1.98]–[1.100] yield that $B_1(m,n,a) \to 0$ as $n,m \to \infty$, it means that the condition [1.86] holds. It is clear that the condition [1.90] also holds true, that is

$$B_1(m,a) = \lim_{n\to\infty} \sum_{k=m}^{n} b_{kn}\left(\sum_{j=m}^{k} \sigma_j^2 b_j^2 a_j^2\right)^{\frac{1}{2}}$$

$$= \sum_{k=m}^{\infty} c_k(a_k^{-1} - a_{k+1}^{-1})\left(\sum_{j=m}^{k} \sigma_j^2 b_j^2 a_j^2\right)^{\frac{1}{2}} = \tilde{B}_1(m,a).$$

Inequality [1.97] follows from [1.91]. □

COROLLARY 1.13.– Consider random series [1.78] (process). Assume that either random variables $\xi = \{\xi_k, k = 1,2,\ldots\}$ are uncorrelated or the functions $f = \{f_k(t), t \in T, k = 1,2,\ldots\}$ are orthogonal, $\mathbf{E}\xi_k^2 = \sigma_k^2 > 0$, $\int_T |f_k(t)|^2 \, d\mu(t) = b_k^2 > 0$. Let the conditions

$$\sum_{k=1}^{\infty} \sigma_k^2 b_k^2 < \infty, \qquad [1.101]$$

$$\sum_{k=1}^{\infty} c_k \frac{\sigma_k^2 b_k^2}{\left(\sum_{s=k}^{\infty} \sigma_s^2 b_s^2\right)^{\frac{1}{2}}} < \infty \qquad [1.102]$$

be satisfied. Then, stochastic process $S(t)$ almost surely belongs to the space $L_U(\Omega)$ and for any $x > \check{B}_1(m)$ the inequality

$$\mathbf{P}\{\|S_m^{\infty}(t)\|_{L_U} > x\}$$
$$\le \exp\left\{\frac{1}{2}\right\} \frac{x}{\check{B}_1(m)} \exp\left\{-\frac{x^2}{2(\check{B}_1(m))^2}\right\} \qquad [1.103]$$

holds, where

$$\check{B}_1(m) = \sum_{k=m}^{\infty} c_k \sigma_k^2 b_k^2 \left[2\left(\sum_{s=k}^{\infty} \sigma_s^2 b_s^2\right)^{-1} - \left(\sum_{s=m}^{\infty} \sigma_s^2 b_s^2\right)^{-1}\right]^{\frac{1}{2}}.$$

PROOF.– Show that the corollary follows from theorem 1.8. Choose in [1.96] the sequence a_k as:

$$a_k = \left(\sum_{s=k}^{\infty} \sigma_s^2 b_s^2\right)^{-1}.$$

Estimate now $\tilde{B}_1(m,a)$ from [1.97] under chosen sequence a. It is easy to see that the following relationships hold true:

$$\sum_{j=m}^{k} \sigma_j^2 b_j^2 a_j^2$$

$$= \sum_{j=m}^{k} \frac{\sigma_j^2 b_j^2}{\left(\sum_{s=j}^{\infty} \sigma_s^2 b_s^2\right)^2} = \sum_{j=m}^{k-1} \frac{\sigma_j^2 b_j^2}{\left(\sum_{s=j}^{\infty} \sigma_s^2 b_s^2\right)^2} + \frac{\sigma_k^2 b_k^2}{\left(\sum_{s=k}^{\infty} \sigma_s^2 b_s^2\right)^2}$$

$$\le \sum_{j=m}^{k-1} \frac{\sum_{s=j}^{\infty} \sigma_s^2 b_s^2 - \sum_{s=j+1}^{\infty} \sigma_s^2 b_s^2}{\left(\sum_{s=j}^{\infty} \sigma_s^2 b_s^2\right)\left(\sum_{s=j+1}^{\infty} \sigma_s^2 b_s^2\right)} + \left(\sum_{s=k}^{\infty} \sigma_s^2 b_s^2\right)^{-1} \qquad [1.104]$$

$$= \sum_{j=m}^{k-1} \left(\left(\sum_{s=j+1}^{\infty} \sigma_s^2 b_s^2 \right)^{-1} - \left(\sum_{s=j}^{\infty} \sigma_s^2 b_s^2 \right)^{-1} \right) + \left(\sum_{s=k}^{\infty} \sigma_s^2 b_s^2 \right)^{-1}$$

$$= 2\left(\sum_{s=k}^{\infty} \sigma_s^2 b_s^2 \right)^{-1} - \left(\sum_{s=m}^{\infty} \sigma_s^2 b_s^2 \right)^{-1}.$$

[1.104] and equality $a_k^{-1} - a_{k+1}^{-1} = \sigma_k^2 b_k^2$ imply that

$$\tilde{B}_1(m,a) \le \check{B}_1(m). \qquad [1.105]$$

From [1.105] and [1.102] it follows that $\tilde{B}_1(1,a) < \infty$, it means that condition [1.96] of theorem 1.8 holds. Inequalities [1.97] and [1.105] provide [1.103], since the function $c(x) = x\exp\{-\frac{x^2}{2}\}$ monotonically decreases as $x > 1$. □

From theorem 1.8 follows corollary 1.14.

COROLLARY 1.14.– Let the assumptions of corollary 1.13 be satisfied. Instead of conditions [1.101] and [1.102] the following condition holds true: suppose that for some sequence $a = \{a_k, k = 1, 2, \ldots\}$, such that $0 \le a_k \le a_{k+1}$, $a_k \to \infty$ as $k \to \infty$,

$$\sum_{k=1}^{\infty} c_k (a_k^{-1} - a_{k+1}^{-1}) < \infty, \qquad [1.106]$$

$$\sum_{j=1}^{\infty} \sigma_j^2 b_j^2 a_j^2 < \infty.$$

Then stochastic process $S(t)$ almost surely belongs to space $L_U(\Omega)$ and for any $x > \check{B}_1(m,a)$ the inequality

$$\mathbf{P}\{\|S_m^\infty(t)\|_{L_U} > x\} \leq \exp\Big\{\frac{1}{2}\Big\}\frac{x}{\check{B}_1(m,a)}\exp\Big\{-\frac{x^2}{2(\check{B}_1(m,a))^2}\Big\} \qquad [1.107]$$

holds, where

$$\check{B}_1(m,a) = \Big(\sum_{j=m}^{\infty}\sigma_j^2 b_j^2 a_j^2\Big)^{\frac{1}{2}}\Big(\sum_{j=m}^{\infty}c_j(a_j^{-1} - a_{j+1}^{-1})\Big).$$

Finding the sequence $a = \{a_k, k = 1, 2, \ldots\}$, which satisfies equation [1.106] will be useful in the next lemma.

LEMMA 1.19.– Assume that there exists a function $c = \{c(u), u \geq 1\}$, such that $c(k) = c_k$ and $c(u)$ monotonically increases. Suppose that $a = \{a(u), u \geq 1\}$ is a function such that $a(u) > 0$, $a(u)$ monotonically increases, $a(u) \to \infty$ as $u \to \infty$ and there exists a derivative $a'(u)$. If the following integral converges

$$\int_1^\infty \frac{c(x)a'(x)}{a^2(x)}\,dx < \infty, \qquad [1.108]$$

then the condition [1.106] is satisfied for the sequence $a = \{a_k, k = 1, 2, \ldots\}$.

PROOF.– The assertion of the lemma follows from such inequalities:

$$\begin{aligned}\sum_{k=1}^{\infty} c_k\Big(\frac{1}{a_k} - \frac{1}{a_{k+1}}\Big) &= \sum_{k=1}^{\infty} c_k \int_k^{k+1} d\Big(-\frac{1}{a(x)}\Big)\\ &\leq \sum_{k=1}^{\infty}\int_k^{k+1} c(x)\,d\Big(-\frac{1}{a(x)}\Big)\\ &= \int_1^\infty c(x)\frac{a'(x)}{a^2(x)}\,dx < \infty.\end{aligned}$$

□

THEOREM 1.9.– Consider random series (process) [1.78]. Let the assumptions of lemma 1.17 hold. Assume that random variables $\xi = \{\xi_k, k = 1, 2, \ldots\}$ are uncorrelated and the functions $f = \{f_k(t), t \in T, k = 1, 2, \ldots\}$ are orthogonal;

$$\mathbf{E}\xi_k^2 = \sigma_k^2 > 0, \qquad \int_T |f_k(t)|^2\,d\mu(t) = b_k^2 > 0.$$

If there exists a sequence $a = \{a_k, k = 1, 2, \ldots\}$ such that $0 \le a_k \le a_{k+1}$, $a_k \to \infty$ as $k \to \infty$, and

$$\sum_{k=1}^{\infty} c_k \left(\frac{1}{a_k} - \frac{1}{a_{k+1}}\right) \left(\sum_{j=1}^{k} |\sigma_j b_j a_j|^{\frac{4}{3}}\right)^{\frac{3}{4}} < \infty, \qquad [1.109]$$

then stochastic process $S(t)$ almost surely belongs to the space $L_U(\Omega)$ and for any $x \ge (\tilde{B}_2(m,a)\tilde{C}_2(m,a))^{\frac{1}{2}}$, the following inequality holds true

$$\begin{aligned} &\mathbf{P}\{\|S_m^{\infty}(t)\|_{L_U} > x\} \\ &\le \frac{(x^2 - \tilde{C}_2(m,a)\tilde{B}_2(m,a) + (\tilde{B}_2(m,a))^2)^{\frac{1}{2}}}{\tilde{B}_2(m,a)} \qquad [1.110] \\ &\quad \times \exp\left\{-\frac{x^2}{2(\tilde{B}_2(m,a))^2} + \frac{\tilde{C}_2(m,a)}{2\tilde{B}_2(m,a)}\right\}, \end{aligned}$$

$$\tilde{B}_2(m,a) = \sum_{k=m}^{\infty} c_k \left(\frac{1}{a_k} - \frac{1}{a_{k+1}}\right) \left(\sum_{j=m}^{k} |\sigma_j b_j a_j|^4\right)^{\frac{1}{4}},$$

$$\tilde{C}_2(m,a) = \sum_{k=m}^{\infty} c_k \left(\frac{1}{a_k} - \frac{1}{a_{k+1}}\right) \left(\sum_{j=m}^{k} |\sigma_j b_j a_j|^{\frac{4}{3}}\right)^{\frac{3}{4}}.$$

PROOF.– The proof of theorem follows from corollary 1.12. Really, it is easy to see that under conditions of theorem

$$J_1(m,k,a) = \sum_{j=m}^{k} \sigma_j^2 b_j^2 a_j^2, \qquad J_2(m,k,a) = \left(\sum_{j=m}^{k} \sigma_j^4 b_j^4 a_j^4\right)^{\frac{1}{2}}.$$

The Hölder inequality yields that

$$\begin{aligned} &\sum_{k=m}^{n} b_{kn} \frac{J_1(m,k,a)}{(J_2(m,k,a))^{\frac{1}{2}}} \\ &\le \sum_{k=m}^{n} b_{kn} \frac{(\sum_{j=m}^{k} |\sigma_j b_j a_j|^4)^{\frac{1}{4}} (\sum_{j=m}^{k} |\sigma_j b_j a_j|^{\frac{4}{3}})^{\frac{3}{4}}}{(J_2(m,k,a))^{\frac{1}{2}}} \\ &= \sum_{k=m}^{n-1} c_k \left(\frac{1}{a_k} - \frac{1}{a_{k+1}}\right) \left(\sum_{j=m}^{k} |\sigma_j b_j a_j|^{\frac{4}{3}}\right)^{\frac{3}{4}} \\ &\quad + \frac{c_n}{a_n} \left(\sum_{j=m}^{k} |\sigma_j b_j a_j|^{\frac{4}{3}}\right)^{\frac{3}{4}}. \end{aligned}$$

From the inequality above and condition [1.109] as in the previous theorem it follows that [1.94] is satisfied. Inequality [1.95] implies [1.110], because of $B_2(m,a) = \tilde{B}_2(m,a)$, and from the Hölder inequality it can be obtained that $C_2(m,a) \leq \tilde{C}_2(m,a)$. □

COROLLARY 1.15.– Suggest that the assumptions of theorem 1.9 are satisfied. Let the following conditions instead of [1.109] hold true: for some sequence $a = \{a_k, k = 1,2,\ldots\}$ such that $0 \leq a_k \leq a_{k+1}$, $a_k \to \infty$ as $k \to \infty$, [1.101] is satisfied, the next series converges

$$\sum_{j=1}^{\infty} |\sigma_j b_j a_j|^{\frac{4}{3}} < \infty. \qquad [1.111]$$

Then, stochastic process $S(t)$ almost surely belongs to the space $L_U(\Omega)$ and for any $x \geq (\check{B}_2(m,a)\check{C}_2(m,a))^{\frac{1}{2}}$ the inequality

$$\mathbf{P}\{\|S_m^{\infty}(t)\|_{L_U} > x\} \leq \frac{(x^2 - \check{C}_2(m,a)\check{B}_2(m,a) + (\check{B}_2(m,a))^2)^{\frac{1}{2}}}{\check{B}_2(m,a)} \times \exp\left\{-\frac{x^2}{2(\check{B}_2(m,a))^2} + \frac{\check{C}_2(m,a)}{2\check{B}_2(m,a)}\right\} \qquad [1.112]$$

holds, where

$$\check{B}_2(m,a) = \left(\sum_{j=m}^{\infty} |\sigma_j b_j a_j|^4\right)^{\frac{1}{4}} \sum_{k=m}^{\infty} c_k \left(\frac{1}{a_k} - \frac{1}{a_{k+1}}\right),$$

$$\check{C}_2(m,a) = \left(\sum_{j=m}^{\infty} |\sigma_j b_j a_j|^{\frac{4}{3}}\right)^{\frac{3}{4}} \sum_{k=m}^{\infty} c_k \left(\frac{1}{a_k} - \frac{1}{a_{k+1}}\right).$$

1.8. Uniform convergence estimates of sub-Gaussian random series

In this section, the conditions and the rate of convergence of sub-Gaussian random series are given.

Let (T,ρ) be separable metric space. U is a σ-algebra of Borelean set on (T,ρ) and $\mu(\cdot)$ is σ-finite measure on (T,U); $C(T)$ is a space of continuous and bounded functions on (T,ρ) with norm $\|f(t)\|_C = \sup_{t\in T}|f(t)|$.

DEFINITION 1.11.– *We will say that the sequence of functions $\{f_k(t), k = 1,2,\ldots\}$ from $C(T)$ belongs to the class B, if the following conditions hold true:*

a) there exists a continuous function $c(t)$, such that $|c(t)| < 1$ and

$$\int_T |c(t)|\, d\mu(t) < \infty,$$

furthermore, for any $\varepsilon > 0$ there exists a compact $K_\varepsilon \subset T$ that outside of this compact inequality $|c(t)| < \varepsilon$ holds.

b) there exist continuous functions $q_n(\delta) \geq 0$, $\delta \in R$, $n = 1, 2, \ldots$, such that for each n and $\delta_1 < \delta_2$ we have $q_n(\delta_1) \leq q_n(\delta_2)$, $q_n(\delta) \to 0$, as $\delta \to 0$, $q_n(\delta) \to \infty$, $n \to \infty$, as $\delta > 0$, $q_{n_1}(\delta) < q_{n_2}(\delta)$ as $n_1 < n_2$, $\delta > 0$.

c) For any sequence of numbers $\{b_n, n = 1, 2, \ldots\}$ for all $t, s \in T$ and all n, the inequality

$$\left| c(t) \sum_{k=1}^{n} b_k f_k(t) - c(s) \sum_{k=1}^{n} b_k f_k(s) \right|$$
$$\leq \left\| c(t) \sum_{k=1}^{n} b_k f_k(t) \right\|_C q_n(\rho(t, s))$$

is satisfied.

REMARK 1.13.– In definition 1.11, sequence $q_n(\delta)$ and function $c(t)$ are the same for any sequence $\{b_k, k = 1, 2, \ldots\}$.

LEMMA 1.20.– Assume that $f = \{f_k(t), k = 1, 2, \ldots\}$ is a sequence from class B, and $R_n(t) = \sum_{k=1}^{n} b_k f_k(t)$. Then, for arbitrary $0 < \theta < 1$ there exists a set $A(\theta, \{b_k\})$ such that $\mu(A(\theta, \{b_k\})) \geq \delta_n(\theta)$, where

$$\delta_n(\theta) = \inf_{t \in T} \mu\left(s\colon\ \rho(t, s) < q_n^{(-1)}(\theta)\right),$$

($q_n^{(-1)}(\theta)$ is inverse function of $q_n(\theta)$) and for $x \in A(\theta, \{b_k\})$ inequality

$$|c(t) R_n(t)| \geq (1 - \theta) \|c(t) R_n(t)\|_C \qquad [1.113]$$

holds.

PROOF.– It follows from the properties of $c(t)$ and $f_k(t)$ that there exists a point t_0 such that $|c(t_0) R_n(t_0)| = \|c(t) R_n(t)\|_C$. To clarify uncertainty, we suggest that $c(t_0) R_n(t_0) > 0$. Then

$$c(t_0) R_n(t_0) - c(t) R_n(t) \leq c(t_0) R_n(t_0) q_n(\rho(t, t_0)).$$

Let $A(\theta, \{b_k\})$ be a set of points t such that $\rho(t, t_0) < q_n^{(-1)}(\theta)$. Then, for $t \in A(\theta, \{b_k\})$ the inequality

$$c(t_0)R_n(t_0) - c(t)R_n(t) \leq c(t_0)R_n(t_0)\theta$$

holds. It means that for $t \in A(\theta, \{b_k\})$ [1.113] carries out and $\mu(A(\theta, \{b_k\})) \geq \delta_n(\theta)$. □

DEFINITION 1.12.– *The measure $\mu(\cdot)$ in definition 1.11 is called admissible if the function $\delta_n(\theta)$ has a property $\delta_n(\theta) \to 0$ as $n \to \infty$, $\theta > 0$, and for any n $\delta_n(\theta) \to 0$ as $\theta \to 0$.*

REMARK 1.14.– It is clear that the function $\delta_n(\theta)$ has such properties: for any n $\delta_n(\theta_1) < \delta_n(\theta_2)$ as $\theta_1 < \theta_2$ and for any θ $\delta_{n_1}(\theta) < \delta_{n_2}(\theta)$ as $n_1 > n_2$.

Consider some examples of sequences from the class B with admissible measure μ.

EXAMPLE 1.11.– Let $T = [-b, b]$, $b > 0$, $\rho(t, s) = |t - s|$, and $\mu(\cdot)$ is Lebesgue measure,

$$f_n(x) = \sum_{k=-n}^{n} \left(a_k \cos \frac{\pi k x}{b} + c_k \sin \frac{\pi k x}{b} \right)$$

is a trigonometric polynom. The sequence of function $f_n(x)$ belongs to class B. Furthermore, $c(t) \equiv 1$, $q_n(\delta) = \frac{\pi n}{b}\delta$, $\delta_n(\theta) = \frac{b}{\pi n}\theta$. Really, since $R_n(x) = \sum_{k=1}^{n} b_k f_k(x)$ is trigonometric polynomial, then from the Taylor formulas and the Bernstein inequality (see [BUL 00]) follows the inequality

$$|R_n(x) - R_n(y)| = |R_n'(x)(x - y)| \leq \frac{n\pi}{b} \|R_n(x)\|_C |x - y|.$$

Hence, $q_n(\delta) = \frac{n\pi}{b}\delta$, $q_n^{(-1)}(\delta) = \frac{b}{n\pi}\delta$ and $\delta_n(\theta) = \frac{b}{n\pi}\theta$.

DEFINITION 1.13.– *A function of complex value z $f(z)$ is called an integer function of exponential type if for any complex z the following inequality holds true*

$$|f(z)| \leq A \exp\{B|z|\}, \qquad [1.114]$$

where the numbers $A > 0$ and $B > 0$ do not depend on z. The type of the function is defined by the formula

$$u = \overline{\lim_{|z| \to \infty}} \frac{\ln |f(z)|}{|z|}. \qquad [1.115]$$

EXAMPLE 1.12.– Let $T = R$, $\rho(t,s) = |t-s|$, $\mu(\cdot)$ is the Lebesgue measure, $f_{u_n}(z)$ is a function of exponential type u_n, bounded on real axis $u_n < u_{n+1}$, $n = 1, 2, \cdots$ (Definition 1.13). A sequence of functions $f_{u_n}(x)$, $n = 1, 2, \cdots$, belongs to the class B, moreover $c(t) \equiv 1$, $q_n(\delta) = u_n\delta$, $\delta_n(\theta) = \frac{1}{u_n}\theta$. Really, since a function $R_n(x) = \sum_{k=1}^{n} b_k f_{u_k}(x)$ is a function of exponential type u_n, then from the Taylor formula and the Bernstein inequality (see [BUL 00]), it follows

$$|R_n(x) - R_n(y)| = |R'_n(x)(x-y)| \leq u_n \|R_n(x)\|_C |x-y|.$$

Hence, $q_n(\delta) = u_n\delta$, $q_n^{(-1)}(\delta) = \frac{\delta}{u_n}$ and $\delta_n(\theta) = \frac{2}{u_n}\theta$.

EXAMPLE 1.13.– Let $T = [0, b]$, $b > 0$, $\rho(t,s) = |t-s|$, $\mu(\cdot)$ is Lebesgue measure, $B(t,s)$, $t, s \in [0,b]$, is continuous symmetric non-negative definite function, $x_k(t)$, $k = 1, 2, \cdots$, are an orthonormal eigenfunctions and λ_k are corresponding eigenvalues of integral equation

$$z(t) = \lambda \int_0^b B(t,s)z(s)\,ds.$$

Then, it can be shown that (see, [KOZ 07a]) the sequence of functions $x_n(t)$, $n = 1, 2, \cdots$, belongs to the class B with $q_n(\delta) = \lambda_n\sqrt{b}\omega_B(\delta)$, $c(t) \equiv 1$, where

$$\omega_B(\delta) = \sup_{|u-v|\leq\delta} \left(\int_0^b \left(B(u,x) - B(v,x)\right)^2 dx \right)^{\frac{1}{2}}.$$

Therefore, $q_n^{(-1)}(\delta) = \omega_B^{(-1)}\left(\frac{\delta}{\lambda_n\sqrt{b}}\right)$ and $\delta_n(\theta) = \omega_B^{(-1)}\left(\frac{\theta}{\lambda_n\sqrt{b}}\right)$.

Consider now the series

$$S(t) = \sum_{k=1}^{\infty} \xi_k f_k(t),$$

where $\xi = \{\xi_k, k = 1, 2, \cdots\}$ is a family of sub-Gaussian random variables, $f = \{f_k, k = 1, 2, \cdots\}$ is a sequence of functions from the class B with admissible measure μ. For $1 \leq m \leq n < \infty$, denote

$$S_m^n(t) = \sum_{k=m}^{n} f_k(t)\xi_k, \quad R_m^n(t) = \sum_{k=m}^{n} b_k f_k(t)\xi_k,$$

$$Z_m^n = \|\tau^2(R_m^n(t))\|_C^{\frac{1}{2}},$$

where $\{b_k, k = 1, 2, \cdots\}$ is some numerical sequence, $\tau(R_m^n(t))$ is sub-Gaussian standard of $R_m^n(t)$.

LEMMA 1.21.– For $n = 1, 2, \ldots$, and $\theta \in (0, 1)$, such that

$$(\delta_n(\theta))^{-1} \int_T |c(t)|\, d\mu(t) \geq 1,$$

for any $y > 0$, the inequality

$$\begin{aligned} &\mathbf{E} \exp\{y \|c(t) R_m^n(t)\|_C\} \\ &\leq \frac{2}{\delta_n(\theta)} \int_T |c(t)|\, d\mu(t) \exp\left\{ \frac{y^2}{2(1-\theta)^2} (Z_m^n)^2 \right\} \end{aligned} \qquad [1.116]$$

holds true.

PROOF.– It follows from lemma 1.20 that with probability 1

$$\begin{aligned} &\delta_n(\theta)\big(\exp\{y\|c(t)R_m^n(t)\|_C\} - 1\big) \\ &\leq \int_{A(\theta,\{b_k\})} \big(\exp\{y\|c(t)R_m^n(t)\|_C\} - 1\big)\, d\mu(t) \\ &\leq \int_{A(\theta,\{b_k\})} \left(\exp\left\{ \frac{y}{1-\theta} |c(t) R_m^n(t)| \right\} - 1 \right) d\mu(t) \\ &\leq \int_T \left(\exp\left\{ \frac{y}{1-\theta} |c(t) R_m^n(t)| \right\} - 1 \right) d\mu(t) \\ &\leq \int_T |c(t)| \left(\exp\left\{ \frac{y}{1-\theta} |R_m^n(t)| \right\} - 1 \right) d\mu(t). \end{aligned}$$

If we take mathematical expectation from both parts and take into account sub-Gaussian random variables

$$\mathbf{E} \exp\{\lambda|\xi|\} \leq 2 \exp\left\{ \frac{\lambda^2 \tau^2(\xi)}{2} \right\},$$

we obtain

$$\begin{aligned} &\mathbf{E} \exp\{y\|c(t)R_m^n(t)\|_C\} \\ &\leq \frac{1}{\delta_n(\theta)} \mathbf{E} \int_T |c(t)| \left(\exp\left\{ \frac{y}{1-\theta} |R_m^n(t)| \right\} - 1 \right) d\mu(t) + 1 \\ &\leq \frac{2}{\delta_n(\theta)} \int_T |c(t)|\, d\mu(t) \exp\left\{ \frac{y^2}{2(1-\theta)^2} (Z_m^n)^2 \right\} \\ &\qquad - \frac{1}{\delta_n(\theta)} \int_T |c(t)|\, d\mu(t) + 1 \\ &\leq \frac{2}{\delta_n(\theta)} \int_T |c(t)|\, d\mu(t) \exp\left\{ \frac{y^2}{2(1-\theta)^2} (Z_m^n)^2 \right\}. \end{aligned}$$

□

LEMMA 1.22.– For arbitrary $y > 0$, $\theta \in (0,1)$ such that

$$(\delta_m(\theta))^{-1} \int_T |c(t)|\, d\mu(t) \geq 1,$$

and for non-decreasing sequence $\{b_k, k = 1, 2, \cdots\}$, $b_k > 0$, the inequality

$$\mathbf{E}\exp\{y\|c(t)S_m^n(t)\|_C\} \qquad [1.117]$$

$$\leq 2\int_T |c(t)|\, d\mu(t) \exp\left\{\frac{y^2}{2(1-\theta)^2}(A_m^n)^2 + \frac{y}{(1-\theta)}D_m^n(\theta)\right\},$$

holds, where

$$A_m^n = \sum_{k=m}^{n-1}(b_k^{-1} - b_{k+1}^{-1})Z_m^k + b_n^{-1}Z_m^n,$$

$$D_m^n(\theta) = 2^{\frac{1}{2}}\left(\sum_{k=m}^{n-1}(b_k^{-1} - b_{k+1}^{-1})Z_m^k|\ln\delta_k(\theta)|^{\frac{1}{2}} + \frac{Z_m^n|\ln\delta_n(\theta)|^{\frac{1}{2}}}{b_n}\right).$$

PROOF.– Consider Abelian transform

$$S_m^n(t) = \sum_{k=m}^{n-1}(b_k^{-1} - b_{k+1}^{-1})R_m^k(t) + b_n^{-1}R_m^n(t).$$

Then

$$\mathbf{E}\exp\{y\|c(t)S_m^n(t)\|_C\} \leq \mathbf{E}\exp\left\{\sum_{k=m}^{n} yd_k\|c(t)R_m^k(t)\|_C\right\},$$

where

$$d_k = \begin{cases} b_k^{-1} - b_{k+1}^{-1}, & k = \overline{m, n-1}; \\ b_n^{-1}, & k = n. \end{cases}$$

From [1.116] and the *Hölder* inequality, for $\{\alpha_k\}$ such that $\sum_{k=m}^{n}\alpha_k^{-1} = 1$, $\alpha_k > 1$, it follows that

$$\mathbf{E}\exp\{y\|c(t)S_m^n(t)\|_C\}$$

$$\leq \prod_{k=m}^{n}\left(\mathbf{E}\exp\{y\alpha_k d_k\|c(t)R_m^k(t)\|_C\}\right)^{\frac{1}{\alpha_k}}$$

$$\leq \prod_{k=m}^{n}\left(\frac{2}{\delta_k(\theta)}\int_T |c(t)|\, d\mu(t)\exp\left\{\frac{y^2\alpha_k^2 d_k^2(Z_m^k)^2}{2(1-\theta)^2}\right\}\right)^{\frac{1}{\alpha_k}}$$

$$\leq 2\int_T |c(t)|\,d\mu(t)\exp\Big\{\sum_{k=m}^{n}\frac{y^2\alpha_k d_k^2(Z_m^k)^2}{2(1-\theta)^2}+\frac{1}{\alpha_k}|\ln\delta_k(\theta)|\Big\}.$$

Note

$$H_k^2=\frac{y^2 d_k^2(Z_m^k)^2}{2(1-\theta)^2},$$

$$\alpha_k=\frac{(|\ln\delta_k(\theta)|+\nu)^{\frac{1}{2}}}{H_k},$$

where $\nu>0$ such that $\sum\limits_{k=m}^{n}\alpha_k^{-1}=1$. Remark that under these conditions

$$1=\sum_{k=m}^{n}\alpha_k^{-1}=\sum_{k=m}^{n}\frac{H_k}{(|\ln\delta_k(\theta)|+\nu)^{\frac{1}{2}}}\leq\sum_{k=m}^{n}\frac{H_k}{\nu^{\frac{1}{2}}},$$

which means that $\nu\leq\big(\sum_{k=m}^{n}H_k\big)^2$. Hence,

$$\begin{aligned}
&\mathbf{E}\exp\{y\|c(t)S_m^n(t)\|_C\}\\
&\leq 2\int_T |c(t)|\,d\mu(t)\\
&\quad\times\exp\Big\{\sum_{k=m}^{n}\Big(\frac{(|\ln\delta_k(\theta)|+\nu)^{\frac{1}{2}}}{H_k}H_k^2+\frac{H_k|\ln\delta_k(\theta)|}{(\nu+|\ln\delta_k(\theta)|)^{\frac{1}{2}}}\Big)\Big\}\\
&=2\int_T |c(t)|\,d\mu(t)\\
&\quad\times\exp\Big\{\sum_{k=m}^{n}\frac{\nu H_k}{(\nu+|\ln\delta_k(\theta)|)^{\frac{1}{2}}}+2\sum_{k=m}^{n}\frac{H_k|\ln\delta_k(\theta)|}{(\nu+|\ln\delta_k(\theta)|)^{\frac{1}{2}}}\Big\}\\
&\leq 2\int_T |c(t)|\,d\mu(t)\exp\Big\{\Big(\sum_{k=m}^{n}H_k\Big)^2+2\sum_{k=m}^{n}H_k|\ln\delta_k(\theta)|^{\frac{1}{2}}\Big\}.
\end{aligned}$$

□

THEOREM 1.10.– If there exists a sequence $\{b_k, k=1,2,\cdots\}$, $b_k>0$, $b_k\leq b_{k+1}$, $b_k\to\infty$ as $k\to\infty$, that satisfies conditions: for any $m\geq 1$, $s\leq m$, $0<\theta<1$

$$\sum_{k=m}^{\infty}(b_k^{-1}-b_{k+1}^{-1})Z_s^k|\ln\delta_k(\theta)|^{\frac{1}{2}}<\infty,$$

$$\frac{Z_s^n|\ln\delta_n(\theta)|^{\frac{1}{2}}}{b_n}\to 0\quad\text{as}\quad n\to\infty,$$

then $\|c(t)(S(t) - S_1^n(t))\|_C \to 0$ as $n \to \infty$ in probability and for $x \geq \frac{D_m(\theta)}{1-\theta}$ and θ, such that

$$(\delta_n(\theta))^{-1} \int_T |c(t)|\, d\mu(t) \geq 1,$$

we have

$$\mathbf{P}\{\|c(t)S_m^\infty(t)\|_C > x\} \qquad [1.118]$$
$$\leq 2\int_T |c(t)|\, d\mu(t) \exp\left\{-\frac{1}{2A_m^2}\left(x - \frac{D_m(\theta)}{1-\theta}\right)^2 (1-\theta)^2\right\},$$

where

$$A_m = \sum_{k=m}^{\infty} (b_k^{-1} - b_{k+1}^{-1}) Z_m^k,$$

$$D_m(\theta) = 2^{\frac{1}{2}} \sum_{k=m}^{\infty} (b_k^{-1} - b_{k+1}^{-1}) Z_m^k |\ln \delta_k(\theta)|^{\frac{1}{2}}.$$

PROOF.– By the Chebyshev inequality and lemma 1.117 for all $y > 0$, the following relationship is obtained

$$\mathbf{P}\{\|c(t)S_m^n(t)\|_C > x\} \leq \frac{\mathbf{E}\exp\{y\|c(t)S_m^n(t)\|_C\}}{\exp\{yx\}}$$
$$\leq 2\int_T |c(t)|\, d\mu(t) \exp\left\{\frac{y^2}{2(1-\theta)^2}(A_m^n)^2 + y\frac{D_m^n(\theta)}{1-\theta} - yx\right\},$$

where A_m^n, $D_m^n(\theta)$ is defined in [1.117]. For $x \geq \frac{D_m^n(\theta)}{1-\theta}$, we can choose

$$y = \frac{1}{(A_m^n)^2}\left(x - \frac{D_m^n(\theta)}{1-\theta}\right)(1-\theta)^2.$$

Then, we have

$$\mathbf{P}\{\|c(t)S_m^n(t)\|_C > x\}$$
$$\leq 2\int_T |c(t)|\, d\mu(t) \exp\left\{-\frac{1}{2(A_m^n)^2}\left(x - \frac{D_m^n(\theta)}{1-\theta}\right)^2 (1-\theta)^2\right\} \to 0,$$

as $m \to \infty$, $n \to \infty$. Since $A_m^n \to 0$ and $D_m^n(\theta) \to 0$ as $m, n \to \infty$, then $\|c(t)S_m^n(t)\|_C \to 0$ as $m, n \to \infty$ in probability, therefore $\|c(t)(S(t) - S_1^n(t)\|_C \to 0$ in probability as $n \to \infty$. The estimation of convergence rate is obtained from the above inequality, if the limit is taken as $n \to \infty$. □

Let $r(u) > 0$, $u > 1$, be a monotonically non-decreasing function such that $r(\exp\{u\})$ as $u > 1$ convex, for example, $r(u) = u^\alpha$, $\alpha > 0$; $r(u) = (\ln u)^\alpha$, $\alpha > 1$.

LEMMA 1.23.– For any $y > 0$, non-decreasing sequence $\{b_k, k = 1, 2, \cdots\}$, such that $b_k > 0$, $b_k \to \infty$ as $k \to \infty$ and such $\theta \in (0,1)$, that $\delta_k(\theta) < 1$ and

$$\frac{1}{\delta_m(\theta)} \int_T |c(t)|\, d\mu(t) > 1$$

the inequality

$$\begin{aligned} &\mathbf{E} \exp\{y \|c(t) S_m^n(t)\|_C\} \\ &\leq 2 \int_T |c(t)|\, d\mu(t) \\ &\times \exp\left\{ \frac{y^2}{2(1-\theta)^2} \left(\left(\sum_{k=m}^{n-1} \left(\frac{1}{b_k} - \frac{1}{b_{k+1}} \right) \right) Z_m^k + \frac{1}{b_n} Z_m^n \right)^2 \right\} \\ &\times r^{(-1)} \left(\frac{\sum_{k=m}^{n-1} \left(\frac{1}{b_k} - \frac{1}{b_{k+1}} \right) Z_m^k r(\delta_k^{-1}(\theta)) + b_n^{-1} Z_m^n r(\delta_n^{-1}(\theta))}{\sum_{k=m}^{n-1} (b_k^{-1} - b_{k+1}^{-1}) Z_m^k + \frac{1}{b_n} Z_m^n} \right), \end{aligned} \qquad [1.119]$$

holds, where $r^{(-1)}(u)$ in an inverse function of $r(u)$.

PROOF.– Similarly to the proof of lemma 1.22 for any $\alpha_k > 0$, $k = m, m+1, \ldots, n$, such that $\sum\limits_{k=m}^{n} \alpha_k^{-1} = 1$, and $y > 0$ we obtain

$$\begin{aligned} &\mathbf{E} \exp\{y \|c(t) S_m^n(t)\|_C\} \\ &\leq 2 \int_T |c(t)|\, d\mu(t) \exp\left\{ \sum_{k=m}^{n} \left(\frac{y^2 \alpha_k d_k^2 (Z_m^k)^2}{2(1-\theta)^2} + \frac{1}{\alpha_k} |\ln \delta_k(\theta)| \right) \right\}. \end{aligned} \qquad [1.120]$$

$$\begin{aligned} \text{Since} \quad &\exp\left\{ \sum_{k=m}^{n} \alpha_k^{-1} |\ln \delta_k(\theta)| \right\} \\ &= r^{(-1)} \left[r \left(\exp\left\{ \sum_{k=m}^{n} \alpha_k^{-1} |\ln \delta_k(\theta)| \right\} \right) \right] \\ &\leq r^{(-1)} \left(\sum_{k=m}^{n} \alpha_k^{-1} r\{\delta_k^{-1}(\theta)\} \right), \end{aligned}$$

then from [1.120] inequality

$$\mathbf{E}\exp\{y\|c(t)S_m^n(t)\|_C\}$$
$$\leq 2\int_T |c(t)|\, d\mu(t)\exp\Big\{\sum_{k=m}^{n}\alpha_k H_k^2\Big\} r^{(-1)}\Big(\sum_{k=m}^{n}\alpha_k^{-1} r(\delta_k^{-1}(\theta))\Big)$$

is obtained,

$$H_k^2 = \frac{y^2 d_k^2 (Z_m^k)^2}{2(1-\theta)^2}.$$

If we put $\alpha_k = H_k^{-1}(\sum_{k=m}^{n} H_k)$, then the assertion of lemma is proved. □

THEOREM 1.11.– If there exists the sequence $\{b_k, k = 1, 2, \cdots\}$, $b_k > 0$, $b_k \leq b_{k+1}$, $b_k \to \infty$ as $k \to \infty$, and the next conditions are satisfied: for arbitrary $m \geq 1$, $s \leq m$, $\theta \in (0,1)$

$$\sum_{k=m}^{\infty}(b_k^{-1} - b_{k+1}^{-1}) Z_s^k r(\delta_k^{-1}(\theta)) < \infty,$$

$$\frac{Z_s^n r(\delta_n^{-1}(\theta))}{b_n} \to 0 \quad \text{as} \quad n \to \infty,$$

then $\|c(t)(S(t) - S_1^n(t))\|_C \to 0$ as $n \to \infty$ in probability. As $x > 0$ and θ such that $\delta_m(\theta) < 1$ and

$$\frac{1}{\delta_m(\theta)}\int_T |c(t)|\, d\mu(t) > 1,$$

the following inequality is fulfilled

$$\mathbf{P}\{\|c(t)S_m^\infty(t)\|_C > x\}$$
$$\leq 2\int_T |c(t)|\, d\mu(t)\exp\Big\{-\frac{x^2(1-\theta)^2}{2A_m^2}\Big\} \qquad [1.121]$$
$$\times\, r^{(-1)}\Big(\frac{1}{A_m}\sum_{k=m}^{\infty}(b_k^{-1} - b_{k+1}^{-1}) Z_m^k r(\delta_k^{-1}(\theta))\Big),$$

$$A_m = \sum_{k=m}^{n-1}(b_k^{-1} - b_{k+1}^{-1}) Z_m^k.$$

PROOF.– If the conditions of theorem 1.11 are satisfied, then the conditions of theorem 1.10 also hold true, that is the assertion about the convergence carries out.

Similarly to the proving of [1.118], the inequality [1.121] follows from [1.119] and the Chebyshev inequality if we put

$$y = \frac{x(1-\theta)^2}{\left(\sum_{k=m}^{n} d_k Z_m^k\right)^2}$$

and take a limit as $n \to \infty$. □

1.9. Convergence estimate of strictly sub-Gaussian random series in C(T)

In this section, the results of the previous section are improved for strictly sub-Gaussian stochastic series. Consider now T as R^n or a rectangle on R^n, $\mu(t)$ is a Lebesgue measure.

Let $C(T)$ be a space of continuous and bounded functions with a norm

$$\|f(t)\|_C = \sup_{t\in T} |f(t)|.$$

Consider a series

$$S(t) = \sum_{k=1}^{\infty} f_k(t)\xi_k,$$

where $\xi = \{\xi_k, k = 1, 2, \ldots\}$ is a family of strictly sub-Gaussian non-correlated random variables, $f = \{f_k, k = 1, 2, \ldots\}$ is a sequence of the functions from class B and $\mu(\cdot)$ is a Lebesgue measure. For function f, there exists constant $Q > 0$ such that for all $k = 1, 2, \ldots$

$$\sup_{t\in T} |f_k^2(t)| \leq Q^2.$$

Denote for $1 \leq m \leq n < \infty$

$$S_m^n(t) = \sum_{k=m}^{n} f_k(t)\xi_k, \quad R_m^n(t) = \sum_{k=m}^{n} b_k f_k(t)\xi_k,$$

$$Z_m^n = \left\|\tau(R_m^n(t))\right\|_C,$$

where $\{b_k, k = 1, 2, \ldots\}$, $b_k > 0$, is some numeric sequence.

REMARK 1.15.– For such series, the following estimates hold true

$$\mathbf{E}(S_m^n(t))^2 = \mathbf{E}\Big(\sum_{k=m}^{n} f_k(t)\xi_k\Big)^2 \leq Q^2 \sum_{k=m}^{n} \sigma_k^2,$$

$$(Z_m^n)^2 \leq Q^2 \sum_{k=m}^{n} b_k^2 \sigma_k^2, \qquad [1.122]$$

where $\sigma_k^2 = \mathbf{E}\xi_k^2$.

REMARK 1.16.– For simplicity, consider the case when $Q = 1$. And

$$\frac{1}{\delta_k(\theta)} \int_T |c(t)|\, dt > 1, \qquad k = 1, 2, \ldots,$$

The obtained estimates can be easily rewritten in the case of $Q \neq 1$ that will be carried out in examples.

THEOREM 1.12.– Let there exist the non-decreasing sequence $\{b_k, k = 1, 2, \ldots\}$, $b_k > 0$, $b_k \to \infty$ as $k \to \infty$, and the following condition is fulfilled

$$\sum_{k=1}^{\infty} (b_k^{-1} - b_{k+1}^{-1}) |\ln(\delta_k(\theta))|^{\frac{1}{2}} \Big(\sum_{s=m}^{k} b_s^2 \sigma_s^2\Big)^{\frac{1}{2}} < \infty. \qquad [1.123]$$

Then, $\|c(t)(S(t) - S_1^m(t))\|_C \to 0$ in probability as $m \to \infty$ and for any $y > 0$, $\theta \in (0, 1)$, such that

$$(\delta_m(\theta))^{-1} \int_T |c(t)|\, dt \geq 1,$$

the inequality

$$\mathbf{E}\exp\{y\|c(t)S_m^{\infty}(t)\|_C\} \qquad [1.124]$$

$$\leq 2 \int_T |c(t)|\, dt \exp\Big\{\frac{y^2}{2(1-\theta)^2} A_m^2 + \frac{y\sqrt{2}}{(1-\theta)} D_m\Big\}$$

is obtained, where

$$A_m = \sum_{k=m}^{\infty} (b_k^{-1} - b_{k+1}^{-1}) \Big(\sum_{s=m}^{k} b_s^2 \sigma_s^2\Big)^{\frac{1}{2}},$$

$$D_m = \sum_{k=m}^{\infty} (b_k^{-1} - b_{k+1}^{-1}) |\ln(\delta_k(\theta))|^{\frac{1}{2}} \Big(\sum_{s=m}^{k} b_s^2 \sigma_s^2\Big)^{\frac{1}{2}}.$$

PROOF.– It is easy to show that the second condition of theorem 1.10 follows from [1.123]

$$
\begin{aligned}
&b_n^{-1} Z_m^n \left|\ln(\delta_k(\theta))\right|^{\frac{1}{2}} \\
&\leq b_n^{-1}\left(\sum_{s=m}^{n} b_s^2\sigma_s^2\right)^{\frac{1}{2}} \left|\ln(\delta_n(\theta))\right|^{\frac{1}{2}} \\
&\leq \left(\sum_{s=m}^{n} b_s^2\sigma_s^2\right)^{\frac{1}{2}} \sum_{k=n}^{\infty}(b_k^{-1}-b_{k+1}^{-1})\left|\ln(\delta_n(\theta))\right|^{\frac{1}{2}} \\
&\leq \sum_{k=n}^{\infty}\left(\sum_{s=m}^{k} b_s^2\sigma_s^2\right)^{\frac{1}{2}} \left|\ln(\delta_k(\theta))\right|^{\frac{1}{2}}(b_k^{-1}-b_{k+1}^{-1}) \to 0
\end{aligned}
$$

as $n \to \infty$. Hence, uniform convergence in probability of the series $c(t)S(t)$ follows from theorem 1.10. From [1.117] and [1.122] it follows that

$$
\begin{aligned}
&\mathbf{E}\exp\{y\|c(t)S_m^n(t)\|_C\} \\
&\leq 2\int_T |c(t)|\,dt \exp\left\{\frac{y^2}{2(1-\theta)^2}(A_m^n)^2 + \frac{y\sqrt{2}}{(1-\theta)}D_m^n\right\},
\end{aligned}
$$

$$
A_m^n = \sum_{k=m}^{n-1}(b_k^{-1}-b_{k+1}^{-1})\left(\sum_{s=m}^{k} b_s^2\sigma_s^2\right)^{\frac{1}{2}} + b_n^{-1}\left(\sum_{s=m}^{n} b_s^2\sigma_s^2\right)^{\frac{1}{2}},
$$

$$
\begin{aligned}
D_m^n &= \sum_{k=m}^{n-1}(b_k^{-1}-b_{k+1}^{-1})\left|\ln(\delta_n(\theta))\right|^{\frac{1}{2}}\left(\sum_{s=m}^{k} b_s^2\sigma_s^2\right)^{\frac{1}{2}} \\
&\quad + b_n^{-1}\left(\sum_{s=m}^{n} b_s^2\sigma_s^2\right)^{\frac{1}{2}}\left|\ln(\delta_n(\theta))\right|^{\frac{1}{2}}
\end{aligned}
$$

Taking a limit as $n \to \infty$, we have [1.124]. □

LEMMA 1.24.– If $\xi = \{\xi_k, k = 1, 2, \ldots\}$ is a family of strictly sub-Gaussian non-correlated random variables and for $\beta \in (0, 1/2)$, the condition

$$
\sum_{k=1}^{\infty}\left|\ln(\delta_k(\theta))\right|^{\frac{1}{2}}\sigma_k^2\left(\sum_{s=k}^{\infty}\sigma_k^2\right)^{\beta-1} < \infty \qquad [1.125]
$$

is satisfied, then the assumption of theorem 1.12 holds and for $y \geq 1$, $\theta \in (0, 1)$ such that

$$
(\delta_m(\theta))^{-1}\int_T |c(t)|\,dt \geq 1,
$$

the estimate

$$\mathbf{E}\exp\left\{y\|c(t)S_m^{\infty}(t)\|_C\left(\sum_{s=m}^{\infty}\sigma_s^2\right)^{-\frac{1}{2}}\right\}$$

$$\leq 2\int_T |c(t)|\,dt \quad [1.126]$$

$$\times\exp\left\{\frac{y^2}{2(1-\theta)^2}+y^{\frac{4\beta+1}{2\beta+1}}\left(\frac{\pi}{\sqrt{2}(1-\theta)^2}+\frac{\sqrt{2}F_\beta}{1-\theta}\right)\right.$$

$$\left.+y^{\frac{4\beta}{2\beta+1}}\left(\frac{\pi^2}{4(1-\theta)^2}+\frac{2F_\beta}{1-\theta}\right)\right\},$$

where

$$F_\beta=\sum_{k=m}^{\infty}\left|\ln(\delta_k(\theta))\right|^{\frac{1}{2}}\sigma_k^2\left(\sum_{s=k}^{\infty}\sigma_s^2\right)^{\beta-1}$$

holds true.

PROOF.– For simplicity in proving we suppose that $\sum\limits_{s=m}^{\infty}\sigma_s^2 = 1$. As $\{b_k, k=1,2,\ldots\}$, $b_k>0$, choose the sequence

$$b_k=1+y^{-p}\left(\left(\sum_{s=k}^{\infty}\sigma_s^2\right)^{-1}-1\right),$$

where $p>0$ is some number. If for the sequence condition [1.123] is satisfied, then theorem 1.12 holds true. Let us estimate the right-hand side of [1.124]. From the Minkowski inequality, we have:

$$\left(\sum_{k=m}^{n}b_k^2\sigma_k^2\right)^{\frac{1}{2}}=\left(\sum_{k=m}^{n}\left(1+\frac{1}{y^p}\left(\left(\sum_{s=k}^{\infty}\sigma_s^2\right)^{-1}-1\right)\right)^2\sigma_k^2\right)^{\frac{1}{2}}$$

$$\leq\left(\sum_{k=m}^{n}\sigma_k^2\right)^{\frac{1}{2}}+\frac{1}{y^p}\left(\sum_{k=m}^{n}\left(\left(\sum_{s=k}^{\infty}\sigma_s^2\right)^{-1}-1\right)^2\sigma_k^2\right)^{\frac{1}{2}}$$

$$\leq 1+\frac{1}{y^p}\left|\sum_{k=m}^{n-1}\sigma_k^2\left(\left(\sum_{s=k}^{\infty}\sigma_s^2\right)^{-1}-1\right)^2+\sigma_n^2\left(\left(\sum_{s=n}^{\infty}\sigma_s^2\right)^{-1}-1\right)^2\right|^{\frac{1}{2}}$$

$$\leq 1+\frac{1}{y^p}\left|\sum_{k=m}^{n-1}\int_{\sum_{s=k+1}^{\infty}\sigma_s^2}^{\sum_{s=k}^{\infty}\sigma_s^2}\left(\left(\sum_{s=k}^{\infty}\sigma_s^2\right)^{-1}-1\right)^2dx\right.$$

$$\left.+\frac{\sigma_n^2}{\sum_{s=n}^{\infty}\sigma_s^2}\left(\left(\sum_{s=n}^{\infty}\sigma_s^2\right)^{-1}-1\right)\right|^{\frac{1}{2}}$$

$$\leq 1+\frac{1}{y^p}\left|\sum_{k=m}^{n-1}\int_{\sum_{s=k+1}^{\infty}\sigma_s^2}^{\sum_{s=k}^{\infty}\sigma_s^2}\frac{dx}{x^2}+\left(\sum_{s=n}^{\infty}\sigma_s^2\right)^{-1}-1\right)\right|^{\frac{1}{2}}$$

$$\leq 1+\frac{1}{y^p}\left|\sum_{k=m}^{n-1}\left(\left(\sum_{s=k+1}^{\infty}\sigma_s^2\right)^{-1}-\left(\sum_{s=k}^{\infty}\sigma_s^2\right)^{-1}\right)+\left(\left(\sum_{s=n}^{\infty}\sigma_s^2\right)^{-1}-1\right)\right|^{\frac{1}{2}}$$

$$\leq 1+\frac{1}{y^p}\left|\left(\sum_{s=n}^{\infty}\sigma_s^2\right)^{-1}-\left(\sum_{s=m}^{\infty}\sigma_s^2\right)^{-1}+\left(\left(\sum_{s=n}^{\infty}\sigma_s^2\right)^{-1}-1\right)\right|^{\frac{1}{2}}$$

$$\leq 1+\frac{\sqrt{2}}{y^p}\left(\left(\sum_{s=n}^{\infty}\sigma_s^2\right)^{-1}-1\right)^{\frac{1}{2}}.$$

Then

$$A_m \leq \sum_{k=m}^{\infty}(b_k^{-1}-b_{k+1}^{-1})\left(1+\frac{\sqrt{2}}{y^p}\left(\left(\sum_{s=k}^{\infty}\sigma_s^2\right)^{-1}-1\right)^{\frac{1}{2}}\right)$$

$$= \sum_{k=m}^{\infty}(b_k^{-1}-b_{k+1}^{-1})+\frac{\sqrt{2}}{y^p}\sum_{k=m}^{\infty}\left(\left(\sum_{s=k}^{\infty}\sigma_s^2\right)^{-1}-1\right)^{\frac{1}{2}}$$

$$\times\left[\left(1+\frac{1}{y^p}\left(\left(\sum_{s=k}^{\infty}\sigma_s^2\right)^{-1}-1\right)\right)^{-1}\right.$$

$$\left.-\left(1+\frac{1}{y^p}\left(\left(\sum_{s=k+1}^{\infty}\sigma_s^2\right)^{-1}-1\right)\right)^{-1}\right].$$

Denote

$$E_k=\frac{1}{y^p}\left(\left(\sum_{s=k}^{\infty}\sigma_s^2\right)^{-1}-1\right),$$

then from above inequality follows that

$$A_m \leq 1+\frac{\sqrt{2}}{y^{\frac{p}{2}}}\sum_{k=m}^{\infty}\frac{(E_{k+1}-E_k)E_k^{\frac{1}{2}}}{(1+E_k)(1+E_{k+1})}$$

$$\leq 1+\frac{\sqrt{2}}{y^{\frac{p}{2}}}\sum_{k=m}^{\infty}\int_{E_k}^{E_{k+1}}\frac{E_k^{\frac{1}{2}}\,dx}{(1+E_k)(1+E_{k+1})}.$$

Since as $0<a<b$

$$\int_a^b\frac{dx}{(1+a)(1+b)}=\int_a^b\frac{dx}{(1+x)^2},$$

then

$$A_m \le 1 + \frac{\sqrt{2}}{y^{\frac{p}{2}}} \sum_{k=m}^{\infty} \int_{E_k}^{E_{k+1}} \frac{\sqrt{x}\,dx}{(1+x)^2}$$

$$\le 1 + \left(\frac{2}{y^p}\right)^{\frac{1}{2}} \int_0^{\infty} \frac{\sqrt{x}\,dx}{(1+x)^2} = 1 + \left(\frac{2}{y^p}\right)^{\frac{1}{2}} \frac{\pi}{2}. \qquad [1.127]$$

For D_m, the following inequality holds true

$$D_m \le \frac{\sqrt{2}}{y^p} \sum_{k=m}^{\infty} \left((1+E_k)^{-1} - (1+E_{k+1})^{-1}\right)$$

$$\times \left(\left(\sum_{s=k}^{\infty} \sigma_s^2\right)^{-1} - 1\right)^{\frac{1}{2}} \left|\ln(\delta_k(\theta))\right|^{\frac{1}{2}}$$

$$+ \sum_{k=m}^{\infty} \left((1+E_k)^{-1} - (1+E_{k+1})^{-1}\right) \left|\ln(\delta_k(\theta))\right|^{\frac{1}{2}}$$

$$\le L_1 + L_2,$$

where

$$L_1 = \sum_{k=m}^{\infty} \left((1+E_k)^{-1} - (1+E_{k+1})^{-1}\right) \left|\ln(\delta_k(\theta))\right|^{\frac{1}{2}}$$

$$L_2 = \frac{\sqrt{2}}{y^p} \sum_{k=m}^{\infty} \left((1+E_k)^{-1} - (1+E_{k+1})^{-1}\right)$$

$$\times \left(\left(\sum_{s=k}^{\infty} \sigma_s^2\right)^{-1} - 1\right)^{\frac{1}{2}} \left|\ln(\delta_k(\theta))\right|^{\frac{1}{2}}$$

$$\le L_1 + L_2.$$

Since $\beta \in (0, \frac{1}{2}]$, then

$$L_1 \le \sum_{k=m}^{\infty} \left|\ln(\delta_k(\theta))\right|^{\frac{1}{2}} \frac{E_{k+1} - E_k}{(1+E_k)^{\beta}(1+E_{k+1})}$$

$$\le \sum_{k=m}^{\infty} \left|\ln(\delta_k(\theta))\right|^{\frac{1}{2}}$$

$$\times \frac{y^{p\beta}\sigma_k^2\left(\sum_{s=k}^{\infty}\sigma_s^2\right)^{\beta-1}}{\left(y^p \sum\limits_{s=k}^{\infty}\sigma_s^2 + 1 - \sum\limits_{s=k}^{\infty}\sigma_s^2\right)^{\beta}\left(y^p \sum\limits_{s=k+1}^{\infty}\sigma_s^2 + 1 - \sum\limits_{s=k+1}^{\infty}\sigma_s^2\right)},$$

Since $y \geq 1$, then $y^p \sum_{s=k}^{\infty} \sigma_s^2 + 1 - \sum_{s=k}^{\infty} \sigma_s^2 \geq 1$, therefore,

$$L_1 \leq y^{p\beta} \sum_{k=m}^{\infty} |\ln(\delta_k(\theta))|^{\frac{1}{2}} \sigma_k^2 \left(\sum_{s=k}^{\infty} \sigma_s^2 \right)^{\beta-1}.$$

Similarly,

$$L_2 \leq \frac{\sqrt{2}}{y^p} \sum_{k=m}^{\infty} |\ln(\delta_k(\theta))|^{\frac{1}{2}} \frac{(E_{k+1} - E_k)\left(\sum_{s=k}^{\infty} \sigma_s^2\right)^{-\frac{1}{2}}}{(1+E_k)^{\beta+1/2}(1+E_{k+1})}$$

$$\leq \frac{\sqrt{2}}{y^p} \sum_{k=m}^{\infty} |\ln(\delta_k(\theta))|^{\frac{1}{2}} \frac{y^{p(1/2+\beta)}\sigma_k^2}{\left(\sum_{s=k}^{\infty} \sigma_s^2\right)^{1-\beta}}.$$

Put $p = 2(1+2\beta)^{-1}$ and define

$$F_\beta = \sum_{k=m}^{\infty} |\ln(\delta_n(\theta))|^{\frac{1}{2}} \frac{\sigma_k^2}{\left(\sum_{s=k}^{\infty} \sigma_s^2\right)^{1-\beta}}.$$

Then

$$D_m \leq \left(y^{p\beta} + \sqrt{2} y^{p(\beta-1/2)} \right) F_\beta = \left(y^{\frac{2\beta}{2\beta+1}} + \sqrt{2} y^{\frac{2\beta-1}{2\beta+1}} \right) F_\beta. \qquad [1.128]$$

Substituting [1.127] and [1.128] into [1.124], [1.126] is obtained. □

THEOREM 1.13.– Let $\beta \in (0, 1/2]$, $\theta \in (0,1)$ and m is large such that

$$(\delta_m(\theta))^{-1} \int_T |c(t)|\, dt \geq 1,$$

and condition [1.125] is satisfied, then $\|c(t)(S(t) - S_1^m(t))\|_C \to 0$ as $m \to \infty$ in probability and for $x \geq 2$ the estimate

$$\mathbf{P}\left\{ \|c(t) S_m^\infty(t)\|_C > x \left(\sum_{s=m}^{\infty} \sigma_s^2 \right)^{\frac{1}{2}} \right\}$$

$$\leq 2 \int_T |c(t)|\, dt \exp\left\{ -\frac{x^2}{2} + 1 + \sqrt{2} x^{\frac{4\beta+1}{2\beta+1}} \left(\bar{F}_\beta + \frac{\pi}{2} \right) \right. \qquad [1.129]$$

$$\left. + 2x^{\frac{4\beta}{2\beta+1}} \left(\bar{F}_\beta q_\beta(x) + x^{\frac{1-2\beta}{1+2\beta}} \frac{\pi^2}{8} \right) \right\}$$

hold true, where

$$\bar{F}_\beta = \sum_{k=m}^{\infty} \left| \ln \delta_k \left(1 - \left(1 - \frac{2}{x^2} \right)^{\frac{1}{2}} \right) \right|^{\frac{1}{2}} \sigma_k^2 \left(\sum_{s=k}^{\infty} \sigma_s^2 \right)^{\beta-1},$$

$$q_\beta(x) = \begin{cases} 1, & \beta \in \left[\frac{1}{6}, \frac{1}{2}\right], \\ x^{\frac{1-6\beta}{2(2\beta+1)}}, & \beta \in \left(0, \frac{1}{6}\right). \end{cases}$$

PROOF.– The convergence follows from lemma 1.24. If for $\theta \in (0,1)$ we put $y = x(1-\theta)^2$, then when $x(1-\theta)^2 \geq 1$ by the Chebyshev–Markov inequality and [1.126] we obtain

$$\mathbf{P}\left\{\|c(t)S_m^\infty(t)\|_C > x\left(\sum_{s=m}^\infty \sigma_s^2\right)^{\frac{1}{2}}\right\}$$

$$\leq 2\int_T |c(t)|\,dt \exp\left\{-\frac{x^2(1-\theta)^2}{2}\right.$$
$$+\sqrt{2}x^{\frac{4\beta+1}{2\beta+1}}(1-\theta)^{\frac{2(4\beta+1)}{2\beta+1}}\left(\frac{F_\beta}{1-\theta}+\frac{\pi}{2(1-\theta)^2}\right)$$
$$\left.+2x^{\frac{4\beta}{2\beta+1}}(1-\theta)^{2\frac{4\beta}{2\beta+1}}\left(\frac{F_\beta}{1-\theta}+\frac{\pi^2}{8(1-\theta)^2}\right)\right\} \quad [1.130]$$

$$\leq 2\int_T |c(t)|\,dt \exp\left\{-\frac{x^2(1-\theta)^2}{2}+\sqrt{2}x^{\frac{4\beta+1}{2\beta+1}}\left(F_\beta+\frac{\pi}{2}\right)\right.$$
$$\left.+2x^{\frac{4\beta}{2\beta+1}}\left(F_\beta(1-\theta)^{\frac{6\beta-1}{2\beta+1}}+\frac{\pi^2}{8}(1-\theta)^{\frac{2(2\beta-1)}{2\beta+1}}\right)\right\}.$$

For $x \geq 2$ set $(1-\theta)^2 = 1-2/x^2$, therefore $x(1-\theta)^2 = x-2/x \geq 1$. Obviously, in a case such that $(1-\theta)^2 \geq 1/x$, then

$$(1-\theta)^{\frac{-2(1-2\beta)}{2\beta+1}} \leq x^{\frac{1-2\beta}{2\beta+1}}, \quad [1.131]$$

$$(1-\theta)^{\frac{-2(1-6\beta)}{2(1+2\beta)}} \leq \begin{cases} 1, & 1-6\beta \leq 0, \\ x^{\frac{1-6\beta}{2(1+2\beta)}}, & 1-6\beta > 0. \end{cases} \quad [1.132]$$

Substituting θ, [1.131] and [1.132] into [1.130], we obtain [1.129]. □

Let $r(u) > 0$, $u > 1$, be such monotonically non-decreasing that as $u \geq 0$ the function $r(\exp\{u\})$ is convex.

THEOREM 1.14.– Let there exist non-decreasing sequence $\{b_k, k = 1, 2, \ldots\}$, $b_k > 0$, $b_k \to \infty$ as $k \to \infty$. Suppose that for all $\theta \in (0,1)$ the condition

$$\sum_{k=1}^\infty (b_k^{-1} - b_{k+1}^{-1})\tilde{Z}_1^k r(\delta_k^{-1}(\theta)) < \infty \quad [1.133]$$

is satisfied, where

$$(\tilde{Z}_1^k)^2 = \sum_{s=1}^{k} b_s^2 \sigma_s^2.$$

Then, $\|c(t)(S(t) - S_1^m(t))\|_C \to 0$ as $m \to \infty$ in probability and for such θ that $\delta_m(\theta) < 1$ and

$$(\delta_m(\theta))^{-1} \int_T |c(t)|\, dt \geq 1,$$

the following estimate

$$\mathbf{E} \exp\{y \|c(t) S_m^\infty(t)\|_C\} \leq 2 \int_T |c(t)|\, dt \exp\left\{\frac{y^2}{2(1-\theta)^2} A_m^2\right\} r^{-1}\left(\frac{D_m(r)}{A_m}\right) \qquad [1.134]$$

holds true, where

$$A_m = \sum_{k=m}^{\infty} (b_k^{-1} - b_{k+1}^{-1}) \left(\sum_{s=m}^{k} b_s^2 \sigma_s^2\right)^{\frac{1}{2}},$$

$$D_m(r) = \sum_{k=m}^{\infty} (b_k^{-1} - b_{k+1}^{-1}) r(\delta_k^{-1}(\theta)) \left(\sum_{s=m}^{k} b_s^2 \sigma_s^2\right)^{\frac{1}{2}}.$$

PROOF.– Remark that

$$\begin{aligned} & b_n^{-1} r(\delta_n^{-1}(\theta)) \left(\sum_{s=m}^{n} b_s^2 \sigma_s^2\right)^{\frac{1}{2}} \\ &= \sum_{k=n}^{\infty} (b_k^{-1} - b_{k+1}^{-1}) r(\delta_n^{-1}(\theta)) \left(\sum_{s=m}^{k} b_s^2 \sigma_s^2\right)^{\frac{1}{2}} \\ &\leq \sum_{k=n}^{\infty} (b_k^{-1} - b_{k+1}^{-1}) r(\delta_k^{-1}(\theta)) \left(\sum_{s=m}^{k} b_s^2 \sigma_s^2\right)^{\frac{1}{2}} \to 0 \end{aligned}$$

as $n \to \infty$, it means that under [1.133] the condition of theorem 1.11 holds true. From theorem 1.11 follows that $\|c(t)(S(t) - S_1^m(t))\|_C \to 0$ as $m \to \infty$ in probability. In the case where $n \to \infty$, relationship [1.119] yields [1.134]. □

LEMMA 1.25.– Let for any $\theta \in (0,1)$, $\beta \in (0, 1/2]$ and

$$\sum_{k=1}^{\infty} r(\delta_k^{-1}(\theta)) \sigma_k^2 \left(\sum_{s=k}^{\infty} \sigma_s^2\right)^{\beta - 1} < \infty. \qquad [1.135]$$

Then, the conditions of theorem 1.14 are satisfied and for $y \geq 1$, $\theta \in (0,1)$, such that $\delta_m(\theta) < 1$ and

$$(\delta_m(\theta))^{-1} \int_T |c(t)|\, dt \geq 1,$$

the following inequality holds true

$$\begin{aligned}
&\mathbf{E}\exp\left\{ y \|c(t) S_m^{\infty}(t)\|_C \left(\sum_{s=m}^{\infty} \sigma_s^2\right)^{-\frac{1}{2}} \right\} \\
&\leq 2 \int_T |c(t)|\, dt \qquad [1.136] \\
&\times \exp\left\{ \frac{y^2}{2(1-\theta)^2} \left(1 + \frac{\pi}{\sqrt{2}y^2}\right)^2 \right\} r^{(-1)}(G_\beta (y^{4\beta} + \sqrt{2})),
\end{aligned}$$

$$G_\beta = \sum_{k=m}^{\infty} r(\delta_k^{-1}(\theta)) \sigma_k^2 \left(\sum_{s=k}^{\infty} \sigma_s^2\right)^{\beta - 1}.$$

PROOF.– Without loss of generality suppose that $\sum_{k=m}^{\infty} \sigma_k^2 = 1$. As a sequence $\{b_k, k = 1, 2, \ldots\}$ consider

$$b_k = 1 + \frac{1}{y^4}\left(\left(\sum_{s=k}^{\infty} \sigma_s^2\right)^{-1} - 1 \right).$$

As in the proving of lemma 1.24, we obtain

$$\sum_{k=m}^{\infty} (b_k^{-1} - b_{k+1}^{-1}) \left(\sum_{s=m}^{k} b_s^2 \sigma_s^2\right)^{\frac{1}{2}} \leq 1 + \frac{\pi}{\sqrt{2}y^2},$$

$$\begin{aligned}
&\sum_{k=m}^{\infty} (b_k^{-1} - b_{k+1}^{-1}) r(\delta_k^{-1}(\theta)) \left(\sum_{s=m}^{k} b_s^2 \sigma_s^2\right)^{\frac{1}{2}} \qquad [1.137] \\
&\leq \left(y^{4\beta} + \sqrt{2} y^{4(\beta - \frac{1}{2})}\right) G_\beta < \infty.
\end{aligned}$$

Hence, the condition [1.133] is fulfilled, which means that theorem 1.14 holds true. The estimate [1.136] follows from [1.134], taking into account [1.137] and $y^{4(\beta-\frac{1}{2})} \leq 1$ as $y \geq 1$, and inequality

$$\sum_{k=m}^{\infty} (b_k^{-1} - b_{k+1}^{-1}) \left(\sum_{s=m}^{k} b_s^2 \sigma_s^2\right)^{\frac{1}{2}} \geq \sum_{s=m}^{k} \sigma_s^2 = 1.$$

□

THEOREM 1.15.– Let for any $\theta \in (0,1)$, such m that $\delta_m(\theta) < 1$ and

$$(\delta_m(\theta))^{-1} \int_T |c(t)|\, dt \geq 1,$$

$\beta \in (0, 1/2]$, the condition [1.135] is satisfied. Then

$$\|c(t)(S(t) - S_1^n(t))\|_C \to 0$$

as $n \to \infty$ in probability and for $x \geq 2$, the estimate

$$\mathbf{P}\Big\{\|c(t)S_m^\infty(t)\|_C > x\Big(\sum_{s=m}^{k} \sigma_s^2\Big)^{\frac{1}{2}}\Big\}$$

$$\leq 2\int_T |c(t)|\, dt \qquad [1.138]$$

$$\times \exp\Big\{-\frac{x^2}{2} + \Big(1 + \frac{\pi}{\sqrt{2}}\Big)^2\Big\} r^{(-1)}(\bar{G}_\beta(x^{4\beta} + \sqrt{2}))$$

holds true, where

$$\bar{G}_\beta = \sum_{k=m}^{\infty} \Big| r\Big(\delta_k^{-1}\Big(1 - \Big(1 - \frac{2}{x^2}\Big)^{\frac{1}{2}}\Big)\Big| \sigma_k^2 \Big(\sum_{s=k}^{\infty} \sigma_s^2\Big)^{\beta - 1}.$$

PROOF.– The convergence follows from lemma 1.25. From the Chebyshev–Markov inequality

$$\mathbf{P}\{\|c(t)S_m^\infty(t)\|_C > x\} \leq \frac{\mathbf{E}\exp\{y\|c(t)S_m^\infty(t)\|_C\}}{\exp\{yx\}}$$

and [1.136] follows the estimate [1.138], if for x such that $x(1-\theta)^2 \geq 1$, put $y = x(1-\theta)^2$, and for $x \geq 2$ put $(1-\theta)^2 = 1 - 2/x^2$. In this case, $x(1-\theta)^2 = x - 2/x \geq 1$. □

REMARK 1.17.– The obtained results can be carried over the case of the random series on finite interval. Really, let there exist

$$\inf_{|t|\leq T} |c(t)| = |c(\gamma)| > 0.$$

Then

$$\mathbf{P}\{\sup_{|t|\leq T} |S_m^\infty(t)| > x\} \leq \mathbf{P}\{\|c(t)S_m^\infty(t)\|_C > x|c(\gamma)|\}.$$

EXAMPLE 1.14.– Consider the series, $T = R$, $t \in T$,

$$\zeta(t) = \sum_{k=1}^{\infty} (\xi_k \cos(\lambda_k t) + \eta_k \sin(\lambda_k t)),$$

where $\{\xi_k, \eta_k, k = 1, 2, \ldots\}$ are independent strictly sub-Gaussian random variables, $\mathbf{E}\xi_k = \mathbf{E}\eta_k = 0$, $\mathbf{E}\xi_k^2 = \mathbf{E}\eta_k^2 = \sigma_k^2$, $k = 1, 2, \ldots$, $0 < \lambda_k \leq \lambda_{k+1}$, $\lambda_k \to \infty$, $k \to \infty$. It follow from example 1.12, the sequence of functions

$$\{c(t)\cos(\lambda_k t), c(t)\sin(\lambda_k t)\}$$

belongs to the class B, if

$$c(t) = \left(\frac{\sin(\varepsilon t)}{\varepsilon t}\right)^2, \qquad \varepsilon \in (0, 1/2).$$

Therefore, $\delta_n(\theta) = \frac{2\theta}{\lambda_n + 2\varepsilon}$ and $\int\limits_{-\infty}^{\infty} \left(\frac{\sin(\varepsilon t)}{\varepsilon t}\right)^2 dt = \frac{\pi}{\varepsilon}$, that is

$$\frac{1}{\delta_n(\theta)} \int_{-\infty}^{\infty} |c(t)|\, dt = \frac{\pi(\lambda_n + 2\varepsilon)}{2\theta\varepsilon} \geq 1.$$

1.10. The estimate of the norm distribution of L_p-processes

Let $X = \{X(t),\ t \in \mathbf{T}\}$ be a $L_p(\Omega)$ -process, $p \geq 1$. Denote $\rho_x(t, s) = \|X(t) - X(s)\|_{L_p}$. Suppose that the following conditions are fulfilled:

A_1) The process X is restricted in L_p, i.e.

$$\sup_{t \in \mathbf{T}} \|X(t)\|_{L_p} < \infty,$$

A_2) A space $(\mathbf{T}, \rho_x)$ is separable and the process X is separable on $(\mathbf{T}, \rho_x)$. Let $\varepsilon_0 = \sup\limits_{t \in \mathbf{T}} \|X(t)\|_{L_p}$.

By $N(\varepsilon) = N_{\rho_x}(\mathbf{T}, \varepsilon)$ and $H(\varepsilon) = \ln N(\varepsilon)$ denote a metric massiveness and metric entropy of parametric set $\mathbf{T}$ with respect to the pseudometric ρ_x, respectively.

THEOREM 1.16.– [BUL 00] Let an $L_p(\Omega)$-process X satisfy conditions A_1) and A_2). Suppose that $\int\limits_0^{\varepsilon_0} N^{\frac{1}{p}}(\varepsilon) d\varepsilon < \infty$, then $\left(\mathbf{E}\left(\sup\limits_{t \in \mathbf{T}} |X(t)|\right)^p\right)^{\frac{1}{p}} \leq B_p$ and for all $x > 0$

$$P\left\{\sup_{t \in \mathbf{T}} |X(t)| \geq x\right\} \leq \frac{B_p^p}{x^p},$$

where

$$B_p = \inf_{t \in \mathbf{T}} (\mathbf{E}|X(t)|^p)^{\frac{1}{p}} + \inf_{0<\theta<1} \frac{1}{\theta(1-\theta)} \int\limits_0^{\theta\gamma_0} N^{\frac{1}{p}}(\varepsilon) d\varepsilon,$$

where $\gamma_0 = \sup_{t,s\in\mathbf{T}} \rho_x(t,s) = \sup_{t,s\in\mathbf{T}} \|X(t) - X(s)\|_{L_p}$. Since $\gamma_0 \leq 2\varepsilon_0$, then we have the following.

COROLLARY 1.16.– Assume that $L_p(\Omega)$-process X satisfies conditions A_1) and A_2). Suppose that $\int_0^{2\varepsilon_0} N^{\frac{1}{p}}(\varepsilon)d\varepsilon < \infty$, then

$$\left(\mathbf{E}\left(\sup_{t\in\mathbf{T}} |X(t)|\right)^p\right)^{\frac{1}{p}} \leq \widetilde{B}_p$$

and for all $x > 0$

$$P\left\{\sup_{t\in\mathbf{T}} |X(t)| \geq x\right\} \leq \frac{\widetilde{B}_p^p}{x^p},$$

where

$$\widetilde{B}_p = \inf_{t\in\mathbf{T}} (\mathbf{E}|X(t)|^p)^{\frac{1}{p}} + \inf_{0<\theta<1} \frac{1}{\theta(1-\theta)} \int_0^{2\theta\varepsilon_0} N^{\frac{1}{p}}(\varepsilon)d(\varepsilon).$$

2

Simulation of Stochastic Processes Presented in the Form of Series

In this chapter, the results of the first chapter are applied to construct the models of random processes that allow for the representations of either Gaussian or strictly sub-Gaussian series. In section 2.1, the general principles of modeling techniques are considered. Section 2.2 is devoted to the models construction of stochastic processes using their Karhunen–Loéve expansion. The models obtained approximate the processes with a certain reliability and accuracy in the spaces $L_p(T)$ and $C(T)$ and some Orlicz spaces $L_U(T)$, where T is an interval. All the models considered in other sections of this chapter also approximate the processes in the same functional spaces. Section 2.3 deals with the models of stochastic processes applying their representation in the form of a Fourier series. The disadvantage of these models is that the items of these models in contrast to all other models are dependent. In section 2.4, models of stationary processes with a discrete spectrum are discussed. In section 2.5, the models of stationary random processes that allow for representation as a series of independent items are investigated. The models of this chapter are considered in the books of [KOZ 99b] and [KOZ 07a] and papers of [KOZ 88, ZEL 88, RYA 90, RYA 91] and [TRI 91].

2.1. General approaches for model construction of stochastic processes

Let a stochastic process $X = \{X(t), t \in T\}$ be represented in the form of the series

$$X(t) = \sum_{k=1}^{\infty} \xi_k f_k(t),$$

that converges in mean square. We say that $X_M = \{X_M(t), t \in T\}$ is a model of the process X if

$$X_M(t) = \sum_{k=1}^{M} \xi_k f_k(t). \qquad [2.1]$$

Assume that stochastic process X and all X_M, $M = 1, 2, \ldots$ belong to some functional Banach space $A(T)$ with a norm $\|\cdot\|$. Fix two numbers α and δ $(0 < \alpha < 1$, $\delta > 0)$. Model X_M approximates X with reliability $1-\alpha$ and accuracy δ with respect to the norm of the space $A(T)$, if the following inequality holds true

$$\mathbf{P}\{\|X(t) - X_M(t)\| > \delta\} \leq \alpha. \qquad [2.2]$$

Therefore, for model construction it is necessary to find the number M that given δ and α inequality [2.2] is satisfied.

Suppose that the next inequality

$$\mathbf{P}\{\|X(t) - X_M(t)\| > \delta\} \leq W_M(\delta)$$

is established, where $W_M(\delta)$, $\delta > 0$ is a known function that monotonically decreases with respect to M and δ. If M is such number that $W_M(\delta) \leq \alpha$, then for the model $X_{M'}$, $M' \geq M$, inequality [2.2] is fulfilled. Hence, to construct the model X_M that approximates X with given reliability $1-\alpha$ and accuracy δ with respect to the norm of the space $A(T)$, it is enough to find such M (the least is desirable) that the inequality $W_M(\delta) \leq \alpha$ holds true.

If in representation [2.1] ξ_k, $k = 1, \ldots, M$ are independent strictly sub-Gaussian random variables with $\mathbf{E}\xi_k = 0$ and $\mathbf{E}\xi_k^2 = \sigma_k^2$, $k = 1, 2, \ldots, M$, then the simulation of ξ_k, $k = 1, \ldots, M$ provides the construction of M independent strictly sub-Gaussian random variables with $\mathbf{E}\eta_k = 0$, $\mathbf{E}\eta_k^2 = 1$, $k = 1, 2, \ldots, M$. Then, $\xi_k = \sigma_k \eta_k$, $k = 1, 2, \ldots, M$, is a required sequence.

If in the expansion of the process $X(t)$ in series random variables ξ_k are Gaussian (it means that $X(t)$ is centered Gaussian process), then the simulation approach that is given above allows to approximate the process $X(t)$ with given accuracy and reliability. If $X(t)$ is a strictly sub-Gaussian random process and in its representation in the form of series random variables ξ_k are independent with known distribution, then $X_M(t)$ approximates $X(t)$ with given accuracy and reliability. If the process $X(t)$ is strictly sub-Gaussian, but either ξ_k are dependent or their distribution is exactly unknown, then there are many processes that can be constructed. The approach above allows us to construct one of the model of such processes.

To simulate of such strictly sub-Gaussian random variables as ξ_k, the following independent copies of random variables can be used

$$\eta = \left(\frac{12}{n}\right)^{\frac{1}{2}} \sum_{i=1}^{n} \left(\alpha_i - \frac{1}{2}\right), \qquad [2.3]$$

where α_i, $i = 1, \ldots, n$, is a family of uniformly distributed on (0,1) independent random variables that is obtained by one of the generators of random variables [ROS 06, OGO 96]. As $n \to \infty$ an item η weakly converges to Gaussian one, as $n < \infty$ – it is strictly sub-Gaussian random variable, $\mathbf{E}\eta^2 = 1$.

2.2. Karhunen–Loéve expansion technique for simulation of stochastic processes

In this section, the simulation method of stochastic processes is considered that is based on series expansion of the process by the eigenfunctions of some integral equations.

Let $T = [0, b]$, $b > 0$, be an interval in $\mathbb{R}$, $X = \{X(t), t \in T\}$ be continuous in mean square stochastic process, $\mathbf{E}X(t) = 0$, $t \in T$, $B(t,s) = \mathbf{E}X(t)X(s)$, $t, s \in T$ is its correlation function. Clearly, that $B(t,s)$ is nonnegative-definite function. Since the process $X(t)$ is mean square continuous, then the function $B(t,s)$ is continuous on $T \times T$.

Consider an integral equation

$$z(t) = \lambda \int_T B(t,s) z(s)\, ds. \qquad [2.4]$$

It is a well-known fact (e.g. [TRI 60]) that integral equation [2.4] has the greatest countable family of eigenvalues. These numbers are non-negative. Let λ_n^2 be the eigenvalues and $z_n(t)$ be corresponding eigenfunctions of equation [2.4]. Numerate λ_n^2 in the increase order $0 < \lambda_1 \leq \ldots \leq \lambda_n \leq \lambda_{n+1} \leq \ldots$. It is known that $z_n(t)$ are orthogonal functions. That means that for the functions $z_n(t)$ the relationship

$$\int_T z_n(t) z_m(t)\, dt = \delta_m^n$$

holds, where δ_m^n is a Kronecker symbol. Note that the functions $z_n(t)$ are continuous as $t \in T$.

THEOREM 2.1.– A stochastic process $X = \{X(t), t \in T\}$ can be represented in the form of series

$$X(t) = \sum_{n=1}^{\infty} \xi_n z_n(t). \qquad [2.5]$$

Moreover, the series [2.5] converges in mean square, ξ_n are uncorrelated random variables: $\mathbf{E}\xi_n = 0$, $\mathbf{E}\xi_n\xi_m = \delta_m^n \lambda_n^{-2}$.

PROOF.– According to Mercer's theorem [TRI 60], the following representation holds true

$$B(t,s) = \sum_{n=1}^{\infty} \frac{z_n(t)z_n(s)}{\lambda_n^2}, \qquad [2.6]$$

where the series in right-hand side of [2.6] converges uniformly in regard to (t,s) on the set $T \times T$. The statement of theorem follows now from Karhunen theorem (see [GIK 04]). □

REMARK 2.1.– It follows from the Karhunen–Loéve theorem that if $X(t)$ is Gaussian stochastic processes, then all ξ_n in series expansion [2.5] are independent Gaussian random variables.

2.2.1. *Karhunen–Loéve model of strictly sub-Gaussian stochastic processes*

Let in expansion [2.5] ξ_n be independent strictly sub-Gaussian random variables such that $\mathbf{E}\xi_n^2 = \lambda_n^{-2}$. Then, by example 1.7 random process [2.5] is strictly sub-Gaussian with correlation function $B(t,s)$.

DEFINITION 2.1.– *A stochastic process $X_M = \{X_M(t),\ t \in T\}$, where*

$$X_M(t) = \sum_{n=1}^{M} \xi_n z_n(t)$$

is called the Karhunen–Loéve model (KL-model) of the process $X = \{X(t), t \in T\}$.

2.2.2. *Accuracy and reliability of the KL model in* $L_2(T)$

THEOREM 2.2.– A stochastic process X_M is KL model that approximates the process X with reliability $1-\alpha$, $0<\alpha<1$ and accuracy $\delta>0$ in $L_2(T)$, that is

$$\mathbf{P}\left\{\left(\int_0^b (X(t)-X_M(t))^2\,dt\right)^{\frac{1}{2}} > \delta\right\} \le \alpha, \qquad [2.7]$$

if for M the following inequalities are satisfied:

$$\delta^2 > \hat{J}_{(M+1)1},$$

$$\left(\frac{\delta^2-\hat{J}_{(M+1)1}}{\hat{J}_{(M+1)2}}+1\right)^{\frac{1}{2}} \exp\left\{-\frac{\delta^2-\hat{J}_{(M+1)1}}{2\hat{J}_{(M+1)2}}\right\} \le \alpha, \qquad [2.8]$$

where

$$\hat{J}_{(M+1)1} = \sum_{k=M+1}^{\infty} \lambda_k^{-2}, \qquad \hat{J}_{(M+1)2} = \left(\sum_{k=M+1}^{\infty} \lambda_k^{-4}\right)^{\frac{1}{2}}.$$

PROOF.– Since random variables ξ_n are independent and the functions z_n are orthogonal, then the statement of the theorem follows from inequality [1.60] of example 1.8. □

REMARK 2.2.– To obtain more precise estimation, the assertion of corollary 1.8 can be used.

2.2.3. *Accuracy and reliability of the KL model in* $L_p(T)$, $p>0$

THEOREM 2.3.– A stochastic process X_M is KL model that approximates the process X with reliability $1-\alpha$, $0<\alpha<1$ and accuracy $\delta>0$ in $L_p(T)$, that is

$$\mathbf{P}\left\{\left(\int_0^b |X(t)-X_M(t)|^p\,dt\right)^{\frac{1}{p}} > \delta\right\} \le \alpha, \qquad [2.9]$$

if M satisfies inequality

$$2\exp\left\{-\frac{\delta^2}{2\sigma_{M+1}^2 b^{\frac{2}{p}}}\right\} \le \alpha, \qquad [2.10]$$

where

$$\sigma_{M+1}^2 = \sup_{t\in[0,b]} \sum_{k=M+1}^{\infty} \lambda_k^{-2} z_k^2(t)$$

and

$$p^{\frac{1}{2}}\sigma_{M+1}b^{\frac{1}{p}} < \delta.$$

The statement of theorem follows from theorem 1.4 and corollary 1.9.

REMARK 2.3.– If for all $t \in T$ the inequalities $|z_k(t)| \leq d_k$ hold, then in the inequalities that define M we can put

$$\sigma^2_{M+1} = \sum_{k=M+1}^{\infty} \lambda_k^{-2} d_k^2.$$

REMARK 2.4.– In many cases, the value M can be reduced if the results of section 1.7 will be applied. Taking into account that the norms in $L_2(T)$ of the functions $z_k(t)$ can be significantly less than $\sup_{0\leq t\leq b}|z_k(t)|$.

THEOREM 2.4.– Let $a = \{a_k, k = 1, 2, \ldots\}$ be such sequence that $0 \leq a_k \leq a_{k+1}$, $a_k \to \infty$ as $k \to \infty$, $p > 2$. Denote

$$c_{np} = 2 \inf_{N\geq 1}\left(1 + \tilde{\Delta}\left(\frac{2b}{N}\right)\lambda_n\left(\frac{b}{2}\right)^{\frac{1}{2}}\right)\left(\frac{N}{2b}\right)^{\frac{1}{2}-\frac{1}{p}}, \tag{2.11}$$

where

$$\tilde{\Delta}(h) = \sup_{|z_1-z_2|\leq h}\left(\int_{-b}^{b}\left(\tilde{B}(z_1, s) - \tilde{B}(z_2, s)\right)^2 ds\right)^{\frac{1}{2}},$$

$\tilde{B}(t, s)$ is even and $2b$-periodic by t, s function, it is convergent to $B(t, s)$ over $0 \leq t, s \leq b$. If the series

$$\sum_{k=1}^{\infty} c_{kp}\left(\frac{1}{a_k} - \frac{1}{a_{k+1}}\right)\left(\sum_{j=1}^{k}(\lambda_j^{-1}a_j)^{\frac{4}{3}}\right)^{\frac{3}{4}}$$

converges, then stochastic process X_M is KL model that approximates the process X with given reliability $1 - \alpha$, $0 < \alpha < 1$ and accuracy $\delta > 0$ in $L_p(T)$, if M satisfies the inequalities

$$\delta > (\check{B}_2(M + 1, a)\check{C}_2(M + 1, a))^{\frac{1}{2}}, \tag{2.12}$$

$$\frac{(\delta^2 - \check{B}_2(M + 1, a)\check{C}_2(M + 1, a) + (\check{B}_2(M + 1, a))^2)^{\frac{1}{2}}}{\check{B}_2(M + 1, a)} \times \exp\left\{-\frac{\delta^2}{2(\check{B}_2(M + 1, a))^2} + \frac{\check{C}_2(M + 1, a)}{2\check{B}_2(M + 1, a)}\right\} \leq \alpha, \tag{2.13}$$

where

$$\check{B}_2(M+1,a) = \sum_{k=M+1}^{\infty} c_{kp}\left(\frac{1}{a_k} - \frac{1}{a_{k+1}}\right)\left(\sum_{j=M+1}^{k} (\lambda_j^{-1}a_j)^4\right)^{\frac{1}{4}},$$

$$\check{C}_2(M+1,a) = \sum_{k=M+1}^{\infty} c_{kp}\left(\frac{1}{a_k} - \frac{1}{a_{k+1}}\right)\left(\sum_{j=M+1}^{k} (\lambda_j^{-1}a_j)^{\frac{4}{3}}\right)^{\frac{3}{4}}.$$

The statement of the theorem follows from theorem 1.9 and from the fact that the family of the functions $z = \{z_k(t), t \in T = [0,b], k = 1,2,\dots\}$ belongs to the class $D_U(c)$ (Definition 1.10), where $U(x) = |x|^p$, $c_n = c_{np}$ is defined in [2.11].

COROLLARY 2.1.– Let $a = \{a_k, k = 1,2,\dots\}$ be such sequence that $0 \le a_k \le a_{k+1}$, $a_k \to \infty$ as $k \to \infty$, and the series

$$\sum_{k=1}^{\infty} c_{kp}\left(\frac{1}{a_k} - \frac{1}{a_{k+1}}\right) < \infty \qquad [2.14]$$

converges, then stochastic process X_M is the process KL model that approximates X with reliability $1-\alpha$, $0 < \alpha < 1$ and accuracy $\delta > 0$ in $L_p(T)$, if M satisfies inequalities [2.12] and [2.13], where $\check{B}_2(M+1,a)$ and $\check{C}_2(M+1,a)$ are defined as

$$\check{B}_2(M+1,a) = \sum_{k=M+1}^{\infty} c_{kp}\left(\frac{1}{a_k} - \frac{1}{a_{k+1}}\right)\left(\sum_{j=M+1}^{\infty} |\lambda_j^{-1}a_j|^4\right)^{\frac{1}{4}},$$

$$\check{C}_2(M+1,a) = \sum_{k=M+1}^{\infty} c_{kp}\left(\frac{1}{a_k} - \frac{1}{a_{k+1}}\right)\left(\sum_{j=M+1}^{\infty} |\lambda_j^{-1}a_j|^{\frac{4}{3}}\right)^{\frac{3}{4}}. \qquad [2.15]$$

2.2.4. *Accuracy and reliability of the KL model in* $L_U(T)$

THEOREM 2.5.– Let $U = \{U(x), x \in R\}$ be a C-function such that the conditions of theorem 1.5 are satisfied. Then, stochastic process X_M is KL model that approximates the process X with reliability $1-\alpha$, $0 < \alpha < 1$, and accuracy $\delta > 0$ in $L_U(T)$, that is

$$\mathbf{P}\left\{ \|X(t) - X_M(t)\|_{L_U} > \delta \right\} \le \alpha, \qquad [2.16]$$

if for M the following inequalities are fulfilled

$$\delta \ge \hat{b}\sigma_{M+1}(2 + U^{(-1)}(1))^{-2})^{\frac{1}{2}}, \qquad [2.17]$$

$$\exp\left\{\frac{1}{2}\right\}\frac{\delta U^{(-1)}(1)}{\hat{b}\sigma_{M+1}}\exp\left\{-\frac{\delta^2\left(U^{(-1)}(1)\right)^2}{2\sigma_{M+1}^2\hat{b}^2}\right\}\leq\alpha, \qquad [2.18]$$

where

$$\hat{b}=\max(b,1), \qquad \sigma_{M+1}^2=\sup_{t\in T}\sum_{k=M+1}^{\infty}\lambda_k^{-2}f_k^2(t)$$

or if

$$\sup_{t\in T}|f_k(t)|\leq d_k, \grave{o}\hat{\imath} \qquad \sigma_{M+1}^2=\sum_{k=M+1}^{\infty}\lambda_k^{-2}d_k^2.$$

The statement of theorem follows from theorem 1.6 and remark 1.9.

The conditions of the theorem for C-functions from examples 1.9 and 1.10 are satisfied. But for C-function $U_\alpha(x)=\exp\{|x|^\alpha\}-1$, where $\alpha>2$, the conditions of theorem 2.5 are not fulfilled.

The following theorem gives a possibility to consider an essentially wider class of Orlicz spaces. As for the spaces $L_p(T)$, for spaces $L_U(\Omega)$, for which the conditions of theorem 2.5 are satisfied, in some cases the estimates of the following theorem can be better than the estimates in theorem 2.5.

THEOREM 2.6.– Let $a=\{a_k, k=1,2,\ldots\}$ be a sequence such that $0\leq a_k\leq a_{k+1}$, $a_k\to\infty$ as $k\to\infty$. C-function $U(x)$ satisfies the condition: the functions $(U(x))^{\frac{1}{2}}$ and $U(\sqrt{|x|})$ are convex. Let

$$c_n(U)=2\inf_{N\geq 1}\left(1+\tilde{\Delta}\left(\frac{2b}{N}\right)\lambda_n\left(\frac{b}{2}\right)^{\frac{1}{2}}\right) \qquad [2.19]$$
$$\times\left(\frac{N}{2b}\right)^{\frac{1}{2}}\left(U^{(-1)}\left(\frac{N}{2b}\right)\right)^{-1},$$

where $\tilde{\Delta}(h)$ is defined in [2.11]. If the series

$$\sum_{k=1}^{\infty}c_k(U)\left(\frac{1}{a_k}-\frac{1}{a_{k+1}}\right)\left(\sum_{j=1}^{k}(\lambda_j^{-1}a_j)^{\frac{4}{3}}\right)^{\frac{3}{4}}$$

converges, then stochastic process X_M is KL model that approximates the process X with reliability $1-\alpha$, $0<\alpha<1$ and accuracy $\delta>0$ in $L_U(T)$, if M satisfies inequalities [2.12] and [2.13], where

$$\check{B}_2(M+1,a)=\sum_{k=M+1}^{\infty}c_k(U)\left(\frac{1}{a_k}-\frac{1}{a_{k+1}}\right)\left(\sum_{j=M+1}^{k}(\lambda_j^{-1}a_j)^4\right)^{\frac{1}{4}},$$

$$\check{C}_2(M+1,a) = \sum_{k=M+1}^{\infty} c_k(U)\left(\frac{1}{a_k} - \frac{1}{a_{k+1}}\right)\left(\sum_{j=M+1}^{k}\left(\frac{1}{\lambda_j}a_j\right)^{\frac{4}{3}}\right)^{\frac{3}{4}}.$$

The assertion of theorem follows from theorem 1.9 and the fact that the family of functions $z = \{z_k(t), t \in T = [0,b], k = 1,2,\ldots\}$ belongs to the class $D_U(c)$, where $c_n = c_n(U)$ are given in [2.19].

COROLLARY 2.2.– Let $a = \{a_k, k = 1,2,\ldots\}$ be a sequence such that $0 \le a_k \le a_{k+1}$, $a_k \to \infty$ as $k \to \infty$, and the series

$$\sum_{k=1}^{\infty} c_k(U)\left(\frac{1}{a_k} - \frac{1}{a_{k+1}}\right) < \infty \qquad [2.20]$$

converges, then stochastic process X_M is KL model that approximates the process X with reliability $1-\alpha$, $0 < \alpha < 1$, and accuracy $\delta > 0$ ό $L_U(T)$, if for M inequalities [2.12] and [2.13] hold true, where $\check{B}_2(M+1,a)$ and $\check{C}_2(M+1,a)$ are defined as

$$\check{B}_2(M+1,a) = \sum_{k=M+1}^{\infty} c_k(U)\left(\frac{1}{a_k} - \frac{1}{a_{k+1}}\right)\left(\sum_{j=M+1}^{\infty} |\lambda_j^{-1}a_j|^4\right)^{\frac{1}{4}}, \qquad [2.21]$$

$$\check{C}_2(M+1,a) = \sum_{k=M+1}^{\infty} c_k(U)\left(\frac{1}{a_k} - \frac{1}{a_{k+1}}\right)\left(\sum_{j=M+1}^{\infty} |\lambda_j^{-1}a_j|^{\frac{4}{3}}\right)^{\frac{3}{4}}.$$

2.2.5. *Accuracy and reliability of the KL model in C(T)*

Following from example 1.13, the sequence $z_n(t)$ belongs to the class B with

$$c(t) \equiv 1, \qquad \delta_n(\theta) = \omega_B^{(-1)}\left(\frac{\theta}{\lambda_n\sqrt{b}}\right), \qquad T = [0,b],$$

where

$$\omega_B(\delta) = \sup_{|u-v|\le\delta}\left(\int_0^b (B(u,x) - B(v,x))^2\,dx\right)^{\frac{1}{2}}.$$

Let $Q = \sup_{k=\overline{1,\infty}} |z_k(t)|$. Then, from theorem 1.13 follows theorem 2.7.

THEOREM 2.7.– Stochastic process X_M is KL model that approximates the process X with reliability $1 - \alpha$, $0 < \alpha < 1$, and accuracy $\delta > 0$ in $C(T)$, if for some $\beta \in \left(0, \frac{1}{2}\right]$ the number M satisfies conditions:

$$\frac{\delta}{QG_{M+1}} > 2, \qquad [2.22]$$

where

$$G_M = \left(\sum_{s=M}^{\infty} \sigma_s^2\right)^{\frac{1}{2}}.$$

For $\theta = 1 - \sqrt{1 - 2/x^2}$, where $x = \frac{\delta}{QG_{M+1}}$, the condition of remark 1.16 is fulfilled and inequality

$$\begin{aligned} 2b\exp\Bigg\{ -\frac{1}{2}\left(\frac{\delta}{QG_{M+1}}\right)^2 + 1 + \sqrt{2}\left(\frac{\delta}{QG_{M+1}}\right)^{\frac{4\beta+1}{2\beta+1}}\left(\bar{F}_\beta + \frac{\pi}{2}\right) \\ + 2\left(\frac{\delta}{QG_{M+1}}\right)^{\frac{4\beta}{2\beta+1}}\left(\bar{F}_\beta q_\beta\left(\frac{\delta}{QG_{M+1}}\right)\right. \\ \left. + \frac{\pi^2}{8}\left(\frac{\delta}{QG_{M+1}}\right)^{\frac{1-2\beta}{1+2\beta}}\right)\Bigg\} \leq \alpha \end{aligned} \qquad [2.23]$$

holds, where

$$\bar{F}_\beta = \sum_{k=M+1}^{\infty} \left| \ln\left(\omega_B^{(-1)}\left(\frac{1 - (1 - 2Q^2G_{M+1}^2\delta^{-2})^{\frac{1}{2}}}{\lambda_k\sqrt{b}}\right)\right)\right|^{\frac{1}{2}} \frac{\sigma_k^2}{G_k^{2(1-\beta)}};$$

$q_\beta(x)$ is defined in [1.129].

From theorem 1.15 follows:

THEOREM 2.8.– A stochastic process X_M is KL model that approximates the process X with reliability $1 - \alpha$, $0 < \alpha < 1$ and accuracy $\delta > 0$ in $C(T)$, that is

$$\mathbf{P}\left\{\sup_{t\in T} |X(t) - X_M(t)| > \delta\right\} \leq \alpha,$$

if for M the conditions of theorem 2.7 are satisfied, but instead of [2.23] inequality

$$\begin{aligned} 2b\exp\left\{-\frac{1}{2}\left(\frac{\delta}{QG_{M+1}}\right)^2 + \left(1 + \frac{\pi}{\sqrt{2}}\right)^2\right\} \\ \times r^{(-1)}\left(\bar{G}_\beta\left(\left(\frac{\delta}{QG_{M+1}}\right)^{4\beta} + \sqrt{2}\right)\right) \leq \alpha \end{aligned} \qquad [2.24]$$

holds true, where $r(u) > 0,\ u > 1,$ is any monotonically non-decreasing such function that for $u \geq 0$ the function $r(\exp\{u\})$ is convex,

$$\bar{G}_\beta = \sum_{k=M+1}^{\infty} \left| r\left(\left(\omega_B^{(-1)}\left(\frac{1-(1-2Q^2G_{M+1}^2\delta^{-2})^{\frac{1}{2}}}{\lambda_k\sqrt{b}}\right)\right)^{-1}\right)\right| \times \frac{\sigma_k^2}{G_k^{2(1-\beta)}},$$

$$G_k = \left(\sum_{s=k}^{\infty}\sigma_s^2\right)^{\frac{1}{2}}.$$

REMARK 2.5.– In theorem 2.8, a more precise estimation is used than in theorem 2.7. That is why the value M can be lower if theorem 2.8 is used. But the conditions, under which theorem 2.8 can be applied, are harder.

EXAMPLE 2.1.– Consider a KL model of Wiener process that approximates it with accuracy δ in the space $C(T),\ \ T = [0,1]$. Let us remind that Wiener process $W(t),\ t > 0$ is a zero-mean Gaussian one with the correlation function

$$\mathbf{E}W(t)W(s) = \min\{t,s\} = B(t,s).$$

Moreover, the sample path of this process is continuous with probability 1. Consider now an integral equation

$$Z(t) = \lambda \int_0^1 B(t,s)Z(s)ds$$

or substituting the correlation function $B(s,t)$, we have

$$Z(t) = \lambda \int_0^t sZ(s)ds + \lambda t \int_t^1 Z(s)ds.$$

It is easy to show (see [GIK 88]) that the eigenvalues of equality above are

$$\lambda_n^2 = \pi^2(n+\frac{1}{2})^2$$

and corresponding eigenfunctions are

$$\varphi_n(t) = \sqrt{2}\sin(\pi t(n+\frac{1}{2})).$$

Hence, the Kahrunen–Loève expansion of Wiener process has such representation:

$$W(t) = \sqrt{2}\sum_{n=0}^{\infty} \xi_n \frac{\sin((n+\frac{1}{2})\pi t)}{(n+\frac{1}{2})\pi},$$

where ξ_n, $n \geq 0$ are independent Gaussian random variables with mean $\mathbf{E}\xi_n = 0$ and variance $\mathbf{E}\xi_n^2 = 1$. The KL model of Wiener process then will be

$$W_N(t) = \sqrt{2}\sum_{n=0}^{N} \xi_n \frac{\sin((n+\frac{1}{2})\pi t)}{(n+\frac{1}{2})\pi}.$$

Now we find such N that the KL model $W_N(t)$ will approximate Wiener process $W(t)$ with reliability $1-\alpha$ and accuracy δ in the space $C([0,1])$. Let us use theorem 2.8. In our case

$$b = 1, \quad Q = \sqrt{2}, \quad \sigma_k^2 = ((k+\frac{1}{2})\pi)^{-2}$$

and

$$G_k^2 = \sum_{s=k}^{\infty} \sigma_s^2 = \sum_{s=k}^{\infty} \frac{1}{((s+\frac{1}{2})\pi)^2} \geq \sum_{s=k}^{\infty} \frac{1}{\pi^2}\left(\frac{1}{s+\frac{1}{2}} \cdot \frac{1}{s+\frac{3}{2}}\right)$$
$$= \sum_{s=k}^{\infty} \frac{1}{\pi^2}\left(\frac{1}{s+\frac{1}{2}} - \frac{1}{s+\frac{3}{2}}\right) = \frac{1}{\pi^2}\frac{1}{k+\frac{1}{2}}.$$

If $u > v$, then

$$\int_0^1 (B(u,x) - B(v,x))^2\, dx = \int_0^1 (\min(u,x) - \min(v,x))^2\, dx$$
$$= \int_0^v (x-x)^2 dx + \int_v^u (x-v)^2 dx + \int_u^1 (u-v)^2 dx$$
$$\leq \int_0^u (u-v)^2 dx + \int_u^1 (u-v)^2 dx$$
$$= (u-v)^2 u + (u-v)^2(1-u) = (u-v)^2.$$

It follows from the above relationship that the function $\omega_B(\gamma) \leq \gamma$ and $\omega_B^{(-1)}(\gamma) \geq \gamma$. Set the function $r(u) = u^w$, where $w > 0$, $w > 1 - \beta$ and the value $\beta \in (0, \frac{1}{2}]$ is defined in theorem 2.7. Then

$$\bar{G}_\beta \cong \sum_{k=M+1}^{\infty} \left(\frac{\delta^2}{\sqrt{2} G_{M+1}^2 \pi (k + \frac{1}{2})} \right)^w \times \frac{1}{G_k^{2(1-\beta)} \pi^2 (k + \frac{1}{2})}$$

$$\leq \frac{\delta^{2w}}{(\sqrt{2})^w G_{M+1}^{2w} \pi^{2+w}} \sum_{k=M+1}^{\infty} \frac{1}{(k + \frac{1}{2})^{w+1} G_k^{2(1-\beta)}}$$

$$\leq \frac{\delta^{2w} \pi^{2w} (M + \frac{3}{2})^w}{(\sqrt{2})^w \pi^{2+w}} \sum_{k=M+1}^{\infty} \frac{(\pi^2 (k + \frac{1}{2}))^{1-\beta}}{(k + \frac{1}{2})^{w+1}}$$

$$= \frac{\delta^{2w} \pi^{w-2\beta} (M + \frac{3}{2})^w}{2^{\frac{w}{2}}} \sum_{k=M+1}^{\infty} \frac{1}{(k + \frac{1}{2})^{w+\beta}}.$$

Since $w > 1 - \beta$, then it is easy to show that

$$\sum_{k=M+1}^{\infty} \frac{1}{(k + \frac{1}{2})^{w+\beta}} \leq \frac{1}{w + \beta - 1} \frac{1}{(M + \frac{1}{2})^{w+\beta-1}}.$$

Denote

$$\bar{G}_\beta^* = \frac{\delta^{2w} \pi^{w-2\beta} (M + \frac{3}{2})^w}{2^{\frac{w}{2}} (w + \beta - 1)(M + \frac{1}{2})^{w+\beta-1}}. \qquad [2.25]$$

Moreover, inequality [2.24] can be rewritten as

$$2 \exp\left\{ -\frac{1}{2} \frac{\delta^2}{2} (M + \frac{1}{2}) \pi^2 + (1 + \frac{\pi}{\sqrt{2}})^2 \right\}$$

$$\times \left(\bar{G}_\beta^* \left(\left(\frac{\delta}{\sqrt{2}} \pi^2 (M + \frac{3}{2}) \right)^{4\beta} + \sqrt{2} \right) \right)^{\frac{1}{w}} \leq \alpha,$$

where G_β^* is defined in [2.25]. Furthermore, if follows from [2.22] that the condition for $0 < \beta \leq \frac{1}{2}$

$$\delta \pi \sqrt{M + \frac{3}{2}} > 2\sqrt{2}$$

should be satisfied.

In the case $\beta = \frac{1}{2}$ and $w = 1$, the values M dependent on accuracy δ and reliability $1 - \alpha$ are found in environment for statistical computing R and are shown in the next table.

	$\alpha = 0.1$	$\alpha = 0.05$	$\alpha = 0.01$
$\delta = 0.1$	$1,045$	$1,076$	$1,148$
$\delta = 0.06$	$2,965$	$3,051$	$3,250$
$\delta = 0.01$	$114,741$	$117,818$	$124,935$

Table 2.1. *The result of simulation of Wiener process*

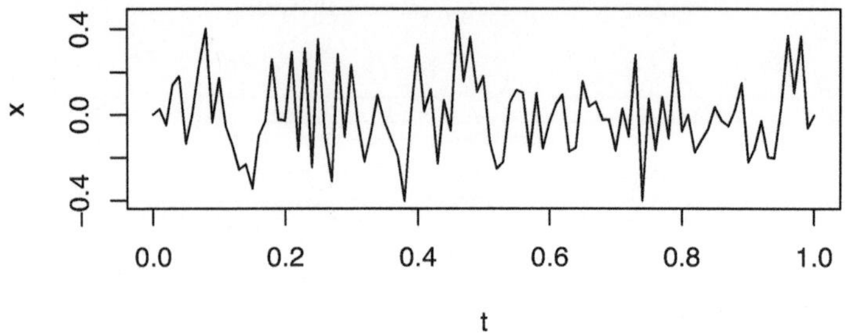

Figure 2.1. *The sample path of the model of Wiener process with accuracy 0.1 and reliability 0.90 in space* $C([0,1])$

2.3. Fourier expansion technique for simulation of stochastic processes

In this section, the simulation method of sub-Gaussian processes is considered that is based on Fourier transform of the processes.

Let $T = [0, b]$ be an interval such that $\mathbb{R}$, $X = \{X(t), t \in T\}$ is mean square continuous stochastic processes, $\mathbf{E}X(t) = 0$, $t \in T$, $B(t,s) = \mathbf{E}X(t)X(s)$, $t, s \in T$ is a correlation function of the process. The function $B(t,s)$ is continuous, therefore, it can be represented in the form of Fourier series, that converges in $L_2\big([0,b] \times [0,b]\big)$

$$B(t,s) = \sum_{n=0}^{\infty} \sum_{m=0}^{\infty} a_{mn} \cos \frac{\pi m t}{b} \cos \frac{\pi n s}{b}, \qquad [2.26]$$

$$a_{mn} = \frac{4}{b^2} r_{mn} \int_0^b \int_0^b B(t,s) \cos \frac{\pi m t}{b} \cos \frac{\pi n s}{b}\, dtds, \qquad [2.27]$$

$$r_{mn} = \begin{cases} \frac{1}{4}, \ m = n = 0, \\ \frac{1}{2}, \ m > 0, n = 0 \quad \text{or} \quad m = 0, n > 0, \\ 1, \ m > 0, n > 0. \end{cases}$$

Since the function $B(t,s)$ is non-negative definite, then $a_{mm} \geq 0$ for all m, so (see, for example, [LOE 60]) X can be written in the form of series, that converges in mean square

$$X(t) = \sum_{n=0}^{\infty} \xi_n \cos \frac{\pi n t}{b}, \qquad [2.28]$$

where ξ_n are random variables such that $\mathbf{E}\xi_n = 0$, $\mathbf{E}\xi_m\xi_n = a_{mn}$. Note that it is easy enough to check the equity $\mathbf{E}X(t)X(s) = B(t,s)$.

2.3.1. *Fourier model of strictly sub-Gaussian stochastic process*

In expansion [2.28] let $\xi = \{\xi_i, i = 0, 1, 2, \ldots\}$ be a strictly independent sub-Gaussian family of random variables. Then, by example 1.7, $X(t)$ is a strictly sub-Gaussian stochastic process with correlation function $B(t,s)$.

DEFINITION 2.2.– *Stochastic process $X_M = \{X_M(t),\, t \in T\}$, where*

$$X_M(t) = \sum_{n=0}^{M} \xi_n \cos \frac{\pi n t}{b}$$

is called the Fourier model (F-model) of the process $X = \{X(t), t \in T\}$.

2.3.2. *Accuracy and reliability of the F-model in* $L_2(T)$

THEOREM 2.9.– Stochastic process X_M is a F-model that approximates the process X with reliability $1-\alpha$, $0 < \alpha < 1$, and accuracy $\delta > 0$ 6 $L_2(T)$, if M satisfies inequalities:

$$\delta > A_{M+1}^{\frac{1}{2}}$$

and

$$\exp\left\{\frac{1}{2}\right\} \frac{\delta}{A_{M+1}^{\frac{1}{2}}} \exp\left\{-\frac{\delta^2}{2A_{M+1}}\right\} \leq \alpha, \qquad [2.29]$$

where $A_{M+1} = \dfrac{b}{2} \displaystyle\sum_{k=M+1}^{\infty} a_{kk}$.

PROOF.– Since the functions $\cos \frac{\pi n t}{b}$, $n = 1, 2, \ldots$ are orthogonal, then the statement of theorem 2.9 follows from corollary 1.7. □

REMARK 2.6.– To obtain more precise estimation, theorem 1.2 can be used.

2.3.3. *Accuracy and reliability of the F-model in* $L_p(T), p > 0$

THEOREM 2.10.– Stochastic process X_M is F-model that approximates the process X with reliability $1-\alpha$, $0<\alpha<1$, and accuracy $\delta > 0$ 6 $L_p(T)$, if M satisfies inequality [2.10], where

$$\sigma^2_{M+1} = \sup_{t\in T} \sum_{k=M+1}^{\infty} \sum_{l=M+1}^{\infty} a_{kl} \cos\frac{\pi kt}{b} \cos\frac{\pi lt}{b}$$

and inequality

$$p^{\frac{1}{2}} \sigma_{M+1} b^{\frac{1}{p}} < \delta.$$

theorem follows from theorem 1.4.

The following theorem gives the estimates that in some cases can be more precise than the estimates in theorem 2.10.

THEOREM 2.11.– Let $a = \{a_k, k = 1,2,\ldots\}$ be such sequence that $0 \le a_k \le a_{k+1}$, $a_k \to \infty$ as $k \to \infty$, $p > 2$. Let the series

$$\sum_{k=1}^{\infty} k^{\frac{1}{2}-\frac{1}{p}} \left(\frac{1}{a_k} - \frac{1}{a_{k+1}}\right) \left(\sum_{j=1}^{k} a_{jj} a_j^2\right)^{\frac{1}{2}} < \infty \qquad [2.30]$$

converge. Then stochastic process X_M is F-model that approximates the process X with reliability $1-\alpha$, $0<\alpha<1$, and accuracy $\delta > 0$ in $L_p(T)$, if M satisfies inequalities

$$\delta > \tilde{B}_1(M+1, a) \qquad [2.31]$$

$$\exp\left\{\frac{1}{2}\right\} \frac{\delta}{\tilde{B}_1(M+1,a)} \exp\left\{-\frac{\delta^2}{2(\tilde{B}_1(M+1,a))^2}\right\} \le \alpha, \qquad [2.32]$$

where

$$\tilde{B}_1(M+1,a) = \left(1+\frac{\pi}{2}\right)\left(\frac{b}{2}\right)^{\frac{1}{p}} \sum_{k=M+1}^{\infty} k^{\frac{1}{2}-\frac{1}{p}} \left(\frac{1}{a_k} - \frac{1}{a_{k+1}}\right) \left(\sum_{j=M+1}^{k} a_{jj} a_j^2\right)^{\frac{1}{2}}$$

The statement of theorem follows from theorem 1.8, since the functions $\cos\frac{\pi nt}{b}$, $n = 0,1,2,\ldots$ are orthogonal, $\int_0^b \cos^2\frac{\pi nt}{b}\,dt = \frac{b}{2}$, and the family of this functions belongs to the class $D_U(c)$, where $U(x) = |x|^p$, $c_n = \left(1+\frac{\pi}{2}\right)\left(\frac{2n}{b}\right)^{\frac{1}{2}-\frac{1}{p}}$.

COROLLARY 2.3.– If the series

$$\sum_{k=1}^{\infty} k^{\frac{1}{2}-\frac{1}{p}} a_{kk} \left(\sum_{s=k}^{\infty} a_{ss}^2\right)^{-\frac{1}{2}} < \infty \qquad [2.33]$$

converges, then stochastic process X_M is F-model that approximates the process X with reliability $1-\alpha$, $0<\alpha<1$ and accuracy $\delta>0$ in $L_p(T)$, if M satisfies inequalities [2.31] and [2.32], where

$$\begin{aligned}\tilde{B}_1(M+1,a) &= \tilde{B}_1(M+1)\\ &= \left(1+\frac{\pi}{2}\right)\left(\frac{b}{2}\right)^{\frac{1}{p}}\\ &\times \sum_{k=M+1}^{\infty} k^{\frac{1}{2}-\frac{1}{p}} a_{kk}\left[2\left(\sum_{s=M+1}^{\infty} a_{ss}^2\right)^{-1} - \left(\sum_{s=M+1}^{\infty} a_{ss}^2\right)^{-1}\right]^{\frac{1}{2}}.\end{aligned} \qquad [2.34]$$

The assertion of corollary follows from corollary 1.13 and theorem 2.11.

COROLLARY 2.4.– Let for some sequence $a=\{a_k, k=1,2,\ldots\}$, such that $0 \le a_k \le a_{k+1}$, $a_k \to \infty$ as $k\to\infty$, and

$$\sum_{k=1}^{\infty} k^{\frac{1}{2}-\frac{1}{p}} \left(\frac{1}{a_k}-\frac{1}{a_{k+1}}\right) < \infty, \qquad [2.35]$$

$$\sum_{j=1}^{\infty} a_{jj} a_j^2 < \infty. \qquad [2.36]$$

Then, the statement of theorem 2.11 holds true, if M satisfies inequalities [2.31] and [2.32], where

$$\begin{aligned}\check{B}_1(M+1,a) &= \left(1+\frac{\pi}{2}\right)\left(\frac{b}{2}\right)^{\frac{1}{p}}\\ &\times \sum_{k=M+1}^{\infty} k^{\frac{1}{2}-\frac{1}{p}}\left(\frac{1}{a_k}-\frac{1}{a_{k+1}}\right)\left(\sum_{j=M+1}^{\infty} a_{jj}a_j^2\right)^{\frac{1}{2}}.\end{aligned} \qquad [2.37]$$

Corollary 2.4 follows from corollary 1.14 and theorem 2.11.

REMARK 2.7.– Lemma 1.19 yields that condition [2.35] is satisfied by, for instance, such sequence: $a_k = k^\beta$, where β is any number that $\beta > \frac{1}{2} - \frac{1}{p}$. In this case,

$$\sum_{k=M+1}^{\infty} k^{\frac{1}{2}-\frac{1}{p}} \left(\frac{1}{a_k} - \frac{1}{a_{k+1}}\right) \leq \int_{M+1}^{\infty} x^{\frac{1}{2}-\frac{1}{p}}\, d\left(-\frac{1}{x^\beta}\right)$$
$$= \beta\left(\beta - \frac{1}{2} + \frac{1}{p}\right)^{-1} (M+1)^{\frac{1}{2}-\frac{1}{p}-\beta}.$$

Therefore, to occur the statement of corollary 2.4, we can in inequalities [2.31] and [2.32] put

$$\check{B}_1(M+1,a)$$
$$= \beta\left(1+\frac{\pi}{2}\right)\left(\frac{b}{2}\right)^{\frac{1}{p}}\left(\beta - \frac{1}{2} + \frac{1}{p}\right)^{-1} \qquad [2.38]$$
$$\times (M+1)^{\frac{1}{2}-\frac{1}{p}-\beta}\left(\sum_{j=M+1}^{\infty} a_{jj} k^{2\beta}\right)^{\frac{1}{2}},$$

if for some $\beta > \frac{1}{2} - \frac{1}{p}$ the series $\sum_{j=M+1}^{\infty} a_{jj} k^{2\beta}$ converges.

2.3.4. *Accuracy and reliability of the F-model in* $L_U(T)$

THEOREM 2.12.– Let $U = \{U(x), x \in \mathbb{R}\}$ be the C-function for which the condition of theorem 1.5 holds true. Then, stochastic process X_M is the F-model that approximates the process X with reliability $1-\alpha$, $0 < \alpha < 1$, and accuracy $\delta > 0$ б $L_U(T)$, if M satisfies inequalities [2.17] and [2.18], where $\check{b} = \max(b, 1)$,

$$\sigma^2_{M+1} = \sup_{t\in T} \sum_{k=M+1}^{\infty} \sum_{l=M+1}^{\infty} a_{kl} \cos\frac{\pi k t}{b} \cos\frac{\pi l t}{b},$$

$$\text{or} \quad \sigma^2_{M+1} = \sum_{k=M+1}^{\infty} \sum_{l=M+1}^{\infty} |a_{kl}|.$$

The theorem follows from theorem 1.6.

The next theorem gives a possibility to consider an essentially wider class of Orlicz spaces than theorem 2.12. Moreover, in some cases the estimates of the theorem are better than ones from theorem 2.12.

THEOREM 2.13.– Let $a = \{a_k, k = 1, 2, \ldots\}$ be such sequence that $0 \le a_k \le a_{k+1}$, $a_k \to \infty$ as $k \to \infty$. C-function $U(x)$ satisfies a condition: the function $(U(x))^{\frac{1}{2}}$ is convex. If the series

$$\sum_{k=1}^{\infty} \frac{k^{\frac{1}{2}}}{U^{(-1)}\left(\frac{k}{2b}\right)} \left(\frac{1}{a_k} - \frac{1}{a_{k+1}}\right) \left(\sum_{j=1}^{k} a_{jj} a_j^2\right)^{\frac{1}{2}} < \infty \qquad [2.39]$$

converges, then stochastic process X_M is F-model that approximates the process X with reliability $1 - \alpha$, $0 < \alpha < 1$, and accuracy $\delta > 0$ in $L_U(T)$, if for M the inequalities [2.31] and [2.32] hold true, where

$$\tilde{B}_1(M+1, a) \qquad [2.40]$$

$$= \frac{1+2\pi}{2} \sum_{k=M+1}^{\infty} \frac{k^{\frac{1}{2}}}{U^{(-1)}\left(\frac{k}{2b}\right)} \left(\frac{1}{a_k} - \frac{1}{a_{k+1}}\right) \left(\sum_{j=M+1}^{k} a_{jj} a_j^2\right)^{\frac{1}{2}},$$

The statement of the theorem follows from theorem 1.8, since the functions $\cos \frac{\pi n t}{b}$, $n = 0, 1, 2, \ldots$ are orthogonal and the family of this functions belongs to the class $D_U(c)$. In this case

$$c_n = c_n(U) = \frac{(1+2\pi)}{\sqrt{2b}} \frac{n^{\frac{1}{2}}}{U^{(-1)}\left(\frac{n}{2b}\right)}.$$

COROLLARY 2.5.– Let C-function $U(x)$ satisfy condition: the function $(U(x))^{\frac{1}{2}}$ is convex. If the series

$$\sum_{k=1}^{\infty} \frac{k^{\frac{1}{2}}}{U^{(-1)}\left(\frac{k}{2b}\right)} a_{kk} \left(\sum_{s=k}^{\infty} a_{ss}^2\right)^{-\frac{1}{2}} < \infty \qquad [2.41]$$

converges, then the assertion of theorem 2.13 is fulfilled, if for M inequalities [2.31] and [2.32] hold true, where

$$\tilde{B}_1(M+1, a) = \tilde{B}_1(M+1)$$

$$= \frac{1+2\pi}{2} \sum_{k=M+1}^{\infty} \frac{k^{\frac{1}{2}}}{U^{(-1)}\left(\frac{k}{2b}\right)} a_{kk} \left[2\left(\sum_{s=M+1}^{\infty} a_{ss}\right)^{-1} \qquad [2.42]$$

$$- \left(\sum_{s=M+1}^{\infty} a_{ss}\right)^{-1}\right]^{\frac{1}{2}}.$$

Corollary follows from corollary 1.13 and theorem 2.13.

COROLLARY 2.6.– Let $a = \{a_k, k = 1, 2, \ldots\}$ be such sequence that $0 \le a_k \le a_{k+1}$, $a_k \to \infty$ as $k \to \infty$, and

$$\sum_{k=1}^{\infty} \frac{k^{\frac{1}{2}}}{U^{(-1)}\left(\frac{k}{2b}\right)} \left(\frac{1}{a_k} - \frac{1}{a_{k+1}} \right) < \infty. \qquad [2.43]$$

If the series

$$\sum_{j=1}^{\infty} a_{jj} a_j^2 < \infty \qquad [2.44]$$

converges and $U(x)$ is such C-function that the function $(U((x))^{\frac{1}{2}}$ is convex, then the statement of theorem 2.13 holds true, if M satisfies inequalities [2.31] and [2.32], where

$$\check{B}_1(M+1, a) = \frac{1+2\pi}{2} \left(\sum_{j=M+1}^{\infty} a_{jj} a_j^2 \right)^{\frac{1}{2}} \sum_{k=M+1}^{\infty} \frac{k^{\frac{1}{2}}}{U^{(-1)}\left(\frac{k}{2b}\right)} \left(\frac{1}{a_k} - \frac{1}{a_{k+1}} \right)$$

Corollary follows from corollary 1.14 and theorem 2.13.

2.3.5. *Accuracy and reliability of the F-model in* $C(T)$

Denote

$$R_m^n(t) = \sum_{k=m}^{n} b_k \xi_k \cos \frac{\pi k t}{b},$$

where $\bar{b} = \{b_k, k = 1, 2, \ldots\}$ is a certain sequence,

$$Z_m^n = Z_m^n(\bar{b}) = \left\| \left(\mathbf{E}(R_m^n(t))^2 \right)^{\frac{1}{2}} \right\|_C = \sup_{0 \le t \le b} \left(\sum_{k=m}^{n} \sum_{l=m}^{n} a_{kl} b_k b_l \cos \frac{\pi k t}{b} \cos \frac{\pi l t}{b} \right)^{\frac{1}{2}}.$$

THEOREM 2.14.– Let $\bar{b} = \{b_k, k = 1, 2, \ldots\}$, $b_k > 0$, $b_k \le b_{k+1}$, $b_k \to \infty$ as $k \to \infty$, be such sequence that the series

$$\sum_{k=1}^{\infty} (\ln k)^{\frac{1}{2}} \left(\frac{1}{b_k} - \frac{1}{b_{k+1}} \right) < \infty \qquad [2.45]$$

converges. Assume that for any $M \geq 1$

$$\sup_{n \geq M} \sup_{0 \leq t \leq b} \mathbf{E}(R_M^n(t))^2 = Z_M^2(\bar{b}) < \infty. \qquad [2.46]$$

Then, for all $0 < \theta < 1$ and $x > \frac{\tilde{D}_{M+1}(\theta)}{(1-\theta)}$, where

$$\tilde{D}_M(\theta) = \sqrt{2} Z_M(\bar{b}) \sum_{k=M}^{\infty} \left| \ln \frac{\pi k}{b\theta} \right|^{\frac{1}{2}} \left(\frac{1}{b_k} - \frac{1}{b_{k+1}} \right), \qquad [2.47]$$

the following inequality

$$\mathbf{P}\{\|X(t) - X_M(t)\|_C > x\} \qquad [2.48]$$
$$\leq 2b \exp\left\{ -\frac{b_{M+1}^2}{Z_{M+1}^2(\bar{b})} (x(1-\theta) - \tilde{D}_{M+1}(\theta))^2 \right\}$$

holds true.

PROOF.– From example 1.11 follows that sequence of functions $\{\cos \frac{\pi k t}{b}\}$, $k = 1, 2, \ldots$, as $t \in T = [0, b]$, where $\mu(\cdot)$ is a Lebesque measure, belongs to the class B with $c(y) \equiv 1$, $\delta_n(\theta) = \frac{b\theta}{\pi n}$. Hence, theorem 1.10 implies the theorem, since in this case

$$\sum_{k=m}^{\infty} Z_s^k |\ln \delta_k(\theta)|^{\frac{1}{2}} \left(\frac{1}{b_k} - \frac{1}{b_{k+1}} \right)$$
$$\leq Z_s(\bar{b}) \sum_{k=m}^{\infty} \left| \ln \frac{\pi k}{b\theta} \right|^{\frac{1}{2}} \left(\frac{1}{b_k} - \frac{1}{b_{k+1}} \right) < \infty,$$

$$\frac{Z_s^n}{b_n} |\ln \delta_n(\theta)|^{\frac{1}{2}} \leq \frac{Z_s(\bar{b})}{b_n} \left| \ln \frac{\pi n}{b\theta} \right|^{\frac{1}{2}}$$
$$\leq Z_s(\bar{b}) \sum_{k=1}^{\infty} \left| \ln \frac{\pi k}{b\theta} \right|^{\frac{1}{2}} \left(\frac{1}{b_k} - \frac{1}{b_{k+1}} \right) \to 0$$

as $n \to \infty$. Therefore, in notation of theorem 1.10

$$\int_T |c(t)|\, dt = b, \quad A_m \leq Z_m(\bar{b}) b_m^{-1}, \quad D_m(\theta) \leq \tilde{D}_m(\theta).$$

□

COROLLARY 2.7.– If in inequality [2.48], we set $x = yb_{M+1}^{-1} Z_{M+1}(\bar{b})$, then as

$$y > \frac{\tilde{D}_m(\theta) b_{M+1}}{Z_{M+1}(\bar{b})(1-\theta)}$$

inequality

$$\mathbf{P}\left\{ \|X(t) - X_M(t)\|_C > y \frac{Z_{M+1}(\bar{b})}{b_{M+1}} \right\} \tag{2.49}$$

$$\leq 2b \exp\left\{ -\left(y(1-\theta) - \frac{b_{M+1} \tilde{D}_{M+1}(\theta)}{Z_{M+1}(\bar{b})} \right)^2 \right\}$$

is obtained.

If for $y > 1$ we put $\theta = 1/y$, then from corollary 2.7 the following corollary is obtained.

COROLLARY 2.8.– Let the conditions of theorem 2.14 be satisfied, then for $y > 1$ such that $y - \sqrt{2}|\ln y|^{\frac{1}{2}} \geq S_{M+1}$, where

$$S_{M+1} = 1$$

$$+ \sqrt{2} b_{M+1} \sum_{k=M+1}^{\infty} \left(\frac{1}{b_k} - \frac{1}{b_{k+1}} \right) \left| \ln \frac{\pi k}{b} \right|^{\frac{1}{2}} \tag{2.50}$$

inequality

$$\mathbf{P}\left\{ \|X(t) - X_M(t)\|_C > y \frac{Z_{M+1}(\bar{b})}{b_{M+1}} \right\}$$

$$\leq 2b \exp\left\{ -\left(y - 1 - \frac{b_{M+1} \tilde{D}_{M+1}(1/y)}{Z_{M+1}(\bar{b})} \right)^2 \right\} \tag{2.51}$$

holds true.

Corollary 2.8 follows from previous one if prove that for $y > 1$ inequality

$$y \geq \frac{b_{M+1} \tilde{D}_{M+1}(1/y)}{Z_{M+1}(\bar{b})(1 - 1/y)}$$

holds true in the case of $y - \sqrt{2}|\ln y|^{\frac{1}{2}} \geq S_{M+1}$.

From corollary 2.8 follows the next theorem.

THEOREM 2.15.– Stochastic process X_M is F-model that approximates the process X with reliability $1-\alpha$, $0<\alpha<1$, and accuracy $\delta>0$ 6 $C(T)$, if M satisfies such conditions

$$\frac{\delta b_{M+1}}{Z_{M+1}(\bar{b})} - \sqrt{2}\left|\ln\left(\frac{\delta b_{M+1}}{Z_{M+1}(\bar{b})}\right)\right|^{\frac{1}{2}} > S_{M+1},$$

where S_{M+1} is defined in [2.50] and

$$2b\exp\left\{-\left(\frac{\delta b_{M+1}}{Z_{M+1}(\bar{b})} - 1 - \frac{b_{M+1}}{Z_{M+1}(\bar{b})}\tilde{D}_{M+1}\left(\frac{Z_{M+1}(\bar{b})}{\delta b_{M+1}}\right)\right)^2\right\}$$
$$\leq \alpha. \qquad [2.52]$$

REMARK 2.8.– To simplify the calculation in inequality [2.52], we substitute $Z_{M+1}(\bar{b})$ for

$$\tilde{Z}_{M+1}(\bar{b}) = \left(\sum_{k=M+1}^{\infty}\sum_{l=M+1}^{\infty}|a_{kl}|b_k b_l\right)^{\frac{1}{2}},$$

as b_k the sequence $b_k = (\ln k)^{\varepsilon}$ can be chosen, where ε is any number such that $\varepsilon > 1/2$. Note, under rigider conditions the better estimations can be obtained from theorem 1.11.

2.4. Simulation of stationary stochastic process with discrete spectrum

This section is devoted to simulation method of strictly sub-Gaussian stationary process with discrete spectrum.

Let $X = \{X(t), t \in \mathbb{R}\}$ be stationary stochastic process, $\mathbf{E}X(t) = 0$, $t \in \mathbb{R}$, $\mathbf{E}X(t+\tau)X(t) = B(\tau)$, $t, \tau \in \mathbb{R}$.

DEFINITION 2.3.– *Stationary process X has discrete spectrum, if its correlation function $B(\tau)$ is equal to*

$$B(\tau) = \sum_{k=0}^{\infty} b_k^2 \cos\lambda_k\tau, \qquad [2.53]$$

where $b_k^2 > 0$, $\sum_{k=0}^{\infty} b_k^2 < \infty$, and λ_k are such that $0 \leq \lambda_k \leq \lambda_{k+1}$, and $\lambda_k \to \infty$ as $k \to \infty$.

THEOREM 2.16.– A stationary stochastic process with discrete spectrum $X = \{X(t), t \in \mathbb{R}\}$ can be written in the form of series

$$X(t) = \sum_{k=0}^{\infty}(\xi_k \cos \lambda_k t + \eta_k \sin \lambda_k t). \quad [2.54]$$

The series [2.54] converges in mean square, $\mathbf{E}\xi_k = \mathbf{E}\eta_k = 0$, $k = 0, 1, \ldots$, $\mathbf{E}\xi_k\eta_l = 0$, $k, l = 0, 1, \ldots$, $\mathbf{E}\xi_k\xi_l = \delta_k^l b_k^2$, $\mathbf{E}\eta_k\eta_l = \delta_k^l b_k^2$, where δ_k^l is a Kronecker symbol.

PROOF.– Since

$$\mathbf{E}X(t)X(s) = B(t-s) = \sum_{k=0}^{\infty} b_k^2 \cos(\lambda_k(t-s))$$

$$= \sum_{k=0}^{\infty} b_k^2 \cos(\lambda_k t)\cos(\lambda_k s) + \sum_{k=0}^{\infty} b_k^2 \sin(\lambda_k t)\sin(\lambda_k s),$$

then the statement of theorem follows from the Karhunen theorem (see [GIK 04]). □

REMARK 2.9.– If $X(t)$ is Gaussian stochastic process, then all ξ_k, η_k in series expansion [2.54] are independent Gaussian random variables. In the case of non-Gaussian stochastic process $X(t)$, the condition of independency ξ_k, η_k should be provided.

2.4.1. *The model of strictly sub-Gaussian stationary process with discrete spectrum*

In expansion [2.54] let $\xi_n, \eta_n, n = 0, 1, 2, \ldots$ be independent strictly sub-Gaussian random variables, then by example 1.7 a random process X is strictly sub-Gaussian with correlation function $B(\tau)$.

DEFINITION 2.4.– *Stochastic process $X_M = \{X_M(t), t \in T\}$, where T is an interval $[0, b]$,*

$$X_M(t) = \sum_{k=0}^{M}(\xi_k \cos(\lambda_k t) + \eta_k \sin(\lambda_k t))$$

is called the model of strictly sub-Gaussian stationary process $X = \{X(t), t \in T\}$ with discrete spectrum on interval $T = [0, b]$, ($D(T)$-model).

2.4.2. *Accuracy and reliability of the* $D(T)$*-model in* $L_2(T)$

THEOREM 2.17.– Stochastic process X_M is $D(T)$-model that approximates the process X with reliability $1-\alpha$, $0<\alpha<1$ and accuracy $\delta>0$ â $L_2(T)$, if M satisfies inequalities [2.29] and $\delta > A_{M+1}^{\frac{1}{2}}$, where

$$A_{M+1} = b \sum_{k=M+1}^{\infty} b_k^2. \qquad [2.55]$$

PROOF.– The theorem follows from corollary 1.7, since random variables ξ_k and η_k, $k=0,1,\ldots$ are non-correlated, and

$$b_k^2 \int_0^b \cos^2(\lambda_k t)\,dt + b_k^2 \int_0^b \sin^2(\lambda_k t)\,dt = b_k^2 b.$$

□

2.4.3. *Accuracy and reliability of the* $D(T)$*-model in* $L_p(T)$*,* $p>0$

THEOREM 2.18.– A stochastic process X_M is $D(T)$-model that approximates the process X with reliability $1-\alpha$, $0<\alpha<1$ and accuracy $\delta>0$ in $L_p(T)$, if M satisfies inequalities [2.10], where

$$\sigma_{M+1}^2 = \sum_{k=M+1}^{\infty} b_k^2,$$

and

$$p^{\frac{1}{2}} \sigma_{M+1} b^{\frac{1}{p}} < \delta.$$

The statement of theorem follows from theorem 1.4.

REMARK 2.10.– It is clear that in this case inequality [2.10] cannot be improved due to the results of section 1.5.

EXAMPLE 2.2.– Let $X=\{X(t),\, t\in\mathbb{R}\}$ be a stationary Gaussian stochastic process with discrete spectrum with $\mathbf{E}X(t)=0$ and correlation function

$$B(\tau) = \mathbf{E}X(t+\tau)X(t) \sum_{k=0}^{\infty} b_k^2 \cos(\lambda_k \tau).$$

Then, by definition 2.4, the model of stochastic process $X(t)$ has the following representation:

$$X_M(t) = \sum_{k=0}^{M} (\xi_k \cos(\lambda_k t) + \eta_k \sin(\lambda_k t)),$$

where ξ_n, η_n, $n = \overline{0, M}$, are independent zero-mean Gaussian random variables with the second moment

$$\mathbf{E}\xi_k^2 = \mathbf{E}\eta_k^2 = b_k^2.$$

Consider a particular case when $b_k^2 = \frac{1}{k^s}$, where $s > 1$. Set

$$\sigma_{M+1}^2 = \sum_{k=M+1}^{\infty} b_k^2.$$

Then

$$\sigma_{M+1}^2 = \sum_{k=M+1}^{\infty} \frac{1}{k^s} \leq \sum_{k=M+1}^{\infty} \int_{k-1}^{k} \frac{1}{u^s} du = \int_{M}^{\infty} \frac{1}{u^s} du = \frac{1}{s-1} \frac{1}{M^{s-1}}. \qquad [2.56]$$

Let us construct the model of Gaussian process $X(t)$ on the segment $[0, b]$ that approximates this process with reliability $1-\alpha$ and accuracy δ in the space $L_2([0, b])$. It follows from theorem 2.18 that it is enough to choose such M that the inequalities

$$p^{\frac{1}{2}} \sigma_{M+1} b^{\frac{1}{p}} < \delta$$

and

$$2\exp\left\{-\frac{\delta^2}{2\sigma_{M+1}^2 b^{\frac{2}{p}}}\right\} \leq \alpha,$$

are fulfilled. This means that

$$\sigma_{M+1} \leq \frac{\delta}{b^{\frac{1}{p}}} \min\left\{\frac{1}{p^{\frac{1}{2}}}, \frac{1}{\sqrt{2(-\ln\frac{\alpha}{2})}}\right\}.$$

Hence, it follows from [2.56] and inequality above that

$$M \geq \left(\frac{(s-1)\delta}{b^{\frac{1}{p}}} \min\left\{\frac{1}{p^{\frac{1}{2}}}, \frac{1}{\sqrt{2(-\ln\frac{\alpha}{2})}}\right\}\right)^{-1/(s-1)}.$$

Assume that $b = 1$. In the case $s = 2$ and $p = 2$ the values of M dependent on accuracy δ and reliability $1 - \alpha$ are found in environment for statistical computing R and are shown in Table 2.2.

	$\alpha = 0.1$	$\alpha = 0.05$	$\alpha = 0.01$
$\delta = 0.1$	25	28	33
$\delta = 0.06$	41	46	55
$\delta = 0.01$	245	272	326

Table 2.2. *The result of the simulation of stationary Gaussian process with discrete spectrum*

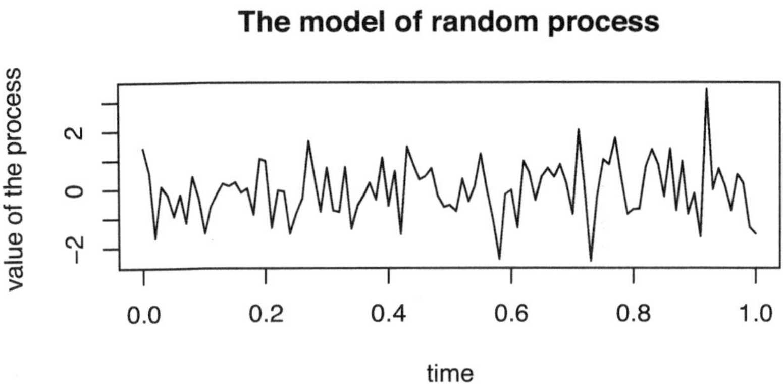

Figure 2.2. *The sample path of the model of Gaussian stationary process with discrete spectrum with accuracy 0.01 and reliability 0.99 in space* $L_2([0,1])$

2.4.4. *Accuracy and reliability of the* $D(T)$*-model in* $L_U(T)$

THEOREM 2.19.– Let $U = \{U(x), x \in \mathbb{R}\}$ be the C-function for which the conditions of theorem 1.5 are satisfied. Then, stochastic process X_M is $D(T)$-model that approximates the process X with reliability $1 - \alpha$, $0 < \alpha < 1$ and accuracy $\delta > 0$ in $L_U(T)$, if M satisfies [2.17] and [2.18], where

$$\sigma_{M+1}^2 = \sum_{k=M+1}^{\infty} b_k^2$$

The statement of theorem follows from theorem 1.6 and remark 1.9. In the next theorem, the estimates of theorem 2.18 cannot be improved, but allows to consider a wider class of Orlicz spaces than in theorem 2.18.

THEOREM 2.20.– Let $a = \{a_k, k = 1, 2, \ldots\}$ be such sequence that $0 \le a_k \le a_{k+1}$, $a_k \to \infty$ as $k \to \infty$. Assume that C-function $U(x)$ is such that the function $(U(x))^{\frac{1}{2}}$ is convex. If the series

$$\sum_{k=1}^{\infty} c_k(U)\left(\frac{1}{a_k} - \frac{1}{a_{k+1}}\right)\left(\sum_{j=1}^{k} a_j^2 b_j^2\right)^{\frac{1}{2}} < \infty, \quad [2.57]$$

converges, where

$$c_k(U) = \inf_{h>0}\left(h^{\frac{1}{2}} U^{(-1)}\left(\frac{1}{h}\right)\right)^{-1}\left(1 + h\left(\lambda_n + \frac{2}{b}\right)\right), \quad [2.58]$$

then stochastic process X_M is $D(T)$-model, $T = [0, b]$, that approximates the process X with reliability $1 - \alpha$, $0 < \alpha < 1$ and accuracy $\delta > 0$ in $L_U(T)$, if M satisfies inequalities

$$G(M+1, a) \ge \delta, \quad [2.59]$$

$$\exp\left\{\frac{1}{2}\right\}\frac{\delta(\sin 1)^2}{G(M+1,a)} \exp\left\{-\frac{\delta^2(\sin 1)^4}{2(G(M+1,a))^2}\right\} \le \alpha, \quad [2.60]$$

where

$$G(M+1, a) = \sum_{k=M+1}^{\infty} c_k(U) b^{\frac{1}{2}} S_4 \left(\frac{1}{a_k} - \frac{1}{a_{k+1}}\right)\left(\sum_{j=M+1}^{k} a_j^2 b_j^2\right)^{\frac{1}{2}},$$

$$S_4 = \left(\int_{-\infty}^{\infty}\left(\frac{\sin u}{u}\right)^4 du\right)^{\frac{1}{2}}.$$

PROOF.– Consider the Orlicz space $L_U(\mathbb{R})$, that is the space of measurable functions on $\mathbb{R}$ with respect to the norm

$$\|f\|_{L_U(\mathbb{R})} = \inf\left\{r : \int_{-\infty}^{\infty} U\left(\frac{f(t)}{r}\right) dt \le 1\right\}.$$

For any $\varepsilon > 0$ define $d_\varepsilon(t) = \left(\frac{\sin t\varepsilon}{t\varepsilon}\right)^2$. Let $f(t)$ be bounded on $\mathbb{R}$ and Borelean function $|f(t)| < A$, $t \in \mathbb{R}$. Then, the function $d_\varepsilon(t) f(t)$ belongs to the space $L_U(\mathbb{R})$. Really, since $|d_\varepsilon(t)| \le 1$, then $U(d_\varepsilon(t) f(t)) \le d_\varepsilon(t) U(f(t))$. Hence,

$$\int_{-\infty}^{\infty} U(d_\varepsilon(t) f(t))\, dt \le \int_{-\infty}^{\infty} d_\varepsilon(t) U(f(t))\, dt \le U(A) \int_{-\infty}^{\infty} d_\varepsilon(t)\, dt < \infty.$$

For any $0 \leq l \leq m$, the function

$$X_l^m(t,\varepsilon) = d_\varepsilon(t)\sum_{k=l}^{m}(\xi_k \cos(\lambda_k t) + \eta_k \sin(\lambda_k t)$$

is a function of exponential type $(\lambda_m + 2\varepsilon)$, bounded on a real axis. Therefore, it follows that the sequence of functions $\{d_\varepsilon(t)\cos(\lambda_k t), d_\varepsilon(t)\sin(\lambda_k t), k = 0, 1, \ldots\}$ belongs to the class $D_U(c)$ from the space $L_U(\mathbb{R})$, where $c_k = c_k(U)$, $\varepsilon = b^{-1}$(see [2.58]). Since random variables ξ_k, η_k are uncorrelated and

$$\int_{-\infty}^{\infty} d_\varepsilon^2(t)\cos^2(\lambda_k t)\,dt + \int_{-\infty}^{\infty} d_\varepsilon^2(t)\sin^2(\lambda_k t)\,dt$$
$$= \int_{-\infty}^{\infty}\left(\frac{\sin \varepsilon t}{\varepsilon t}\right)^4 dt = \frac{1}{\varepsilon}\int_{-\infty}^{\infty}\left(\frac{\sin u}{u}\right)^4 du,$$

then from theorem 1.8 follows that as $x > G(m,a)$ an inequality

$$\mathbf{P}\left\{\left\|d_\varepsilon(t)\sum_{k=m}^{\infty}(\xi_k \cos(\lambda_k t) + \eta_k \sin(\lambda_k t)\right\|_{L_U(\mathbb{R})} > x\right\}$$
$$\leq \exp\left\{\frac{1}{2}\right\}\frac{x}{G(M,a)}\exp\left\{-\frac{x^2}{2(G(M,a))^2}\right\} \qquad [2.61]$$

holds true, where $G(M,a)$ is defined in [2.60]. Since as $0 \leq t \leq \frac{1}{\varepsilon}$ the inequality holds $d_\varepsilon(t) > (\sin 1)^2$, then for any bounded on $\mathbb{R}$ function $f(t)$ and for any $r > 0$ the following inequalities are fulfilled

$$\int_0^{\frac{1}{\varepsilon}} U\left(\frac{f(t)}{r}\right) dt \leq \int_0^{\frac{1}{\varepsilon}} U\left(\frac{d_\varepsilon(t)f(t)}{(\sin 1)^2 r}\right) dt$$
$$\leq \int_{-\infty}^{\infty} U\left(\frac{d_\varepsilon(t)f(t)}{(\sin 1)^2 r}\right) dt$$

If, in the inequality above, $r = (\sin 1)^{-2}\|d_\varepsilon(t)f(t)\|_{L_U(\mathbb{R})}$, then the relationship

$$\|f(t)\|_{L_U(T)} \leq (\sin 1)^{-2}\|d_\varepsilon(t)f(t)\|_{L_U(\mathbb{R})}$$

is obtained. If in the last inequality $\varepsilon = b^{-1}$ is considered, then for any $x > 0$ we obtain

$$\mathbf{P}\{\|X - X_M\|_{L_U(T)} > x\}$$
$$\leq \mathbf{P}\left\{\left\|d_{b^{-1}}(t)\sum_{k=M+1}^{\infty}(\xi_k \cos(\lambda_k t) + \eta_k \sin(\lambda_k t)\right\|_{L_U(\mathbb{R})} > x(\sin 1)^2\right\}.$$

From the above inequality and [2.61] follows the statement of the theorem. □

COROLLARY 2.9.– If C-function $U(x)$ satisfies the condition of theorem 2.20 and the series

$$\sum_{k=1}^{\infty} c_k(U) b_k^2 \left(\sum_{s=k}^{\infty} b_k^2 \right)^{-\frac{1}{2}} < \infty \qquad [2.62]$$

converges, where $c_k(U)$ is defined in [2.58], then the statement of theorem 2.20 holds, if for M inequalities [2.59] and [2.60] are fulfilled, where

$$G(M+1, a) = G(M+1)$$
$$= \sqrt{2} b^{\frac{1}{2}} S_4 \sum_{k=M+1}^{\infty} c_k(U) b_k^2 \left(\sum_{s=k}^{\infty} b_k^2 \right)^{-\frac{1}{2}}. \qquad [2.63]$$

Corollary 2.9 follows from corollary 1.13 and theorem 2.20.

COROLLARY 2.10.– Let $a = \{a_k, k = 1, 2, \ldots\}$ be such sequence that $0 \leq a_k \leq a_{k+1}$, $a_k \to \infty$ as $k \to \infty$, and

$$\sum_{k=1}^{\infty} c_k(U) \left(\frac{1}{a_k} - \frac{1}{a_{k+1}} \right) < \infty. \qquad [2.64]$$

If the series converges

$$\sum_{j=1}^{\infty} a_j^2 b_j^2 < \infty, \qquad [2.65]$$

and C-function $U(x)$ satisfies the conditions of theorem 2.20, then the statement of theorem 2.20 holds true, is for M inequalities [2.59] and [2.60] are fulfilled, where

$$G(M+1, a)$$
$$= b^{\frac{1}{2}} S_4 \left(\sum_{j=M+1}^{\infty} a_j^2 b_j^2 \right)^{\frac{1}{2}} \sum_{j=M+1}^{\infty} c_k(U) \left(\frac{1}{a_k} - \frac{1}{a_{k+1}} \right). \qquad [2.66]$$

Corollary follows from corollary 1.14 and theorem 2.20.

REMARK 2.11.– If in [2.58], we set $h = \left(\lambda_n + \frac{2}{b}\right)^{-1}$, then we obtain that as $c_n(U)$ the following sequence can be considered

$$c_n(U) = \frac{2\left(\lambda_n + \frac{2}{b}\right)^{\frac{1}{2}}}{U^{(-1)}\left(\lambda_n + \frac{2}{b}\right)}.$$

2.4.5. *Accuracy and reliability of the* $D(T)$*-model in* $C(T)$

No loss of generality suggests that $b \geq 2$, $(T = [0,b])$. Following from example 1.14, the sequence of functions $\{c(t)\cos\lambda_k t, c(t)\sin\lambda_k t\}$, $k = 0,1,2,\ldots$, where

$$c(t) = \left(\frac{\sin\varepsilon t}{\varepsilon t}\right)^2, \quad \varepsilon \in (0,1/2], \quad \left(\int_{-\infty}^{\infty} c(t)\,dt = \frac{\pi}{\varepsilon}\right),$$

belongs to the class B, where $\delta_n(\theta) = \frac{2\theta}{\lambda_n + 2\varepsilon}$. If $\varepsilon = 1/b$, then, taking into account remark 1.17 $\left(\inf_{|t|<b}|c(t)| = (\sin 1)^2\right)$, according to theorem 1.13 the following theorem is obtained.

THEOREM 2.21.– Stochastic process X_M is $D(T)$-model that approximates the process X with reliability $1-\alpha$, $0<\alpha<1$ and accuracy $\delta > 0$ in $C(T)$, if for any $\beta \in (0,1/2]$ M satisfies conditions

$$\delta \geq 2(\sin 1)^{-2} G_{M+1}, \qquad [2.67]$$

where

$$G_M = \left(\sum_{s=M}^{\infty} \sigma_s^2\right)^{\frac{1}{2}},$$

$$\begin{aligned} 2\pi b \exp\Bigg\{ & -\frac{\delta^2(\sin 1)^4}{2G_{M+1}^2} + 1 + \sqrt{2}\left(\frac{\delta(\sin 1)^2}{G_{M+1}}\right)^{\frac{4\beta+1}{2\beta+1}}\left(\bar{F}_\beta + \frac{\pi}{2}\right) \\ & + 2\left(\frac{\delta(\sin 1)^2}{G_{M+1}}\right)^{\frac{4\beta}{2\beta+1}}\left(\bar{F}_\beta q_\beta\left(\frac{\delta(\sin 1)^2}{G_{M+1}}\right)\right) \\ & + \left(\frac{\delta(\sin 1)^2}{G_{M+1}}\right)^{\frac{1-2\beta}{1+2\beta}}\frac{\pi^2}{8}\Bigg)\Bigg\} \leq \alpha, \end{aligned} \qquad [2.68]$$

$$\text{äå} \quad \bar{F}_\beta = \sum_{k=M+1}^{\infty}\left|\ln\frac{\lambda_k + 2b^{-1}}{2(1-(1-2x^{-2})^{\frac{1}{2}})}\right|^{\frac{1}{2}} \sigma_k^2 G_k^{2(\beta-1)},$$

$$x = \frac{\delta(\sin 1)^2}{G_{M+1}} \geq 2,$$

$q_\beta(x)$ is from [1.129].

Similarly to theorem 1.15, the next theorem can be proved.

THEOREM 2.22.– Stochastic process X_M is $D(T)$-model that approximates the process X with reliability $1-\alpha$, $0<\alpha<1$, and accuracy $\delta>0$ in $C(T)$, $T=[0,b]$, $b>2$, if the conditions of theorem 2.21 hold true, but instead of [2.68] inequality

$$2\pi b \exp\left\{-\frac{1}{2}\left(\frac{\delta(\sin 1)^2}{G_{M+1}}\right)^2+\left(1+\frac{\pi}{\sqrt{2}}\right)^2\right\}$$

[2.69]

$$\times\, r^{(-1)}\left(\bar{G}_\beta\left(\frac{\delta(\sin 1)^2}{G_{M+1}}\right)^{4\beta}+\sqrt{2}\right)\leq\alpha$$

is fulfilled, where

$$\bar{G}_\beta=\sum_{k=M+1}^{\infty}\left|r\left(\frac{\lambda_n+2b^{-1}}{2(1-(1-2x^{-2})^{\frac{1}{2}})}\right)\right|\sigma_k^2G_k^{2(\beta-1)},$$

$$x=\frac{\delta(\sin 1)^2}{G_{M+1}}.$$

2.5. Application of Fourier expansion to simulation of stationary stochastic processes

Let $Y=\{Y(t),t\in\mathbb{R}\}$ be continuous in a mean square stationary stochastic process, $\mathbf{E}Y(t)=0$, $t\in\mathbb{R}$, $\mathbf{E}Y(t+\tau)Y(t)=B(\tau)$, $t,\tau\in\mathbb{R}$. Note that on interval $[0,2b]$ the correlation function $B(\tau)$ of the process Y can be expanded in Fourier series

$$B(\tau)=\sum_{k=0}^{\infty}g_k^2\cos\frac{\pi k\tau}{2b} \qquad [2.70]$$

and the coefficients

$$g_k^2=\frac{\gamma_k}{b}\int_0^{2b}B(\tau)\cos\frac{\pi k\tau}{2b}\,d\tau$$

are non-negative ($\gamma_k=1$, $k\geq 1$, $\gamma_0=\frac{1}{2}$).

Consider a stochastic process

$$X(t)=\sum_{k=0}^{\infty}\left(\xi_k\cos\frac{\pi kt}{2b}+\eta_k\sin\frac{\pi kt}{2b}\right), \qquad [2.71]$$

where ξ_k, η_k, $k = 0, 1, \ldots$ are random variables such that $\mathbf{E}\eta_k = \mathbf{E}\xi_k = 0$, $k = 0, 1, 2, \ldots$, $\mathbf{E}\eta_k^2 = \mathbf{E}\xi_k^2 = g_k^2$.

It's easy to verify that $X = \{X(t), t \in \mathbb{R}\}$ is stationary stochastic process. If this process is considered on $T = [0, b]$, then its correlation function $\mathbf{E}X(t+\tau)X(t)$, $\tau \in [-b, b]$ coincides with $B(\tau)$. Hence, this process can be used for model construction of the process Y as $t \in T$. It is clear that out of this interval, the correlation functions of the processes X and Y can be different.

EXAMPLE 2.3.– If the correlation function $B(\tau)$ is convex on $[0, 2b]$, then from the Hardi theorem [HAR 66] follows that Fourier coefficients of the function g_k^2 are non-negative. For instance, the correlation functions $B(\tau) = A\exp\{-\beta|\tau|^\delta\}$, $0 < \delta \leq 1$, $A > 0$, $B > 0$, are convex for all $\tau > 0$.

2.5.1. *The model of a stationary process in which a correlation function can be represented in the form of a Fourier series with positive coefficients*

Following expansion [2.71], let ξ_k, η_k, $k = 0, 1, \ldots$ be independent strictly sub-Gaussian random variables. Then, by example 1.7, a stochastic process X is strictly a sub-Gaussian stationary process.

If $X(t)$ is a Gaussian stochastic process, then ξ_k, η_k, $k = 0, 1, \ldots$ are independent Gaussian random variables.

DEFINITION 2.5.– *Stochastic process $X_M = \{X_M(t), t \in T\}$, where $T = [0, b]$,*

$$X_M(t) = \sum_{k=0}^{M} \left(\xi_k \cos \frac{\pi k t}{2b} + \eta_k \sin \frac{\pi k t}{2b} \right)$$

is called Fourier model of stationary process $X = \{X(t), t \in T\}$ with correlation function $B(\tau)$, $\tau \in [-b, b]$ (FS-model).

REMARK 2.12.– Since stochastic process X, defined in [2.71], is a stochastic stationary process with discrete spectrum $b_k^2 = g_k^2$, $\lambda_k = \frac{\pi k}{2b}$, $k = 0, 1, \ldots$, then to simulate these processes the results of section 2.4 can be used.

Show, for example, how to construct FS-model of the process X that approximates the process with given accuracy and reliability in $C(T)$. From example [1.11] follows that the sequence of functions

$$\left\{ \cos \frac{\pi k t}{2b}, \sin \frac{\pi k t}{2b} \right\}, \quad k = 0, 1, 2, \ldots, \quad T \in [0, b], \quad t \in T,$$

belongs to the class B from $c(t) \equiv 1$, $\delta_n(\theta) = \frac{b\theta}{n\pi}$. One can easily show that the restriction of remark 1.16 holds true. Really, $Q = 1$

$$\text{and} \quad \int_T |c(t)|\, dt(\delta_k(\theta))^{-1} = \frac{b\pi k}{b\theta} = \frac{\pi k}{\theta} > 1, \quad k = 1, 2, \ldots.$$

Hence, from theorem 1.13 follows such assertion.

THEOREM 2.23.– Stochastic process X_M is FS-model that approximates the process X with reliability $1-\alpha$, $0<\alpha<1$, and accuracy $\delta>0$ in $C(T)$, if M satisfies the conditions

$$\delta > 2G_{M+1}, \qquad [2.72]$$

where

$$G_M = \left(\sum_{s=M}^{\infty} g_k^2\right)^{\frac{1}{2}},$$

and for some $\beta \in \left[0, \frac{1}{2}\right]$ inequality [2.23] holds true as

$$\bar{F}_\beta = \sum_{k=M+1}^{\infty} \left| \ln \frac{b}{k\pi}\left(1 - \left(1 - \frac{2G_{M+1}^2}{\delta^2}\right)^{\frac{1}{2}}\right)\right|^{\frac{1}{2}} \sigma_k^2 G_k^{2(\beta-1)}. \qquad [2.73]$$

From theorem 1.15 follows such theorem.

THEOREM 2.24.– Stochastic process X_M is FS-model that approximates the process X with reliability $1-\alpha$, $0<\alpha<1$ and accuracy $\delta>0$ â $C(T)$, if for M the conditions of theorem 2.23 are satisfied, but instead [2.23] inequality [2.24] holds true, where

$$\bar{G}_M = \sum_{k=M+1}^{\infty} \left| r\left(\frac{k\pi}{b}\left(1 - \left(1 - \frac{2G_{M+1}^2}{\delta^2}\right)^{\frac{1}{2}}\right)^{-1}\right)\right| \sigma_k^2 G_k^{2(\beta-1)}.$$

3

Simulation of Gaussian Stochastic Processes with Respect to Output Processes of the System

In many applied areas that use the theory of stochastic processes, the problem arises to construct the model of a stochastic process, that is considered as an input process to some system or filter, with respect to the output process. We are interested in the model that approximates a Gaussian stochastic process with respect to the output process with predetermined accuracy and reliability in Banach space $C(T)$. In this case, at first we estimate the probability that a certain stochastic vector process

$$\vec{X}^T(t) = (X_1(t), X_2(t), \ldots, X_d(t))$$

leaves some region on some interval of time. For example,

$$\sup_{t \in T} \vec{X}^T(t) A(t) \vec{X}(t) > \varepsilon,$$

where ε is a sufficiently large number, (T, ρ) is a metric space, $\vec{X} = (\vec{X}(t), t \in T)$ is a process that generates the system and $A(t)$ is a matrix (in most cases positive semidefinite). The process $\vec{X}(t)$ will be considered as Gaussian due to the central limit theorem. Thus, the problem arises to estimate the probability

$$P\left\{\sup_{t \in T} \vec{X}^T(t) A(t) \vec{X}(t) > \varepsilon\right\},$$

or the probability

$$P\left\{\sup_{t \in T} |\vec{X}^T(t) A(t) \vec{X}(t) - \mathbb{E}\vec{X}^T(t) A(t) \vec{X}(t)| > \varepsilon\right\},$$

where $\vec{X}(t)$ is a Gaussian vector process and $A(t)$ is a symmetric matrix. The process $\vec{X}(t)$ is considered as centered one. In this chapter, the estimates of large deviation probability for square-Gaussian stochastic processes are established. An exact definition of the class of these processes is proposed in section 3.2.

Distribution properties of supremum of stochastic process were investigated by many authors. They investigate the problem of existence of moments and exponential moments of distribution of supremum of the process, estimates of probability $P\left\{\sup_{0\leq t\leq T}|X(t)| > \varepsilon\right\}$, distribution of the number exceeding a certain level, etc. For more details, readers can refer to books and papers by Cramér and Leadbetter [CRA 67], Lindgren [LIN 71], Dudley [DUD 73], Fernique [FER 75], Nanopoulos and Nobelis [NAN 76], Kôno [KÔN 80], Kozachenko [KOZ 85b, KOZ 85c, KOZ 99a], Piterbarg [PIT 96] and Ostrovs'kij [OST 90]. In [KOZ 99a], the estimates for large deviations of supremum of square-Gaussian stochastic processes were obtained.

The structure of this chapter is as follows. The chapter consists of eight sections. In section 3.1, we obtain the inequalities for exponential moments

$$\mathbb{E}\exp\left\{\frac{s}{\sqrt{2}}\,\frac{\vec{\xi}^T A\vec{\xi} - \mathbb{E}\vec{\xi}^T A\vec{\xi}}{\left(\mathrm{Var}\,\vec{\xi}^T A\vec{\xi}\right)^{1/2}}\right\}$$

where $\vec{\xi}$ is a Gaussian centered random vector and A is a symmetric matrix. Estimates depend only on the eigenvalues of the matrix $B^{1/2}AB^{1/2}$, where $B = \mathrm{cov}\vec{\xi}$. Similar estimates are obtained for the mean square limits of sequences of quadratic forms of jointly Gaussian random variables $\vec{\xi}_n^T A_n\vec{\xi}_n$ as $n \to \infty$. In section 3.2, the definition of the space of square-Gaussian random variables and the definition of square-Gaussian stochastic process are given. In section 3.3, estimates of probability of large deviation of supremum of a square-Gaussian stochastic process on a compact metric space are studied. Similar estimates are discussed in [KOZ 99a, KOZ 07a].

In section 3.4, the estimations of distribution of supremum of square-Gaussian stochastic processes in the space $[0,T]^d$ are obtained. The results of this section are used in section 3.5 to construct the model of a Gaussian process, that is considered as input process on some system or filter, with respect to the output process in Banach space $C(T)$ with given accuracy and reliability. Section 3.6 deals with a stationary Gaussian stochastic process with discrete spectrum. Theorems for the simulation of these processes, that are considered as an input process on some filter with respect to output process in space $C(T)$, are also proven. A particular case is also considered when the system output process is a derivative of the initial one.

Similar results of sections 3.4–3.6 are obtained in [KOZ 03, KOZ 06b, KOZ 07a, ROZ 07, ROZ 08, ROZ 09a].

In section 3.7, Gaussian stochastic fields are considered. The conditions for simulation of these fields, that are considered as input process on some filter with respect to output process in space $C(T)$, are also given. Section 3.7 is based on the study of [KOZ 04a].

3.1. The inequalities for the exponential moments of the quadratic forms of Gaussian random variables

In this section, we prove some inequalities for the exponential moments of the quadratic forms of jointly Gaussian random variables. These inequalities will be applied in the following sections.

LEMMA 3.1.– Let $\xi_1, \xi_2, \ldots, \xi_n, \xi_{n+1}, \xi_{n+2}, \ldots, \xi_{n+m}$ be independent Gaussian random variables; $n = 0,1,2,\ldots$; $m = 0,1,2,\ldots$; $m+n>0$; and $\mathbf{E}\xi_k = 0$, $\mathbf{E}\xi_k^2 = \sigma_k^2 > 0$, $k = 1,2,\ldots$ and let

$$\delta_{N+1}^{+} = \left(\frac{\sum_{i=1}^{n} \sigma_i^{2(N+1)}}{\sum_{i=1}^{n+m} \sigma_i^{2(N+1)}} \right)^{\frac{1}{N+1}},$$

$$\delta_{N+1}^{-} = \left(\frac{\sum_{i=n+1}^{n+m} \sigma_i^{2(N+1)}}{\sum_{i=1}^{n+m} \sigma_i^{2(N+1)}} \right)^{\frac{1}{N+1}}. \qquad [3.1]$$

Then, for integer $N = 1,2,\ldots$ and real s such that $0 \le \delta_{N+1}^{+} s < 1$, the following inequality holds true

$$\mathbf{E}\exp\left\{ \frac{s\left(\sum_{i=1}^{n} \xi_i^2 - \sum_{i=n+1}^{n+m} \xi_i^2\right)}{2\left(\sum_{i=1}^{n+m} \sigma_i^{2(N+1)}\right)^{\frac{1}{N+1}}} \right\} \le$$

$$\exp\left\{ \frac{1}{2} \sum_{k=1}^{N} \frac{s^k \left(\sum_{i=1}^{n} \sigma_i^{2k} + (-1)^k \sum_{i=n+1}^{n+m} \sigma_i^{2k}\right)}{k\left(\sum_{i=1}^{n+m} \sigma_i^{2(N+1)}\right)^{\frac{k}{N+1}}} \right\} \times$$

$$\exp\left\{ \frac{1}{2} \sum_{k=N+1}^{\infty} \frac{\left(s\delta_{N+1}^{+}\right)^k}{k} + \frac{1}{2} \frac{\left(s\delta_{N+1}^{-}\right)^{N+1}}{N+1} \chi(N+1) \right\}, \qquad [3.2]$$

where $\chi(N+1) = \frac{(-1)^{N+1}+1}{2}$.

PROOF.– We will prove the lemma in the case when $m > 0$ and $n > 0$. In the case of $m = 0$ or $n = 0$, the proof is analogous.

Let r be a number such that $r > 2\sigma_i^2$ for $i = 1, 2, \ldots, n$. Then, from the equality

$$\mathbf{E}\exp\left\{\frac{u\xi_i^2}{2\sigma_i^2}\right\} = (1-u)^{-\frac{1}{2}},$$

which holds true for all $u < 1$, it follows the equality

$$I = \mathbf{E}\exp\left\{\frac{1}{r}\left(\sum_{i=1}^{n}\xi_i^2 - \sum_{i=n+1}^{n+m}\xi_i^2\right)\right\}$$

$$= \prod_{i=1}^{n}\left(1 - \frac{2\sigma_i^2}{r}\right)^{-\frac{1}{2}} \prod_{i=n+1}^{n+m}\left(1 + \frac{2\sigma_i^2}{r}\right)^{-\frac{1}{2}}. \qquad [3.3]$$

Therefore,

$$\ln I = -\frac{1}{2}\sum_{i=1}^{n}\ln\left(1 - \frac{2\sigma_i^2}{r}\right) - \frac{1}{2}\sum_{i=n+1}^{n+m}\ln\left(1 + \frac{2\sigma_i^2}{r}\right)$$

$$= \frac{1}{2}\sum_{k=1}^{\infty}\frac{2^k}{kr^k}\sum_{i=1}^{n}\sigma_i^{2k} - \frac{1}{2}\sum_{i=n+1}^{n+m}\ln\left(1 + \frac{2\sigma_i^2}{r}\right).$$

It follows from the last inequality, then

$$I = \exp\left\{\frac{1}{2}\sum_{k=1}^{N}\frac{2^k}{kr^k}\sum_{i=1}^{n}\sigma_i^{2k} + \frac{1}{2}\sum_{k=1}^{N}\frac{2^k}{kr^k}(-1)^k\sum_{i=n+1}^{n+m}\sigma_i^{2k}\right\}$$

$$\times \exp\left\{\frac{1}{2}\sum_{k=N+1}^{\infty}\frac{2^k}{kr^k}\sum_{i=1}^{n}\sigma_i^{2k} + \frac{1}{2}\sum_{k=1}^{N}\frac{2^k(-1)^{k+1}}{kr^k}\sum_{i=n+1}^{n+m}\sigma_i^{2k} - \frac{1}{2}\sum_{i=n+1}^{n+m}\ln\left(1 + \frac{2\sigma_i^2}{r}\right)\right\}. \qquad [3.4]$$

Since for all $x \geq 0$ and all odd integer N, the inequality holds true

$$\sum_{k=1}^{N}(-1)^{k+1}\frac{x^k}{k} - \ln(1+x) \leq \frac{x^{N+1}}{N+1}, \qquad [3.5]$$

and for all $x \geq 0$ and all even integer N holds true the inequality

$$\sum_{k=1}^{N}(-1)^{k+1}\frac{x^k}{k} - \ln(1+x) \leq 0, \qquad [3.6]$$

then from [3.4]–[3.6] follows that

$$I \leq \exp\left\{\frac{1}{2}\sum_{k=1}^{N}\frac{2^k}{kr^k}\sum_{i=1}^{n}\sigma_i^{2k} + \frac{1}{2}\sum_{k=1}^{N}\frac{2^k}{kr^k}(-1)^k\sum_{i=n+1}^{n+m}\sigma_i^{2k}\right\}$$

$$\times \exp\left\{\frac{1}{2}\sum_{k=N+1}^{\infty}\frac{2^k}{kr^k}\sum_{i=1}^{n}\sigma_i^{2k} + \frac{1}{2}\sum_{i=n+1}^{n+m}\frac{\sigma_i^{2(N+1)}2^{N+1}}{(N+1)r^{N+1}}\,\chi(N+1)\right\}. \qquad [3.7]$$

If in [3.7] we put

$$r = \frac{2}{s}\left(\sum_{i=1}^{m+n}\sigma_i^{2(N+1)}\right)^{\frac{1}{N+1}},$$

where s is an arbitrary number such that $0 \leq s$ and $s\delta_{N+1}^{+} < 1$, then we will have the inequality

$$\mathbf{E}\exp\left\{\frac{s\left(\sum\limits_{i=1}^{n}\xi_i^2 - \sum\limits_{i=n+1}^{n+m}\xi_i^2\right)}{2\left(\sum\limits_{i=1}^{n+m}\sigma_i^{2(N+1)}\right)^{\frac{1}{N+1}}}\right\}$$

$$\leq \exp\left\{\frac{1}{2}\sum_{k=1}^{N}\frac{s^k\left(\sum\limits_{i=1}^{n}\sigma_i^{2k} + (-1)^k\sum\limits_{i=n+1}^{n+m}\sigma_i^{2k}\right)}{k\left(\sum\limits_{i=1}^{n+m}\sigma_i^{2(N+1)}\right)^{\frac{k}{N+1}}}\right\}$$

$$\times \exp\left\{\frac{1}{2}\sum_{k=N+1}^{\infty}\frac{s^k\sum\limits_{i=1}^{n}\sigma_i^{2k}}{k\left(\sum\limits_{i=1}^{n+m}\sigma_i^{2(N+1)}\right)^{\frac{k}{N+1}}} + \frac{1}{2}\frac{s^{N+1}\sum\limits_{i=n+1}^{n+m}\sigma_i^{2(N+1)}}{(N+1)\sum\limits_{i=1}^{n+m}\sigma_i^{2(N+1)}}\chi(N+1)\right\}. \qquad [3.8]$$

Since for $k \geq N+1$, we have

$$\frac{\sum_{i=1}^{n} \sigma_i^{2k}}{\left(\sum_{i=1}^{n+m} \sigma_i^{2(N+1)}\right)^{\frac{k}{N+1}}} = \left(\delta_{N+1}^{+}\right)^k \frac{\sum_{i=1}^{n} \sigma_i^{2k}}{\left(\sum_{i=1}^{n} \sigma_i^{2(N+1)}\right)^{\frac{k}{N+1}}} \leq \left(\delta_{N+1}^{+}\right)^k,$$

then from [3.8] follows the statement of the lemma. □

LEMMA 3.2.– Let $\vec{\xi}$ be an d-dimensional Gaussian random vector with $\mathbf{E}\vec{\xi} = 0$ and covariance $B = \mathrm{Cov}\,\vec{\xi}$. Let $A = \|a_{ij}\|_{i,j=1}^{d}$ be a real-valued symmetric matrix. Assume that S is an orthogonal matrix that transforms the matrix $B^{1/2}AB^{1/2}$ to the diagonal one $\Lambda = \mathrm{diag}(\lambda_k)_{k=1}^{d}$, which means $S^T B^{1/2}AB^{1/2}S = \Lambda$. If not all $\lambda_k, k = 1, 2, \ldots, d$ are equal to zero, then for all $N = 1, 2, \ldots$ and all s such that $0 \leq s\gamma_{N+1}^{+} < 1$, where

$$\gamma_{N+1}^{+} = \left(\frac{\sum^{+} |\lambda_i|^{N+1}}{\sum_{i=1}^{d} |\lambda_i|^{N+1}}\right)^{\frac{1}{N+1}}, \quad \gamma_{N+1}^{-} = \left(\frac{\sum^{-} |\lambda_i|^{N+1}}{\sum_{i=1}^{d} |\lambda_i|^{N+1}}\right)^{\frac{1}{N+1}},$$

and where $\sum^{-}$ is sum over all negative λ_i while $\sum^{+}$ is sum over all positive λ_i, the following inequality holds true

$$\mathbf{E}\exp\left\{\frac{s}{2}\frac{\vec{\xi}^T A\vec{\xi}}{\left(\sum_{i=1}^{d} |\lambda_i|^{N+1}\right)^{\frac{1}{N+1}}}\right\}$$

$$\leq \exp\left\{\frac{1}{2}\sum_{k=1}^{N} \frac{s^k\left(\sum^{+} |\lambda_i|^k + (-1)^k \sum^{-} |\lambda_i|^k\right)}{k\left(\sum_{i=1}^{d} |\lambda_i|^{N+1}\right)^{\frac{k}{N+1}}}\right\}$$

$$\times \exp\left\{\frac{1}{2}\sum_{k=N+1}^{\infty} \frac{\left(s\gamma_{N+1}^{+}\right)^k}{k} + \frac{\left(s\gamma_{N+1}^{-}\right)^{N+1}}{2(N+1)}\chi(N+1)\right\}, \qquad [3.9]$$

where $\chi(N+1)$ is defined in [3.2]. In the case $-1 < s\gamma_{N+1}^{-} \leq 0$, the following inequality holds true

$$\mathbf{E}\exp\left\{\frac{s}{2}\frac{\vec{\xi}^T A\vec{\xi}}{\left(\sum_{i=1}^{d} |\lambda_i|^{N+1}\right)^{\frac{1}{N+1}}}\right\}$$

$$\leq \exp\left\{\frac{1}{2}\sum_{k=1}^{N} \frac{|s|^k\left(\sum^{-} |\lambda_i|^k + (-1)^k \sum^{+} |\lambda_i|^k\right)}{k\left(\sum_{i=1}^{d} |\lambda_i|^{N+1}\right)^{\frac{k}{N+1}}}\right\}$$

$$\times \exp\left\{\frac{1}{2}\sum_{k=N+1}^{\infty}\frac{\left(|s|\gamma_{N+1}^{-}\right)^{k}}{k}+\frac{\left(|s|\gamma_{N+1}^{+}\right)^{N+1}}{2(N+1)}\chi(N+1)\right\}. \qquad [3.10]$$

PROOF.– Let $\det B > 0$ and let $\vec{\zeta} = S^T(B^{1/2})^{-1}\vec{\xi}$. The vector $\vec{\zeta}$ is a Gaussian random vector with $\mathbf{E}\vec{\zeta} = 0$ and

$$\mathrm{Cov}\,\vec{\zeta} = S^T(B^{1/2})^{-1}\mathrm{Cov}\,\vec{\xi}\,(B^{1/2})^{-1}S = S^T(B^{1/2})^{-1}B(B^{1/2})^{-1}S = I,$$

where I is the identity matrix. Therefore, the components $\zeta_i, i = 1,2,\ldots,d$ of the vector $\vec{\zeta}$ are independent centered Gaussian random variables with variance $\mathbf{E}\vec{\zeta}_i^{\,2} = 1$. Moreover, since $\vec{\xi} = B^{1/2}S\vec{\zeta}$, then $\vec{\xi}^{\,T}A\vec{\xi} = \vec{\zeta}^{\,T}S^TB^{1/2}AB^{1/2}S\vec{\zeta} = \vec{\zeta}^{\,T}\Lambda\vec{\zeta} = \sum_{k=1}^{d}\zeta_k^2\lambda_k$. Therefore, the statement of the lemma in the case $0 \le s\gamma_{N+1}^{+} < 1$ follows from lemma 3.1. In the case $-1 < s\gamma_{N+1}^{-} \le 0$, the statement of the lemma also follows from lemma 3.1, the signs of λ_i change by substituting $-s$ instead of s.

Consider now the case where $\det B = 0$ (that is the matrix B^{-1} does not exist). Let $\vec{\theta}$ be a d-dimensional centered Gaussian vector that does not depend on $\vec{\xi}$, and such that $\mathrm{Cov}\,\vec{\theta} = I$; let ε be an arbitrary positive number. Put $\vec{\xi}_\varepsilon = \vec{\xi} + \varepsilon\vec{\theta}$. Since $\vec{\xi}_\varepsilon$ is a Gaussian centered vector with covariance $B_\varepsilon = \mathrm{Cov}\,\vec{\xi}_\varepsilon = B + \varepsilon^2 I$, then $\det B_\varepsilon > 0$. Therefore, for $\vec{\xi}_\varepsilon$ inequalities [3.9] and [3.10] hold true with the matrix Λ_ε and $S_\varepsilon^T B_\varepsilon^{1/2}AB_\varepsilon^{1/2}S_\varepsilon = \Lambda_\varepsilon = \mathrm{diag}(\lambda_k^\varepsilon)_{k=1}^d$. Since $\lambda_k^\varepsilon \to \lambda_k$ and $\vec{\xi}_\varepsilon \to \vec{\xi}$ as $\varepsilon \to 0$ in mean square, then the statement of the lemma follows from inequalities [3.9] and [3.10] for $\vec{\xi}_\varepsilon$ and the Fatou lemma. □

COROLLARY 3.1.– Suppose that the assumptions of lemma 3.2 are satisfied for all s such that $0 \le s\gamma_2^{+} < 1$ and $-1 < s\gamma_2^{-} \le 0$. Then, the following inequality holds true

$$\mathbf{E}\exp\left\{\frac{s}{\sqrt{2}}\frac{\vec{\xi}^{\,T}A\vec{\xi} - \mathbf{E}\,\vec{\xi}^{\,T}A\vec{\xi}}{\left(\mathbf{Var}\,\vec{\xi}^{\,T}A\vec{\xi}\right)^{1/2}}\right\} \le L(s\gamma_2^{+}, s\gamma_2^{-}), \qquad [3.11]$$

where

$$L(s\gamma_2^{+}, s\gamma_2^{-}) = \begin{cases}(1 - s\gamma_2^{+})^{-1/2}\exp\left\{-\frac{s\gamma_2^{+}}{2} + \frac{(s\gamma_2^{-})^2}{4}\right\}, & s \ge 0;\\ (1 - |s|\gamma_2^{-})^{-1/2}\exp\left\{-\frac{|s|\gamma_2^{-}}{2} + \frac{(s\gamma_2^{+})^2}{4}\right\}, & s \le 0.\end{cases}$$

Moreover

$$\mathbf{Var}\,\vec{\xi}^{\,T}A\vec{\xi} = \sum_{i=1}^{d}\sum_{j=1}^{d}\sum_{k=1}^{d}\sum_{l=1}^{d}a_{ij}a_{kl}\left(\mathbf{E}\xi_i\xi_k\,\mathbf{E}\xi_j\xi_l + \mathbf{E}\xi_i\xi_l\,\mathbf{E}\xi_j\xi_k\right), \qquad [3.12]$$

where a_{ij} are the elements of the matrix A.

PROOF.– Consider the case $s \geq 0$. In this case, the proof is similar. Let us put in inequality [3.9] $N = 1$. If $0 \leq s\gamma_2^+ < 1$, then we will have the inequality

$$\mathbf{E}\exp\left\{\frac{s}{2}\frac{\vec{\xi}^T A\vec{\xi}}{\left(\sum_{i=1}^d \lambda_i^2\right)^{\frac{1}{2}}}\right\}$$

$$\leq \exp\left\{\frac{1}{2}\frac{s\sum_{i=1}^d \lambda_i}{\left(\sum_{i=1}^d \lambda_i^2\right)^{\frac{1}{2}}}\right\}\exp\left\{\frac{1}{2}\sum_{k=2}^{\infty}\frac{\left(s\gamma_2^+\right)^k}{k}+\frac{\left(s\gamma_2^-\right)^2}{4}\right\}.$$

Since

$$\frac{1}{2}\sum_{k=2}^{\infty}\frac{\left(s\gamma_2^+\right)^k}{k} = -\frac{1}{2}\ln(1-s\gamma_2^+) - \frac{s\gamma_2^+}{2},$$

then from the last inequality, we have

$$\mathbf{E}\exp\left\{\frac{s}{2}\frac{\vec{\xi}^T A\vec{\xi} - \sum_{i=1}^d \lambda_i}{\left(\sum_{i=1}^d \lambda_i^2\right)^{\frac{1}{2}}}\right\}$$

$$\leq \frac{1}{\sqrt{1-s\gamma_2^+}}\exp\left\{-\frac{s\gamma_2^+}{2}+\frac{\left(s\gamma_2^-\right)^2}{4}\right\}. \qquad [3.13]$$

Since $\vec{\xi}^T A\vec{\xi} = \vec{\zeta}^T \Lambda\vec{\zeta} = \sum_{k=1}^d \zeta_k^2\lambda_k$, then

$$\mathbf{E}\,\vec{\xi}^T A\vec{\xi} = \mathbf{E}\sum_{k=1}^d \zeta_k^2\lambda_k = \sum_{k=1}^d \lambda_k \qquad [3.14]$$

and

$$\mathbf{Var}\,\vec{\xi}^T A\vec{\xi} = \mathbf{Var}\sum_{k=1}^d \zeta_k^2\lambda_k = \sum_{k=1}^d \lambda_k^2\mathbf{Var}\,\zeta_k^2 = 2\sum_{k=1}^d \lambda_k^2. \qquad [3.15]$$

Inequality [3.11] follows from [3.13]–[3.15]. Let us prove equality [3.12]. Since

$$\mathbf{Var}\,\vec{\xi}^T A\vec{\xi} = \sum_{i=1}^d\sum_{j=1}^d\sum_{k=1}^d\sum_{l=1}^d \left(\mathbf{E}\xi_i\xi_j\xi_k\xi_l a_{ij}a_{kl} - \mathbf{E}\xi_i\xi_j\,\mathbf{E}\xi_k\xi_l a_{ij}a_{kl}\right),$$

then equality [3.12] follows from the Isserlis formula:

$$\mathbf{E}\xi_i\xi_j\xi_k\xi_l = \mathbf{E}\xi_i\xi_j\,\mathbf{E}\xi_k\xi_l + \mathbf{E}\xi_i\xi_k\,\mathbf{E}\xi_j\xi_l + \mathbf{E}\xi_i\xi_l\,\mathbf{E}\xi_j\xi_k. \qquad \square$$

It is not a simple problem in many cases to find the values γ_2^+ and γ_2^-. For this reason, we will give some estimates that do not depend on these parameters.

COROLLARY 3.2.– Assume that the assumptions of lemma 3.2 are satisfied for all s such that $|s| < 1$, the following inequality holds true

$$\mathbf{E}\exp\left\{\frac{s}{\sqrt{2}}\frac{\vec{\xi}^{\,T}A\vec{\xi} - \mathbf{E}\,\vec{\xi}^{\,T}A\vec{\xi}}{\left(\mathbf{Var}\,\vec{\xi}^{\,T}A\vec{\xi}\right)^{1/2}}\right\} \le (1-|s|)^{-1/2}\exp\left\{-\frac{|s|}{2}\right\}. \qquad [3.16]$$

PROOF.– Since $(\gamma_2^+)^2 + (\gamma_2^-)^2 = 1$, then for $0 \le s < 1$

$$L(s\gamma_2^+, s\gamma_2^-) = (1-s\gamma_2^+)^{-1/2}\exp\left\{-\frac{s\gamma_2^+}{2} - \frac{s^2(\gamma_2^+)^2}{4}\right\}\exp\left\{\frac{s^2}{4}\right\}$$
$$\le (1-s)^{-1/2}\exp\left\{-\frac{s}{2}\right\}.$$

Therefore, in the case $0 \le s < 1$ inequality [3.16] follows from inequality [3.11]. In the case $-1 < s \le 0$, the proof is the same as in the previous case. $\square$

COROLLARY 3.3.– Assume that the assumptions of lemma 3.2 are satisfied and let the matrix A be positive semidefinite. Then, for any integer $N > 0$, the next inequalities hold true for $0 \le s < 1$:

$$I_N(s) = \mathbf{E}\exp\left\{\frac{s}{2}\frac{\vec{\xi}^{\,T}A\vec{\xi} - \mathbf{E}\,\vec{\xi}^{\,T}A\vec{\xi}}{(\mathrm{Sp}((AB)^{N+1}))^{\frac{1}{N+1}}}\right\}$$

$$\le Z_N^+\,\exp\left\{\frac{1}{2}\sum_{k=N+1}^{\infty}\frac{s^k}{k}\right\}, \qquad [3.17]$$

where $Z_1^+ = 1$ and for $N > 1$

$$Z_N^+ = \exp\left\{\frac{1}{2}\sum_{k=2}^{N}\frac{s^k\mathrm{Sp}((AB)^k)}{k\,(\mathrm{Sp}((AB)^{N+1}))^{\frac{k}{N+1}}}\right\}.$$

For $-\infty < s \le 0$

$$I_N(s) \le Z_N^-\,\exp\left\{\frac{|s|^{N+1}}{2(N+1)}\,\chi(N+1)\right\}, \qquad [3.18]$$

where $Z_1^- = 1$ and for $N > 1$

$$Z_N^- = \exp\left\{\frac{1}{2}\sum_{k=2}^{N}\frac{|s|^k(-1)^k \mathrm{Sp}((AB)^k)}{k\,(\mathrm{Sp}((AB)^{N+1}))^{\frac{k}{N+1}}}\right\},$$

the value $\chi(N+1)$ is determined in [3.2].

PROOF.– Under assumptions of this corollary, all λ_i are non-negative. Therefore, the sum of negative items is equal to zero $\sum^- = 0$; that implies $\gamma_{N+1}^+ = 1$ and $\gamma_{N+1}^- = 0$. Moreover,

$$\sum_{k=1}^{d}\lambda_i^k = \mathrm{Sp}(\Lambda^k) = \mathrm{Sp}(S^T B^{1/2}(AB)^{k-1}AB^{1/2}S) = \mathrm{Sp}((AB)^k).$$

For these reasons, inequalities [3.17] and [3.18] follow directly from inequalities [3.9] and [3.10]. □

COROLLARY 3.4.– Suppose that the assumptions of lemma 3.2 are satisfied and let the matrix A be positive semidefinite. Then, for all s, $0 \le s < 1$, the following inequality holds true

$$\begin{aligned} I_1(s) =& \mathbf{E}\exp\left\{\frac{s}{\sqrt{2}}\frac{\vec{\xi}^T A\vec{\xi} - \mathbf{E}\vec{\xi}^T A\vec{\xi}}{\left(\mathbf{Var}\,\vec{\xi}^T A\vec{\xi}\right)^{1/2}}\right\} \\ =& \mathbf{E}\exp\left\{\frac{s}{2}\frac{\vec{\xi}^T A\vec{\xi} - \mathbf{E}\vec{\xi}^T A\vec{\xi}}{((\mathrm{Sp}(AB)^2))^{1/2}}\right\} \le (1-s)^{-1/2}\exp\left\{-\frac{s}{2}\right\}, \end{aligned}$$

and for all s, $-\infty < s \le 0$ the inequality holds true

$$I_1(s) \le \exp\left\{\frac{s^2}{2}\right\}.$$

REMARK 3.1.– The assertions of lemma 3.2, as well as corollaries 3.1–3.4, may be proved in the case where the matrix A is not symmetric. In this case, we can write

$$\vec{\xi}^T A\vec{\xi} = \vec{\xi}^T\left(\frac{A+A^T}{2}\right)\vec{\xi}$$

and use the inequalities with the symmetric matrix $(A + A^T)/2$.

REMARK 3.2.– Let $\vec{\xi}_k, k = 1, 2, \ldots, m$ be centered jointly Gaussian random vectors and let A_k be real-valued matrices of the corresponding dimension, and let

$u_k, k = 1, 2, \ldots, m$ be arbitrary numbers. Then, for the random variable $\eta = \sum_{i=1}^{m} u_i \vec{\xi}_i^T A_i \vec{\xi}_i$ and for all s from $|s| < 1$, the following inequality holds true

$$\mathbf{E} \exp\left\{ \frac{s}{\sqrt{2}} \frac{\eta - \mathbf{E}\eta}{(\mathbf{Var}\,\eta)^{1/2}} \right\} \leq (1 - |s|)^{-\frac{1}{2}} \exp\left\{ -\frac{|s|}{2} \right\}. \qquad [3.19]$$

Inequality [3.19] follows from [3.16] since the random variable η may be represented in the form $\eta = \vec{\zeta}^T A \vec{\zeta}$, where ζ is a Gaussian centered random vector such that $\zeta^T = (\vec{\xi}_1^T, \vec{\xi}_2^T, \ldots, \vec{\xi}_m^T)$ and the matrix A is

$$A = \begin{pmatrix} u_1 A_1 & 0 & \ldots & 0 \\ 0 & u_2 A_2 & \ldots & 0 \\ \ldots & \ldots & \ldots & \ldots \\ 0 & 0 & \ldots & u_m A_m \end{pmatrix}.$$

With the help of the symmetric matrix $\frac{A + A^T}{2}$, we can obtain inequalities from lemma 3.2 and corollaries 3.1 and 3.3.

REMARK 3.3.– From inequalities [3.9]–[3.11] and [3.16]–[3.18], with the help of the Chebyshev inequality

$$\mathbf{P}\{\eta > x\} \leq \inf_{s>0} \frac{\mathbf{E}\exp\{s\eta\}}{\exp\{sx\}}$$

we can deduce inequalities for deviations such as

$$\mathbf{P}\{\vec{\xi}^T A \vec{\xi} > x\}, \quad \mathbf{P}\{|\vec{\xi}^T A \vec{\xi} - \mathbf{E}\vec{\xi}^T A \vec{\xi}| > x\}$$

For example, under the conditions of corollary 3.2 for all $x > 0$, the following inequality holds true

$$\mathbf{P}\left\{ \frac{\vec{\xi}^T A \vec{\xi} - \mathbf{E}\vec{\xi}^T A \vec{\xi}}{\left(\mathbf{Var}\,\vec{\xi}^T A \vec{\xi}\right)^{1/2}} > x \right\}$$

$$\leq \exp\left\{ -\frac{sx}{\sqrt{2}} \right\} \inf_{0 \leq s < 1} \mathbf{E}\exp\left\{ \frac{s}{\sqrt{2}} \frac{\vec{\xi}^T A \vec{\xi} - \mathbf{E}\vec{\xi}^T A \vec{\xi}}{\left(\mathbf{Var}\,\vec{\xi}^T A \vec{\xi}\right)^{1/2}} \right\}$$

$$\leq \inf_{0 \leq s < 1} (1 - s)^{-1/2} \exp\left\{ -\frac{s}{2} \right\} \exp\left\{ -\frac{sx}{\sqrt{2}} \right\}$$

$$= \left(1+x\sqrt{2}\right)^{1/2} \exp\left\{-\frac{x}{\sqrt{2}}\right\}. \qquad [3.20]$$

In the same way, we can prove that for all $x > 0$, the following inequality holds

$$\mathbf{P}\left\{\frac{\vec{\xi}^{\,T} A \vec{\xi} - \mathbf{E}\,\vec{\xi}^{\,T} A \vec{\xi}}{\left(\mathbf{Var}\ \vec{\xi}^{\,T} A \vec{\xi}\right)^{1/2}} < -x\right\} \le \left(1+x\sqrt{2}\right)^{1/2} \exp\left\{-\frac{x}{\sqrt{2}}\right\}. \qquad [3.21]$$

From inequalities [3.20] and [3.21] it follows that for all $x > 0$, the next inequality holds true

$$\mathbf{P}\left\{\frac{|\vec{\xi}^{\,T} A \vec{\xi} - \mathbf{E}\,\vec{\xi}^{\,T} A \vec{\xi}|}{\left(\mathbf{Var}\ \vec{\xi}^{\,T} A \vec{\xi}\right)^{1/2}} > x\right\} \le 2\left(1+x\sqrt{2}\right)^{1/2} \exp\left\{-\frac{x}{\sqrt{2}}\right\}. \qquad [3.22]$$

3.2. The space of square-Gaussian random variables and square-Gaussian stochastic processes

In this section, we will give all necessary information concerning the space of square-Gaussian random variables. We will use the definition of this space that was introduced in [KOZ 98].

DEFINITION 3.1.– *Let $\Xi = \{\xi_t,\ t \in \mathbf{T}\}$ be a family of jointly Gaussian random variables, $\mathbf{E}\xi_t = 0$ (for example let ξ_t, $t \in \mathbf{T}$ be a Gaussian stochastic process).*

The space $SG_\Xi(\Omega)$ is called the space of square-Gaussian random variables with respect to Ξ, if any element η from $SG_\Xi(\Omega)$ can be represented in the form

$$\eta = \bar{\xi}^T A \bar{\xi} - \mathbf{E}\bar{\xi}^T A \bar{\xi}, \qquad [3.23]$$

where $\bar{\xi}^T = (\xi_1, \xi_2, \ldots, \xi_n)$, $\xi_k \in \Xi$, $k = 1, \ldots, n$, A is a real-valued matrix, or this element $\eta \in SG_\Xi(\Omega)$ is a mean square limit of a sequence of random variables of the form [3.23]

$$\eta = \text{l.i.m.}_{n\to\infty}(\bar{\xi}_n^T A \bar{\xi}_n - \mathbf{E}\bar{\xi}_n^T A \bar{\xi}_n).$$

DEFINITION 3.2.– *A stochastic process $X = \{X(t),\ t \in \mathbf{T}\}$ is called square-Gaussian if for any $t \in \mathbf{T}$ random variable $X(t)$ belongs to the space $SG_\Xi(\Omega)$.*

LEMMA 3.3.– Let $\eta_1, \eta_2, \dots, \eta_n$ be random variables from the space $SG_\Xi(\Omega)$. Then, for all real s such that $|s| < 1$ and all real $\lambda_1, \dots, \lambda_n$, the following inequality holds true

$$\mathbf{E}\exp\left\{\frac{s}{\sqrt{2}}\frac{\eta}{(\mathbf{Var}\ \eta)^{1/2}}\right\} \leq (1-|s|)^{-1/2}\exp\left\{-\frac{|s|}{2}\right\}, \qquad [3.24]$$

where $\eta = \sum_{i=1}^{n}\lambda_i\eta_i$.

PROOF.– The assertion of lemma 3.3 follows from inequality [3.19] and the Fatou lemma. □

EXAMPLE 3.1.– Consider $\xi_1(t), \xi_2(t), \dots, \xi_n(t)$, $t \in \mathbf{T}$ a family of jointly Gaussian centered stochastic processes and let $A(t)$ be a symmetric matrix. Then

$$X(t) = \bar{\xi}^T(t)A(t)\bar{\xi}(t) - \mathbf{E}\bar{\xi}^T(t)A(t)\bar{\xi}(t),$$

where $\bar{\xi}^T(t) = (\xi_1(t), \xi_2(t), \dots, \xi_n(t))$ is a square-Gaussian stochastic process.

LEMMA 3.4.– Let a stochastic process $X = \{X(t), t \in \mathbf{T}\}$ belongs to $SG_\Xi(\Omega)$. Then, for all real s, $|s| < 1$, and all $t \in \mathbf{T}$, $t_1 \in \mathbf{T}$, $t_2 \in \mathbf{T}$, $t_1 \neq t_2$, the following inequality holds true

$$\mathbf{E}\exp\left\{\frac{sX(t)}{\sqrt{2}(\mathbf{Var}X(t))^{1/2}}\right\} \leq (1-|s|)^{-\frac{1}{2}}\exp\{-\frac{|s|}{2}\}, \qquad [3.25]$$

$$\mathbf{E}\exp\left\{\frac{s(X(t_1)-X(t_2))}{\sqrt{2}(\mathbf{Var}(X(t_1)-X(t_2)))^{1/2}}\right\} \leq (1-|s|)^{-\frac{1}{2}}\exp\{-\frac{|s|}{2}\}. \qquad [3.26]$$

PROOF.– Lemma 3.4 follows from lemma 3.3 and remark 3.2. □

REMARK 3.4.– For some stochastic process from the space $SG_\Xi(\Omega)$, inequality 3.25 may be obtained in more precise form. See, for example, inequalities [3.9]–[3.11], [3.17] and [3.18].

3.3. The distribution of supremums of square-Gaussian stochastic processes

Let $(\mathbf{T}, \rho)$ be a compact metric space with the metric ρ and let $X = \{X(t), t \in \mathbf{T}\}$ be a square-Gaussian stochastic processes.

REMARK 3.5.– All results from this and the following sections are true in the case where ρ is a pseudometric. For any pseudometric ρ, the equality $\rho(t, s) = 0$ does not imply equality $t = s$. This is the difference of a pseudometric from a metric. Note that in many cases stochastic processes are considered on a space (T, ρ), where ρ is a pseudometric. For example, we may consider the pseudometric $\rho(t,s) = (\mathbf{Var}\ (X(t) - X(s)))^{1/2}$, or $\rho(t,s) = ||X(t) - X(s)||$, where $||\cdot||$ is some norm.

Let there exist a monotonically increasing continuous function $\sigma(h)$, $h > 0$, $\sigma(h) \to 0$ as $h \to 0$ and such that the inequality holds true:

$$\sup_{\rho(t,s)\leq h} (\mathbf{Var}(X(t) - X(s)))^{\frac{1}{2}} \leq \sigma(h). \qquad [3.27]$$

Note, that this property has the function

$$\sigma(h) = \sup_{\rho(t,s)\leq h} (\mathbf{Var}\ (X(t) - X(s)))^{1/2},$$

if the process $X(t)$ is continuous in mean square.

Let us also suppose that for some $A^- \geq 1, A^+ \geq 1$ and all s such that $-A^- < s < A^+$ the following inequality holds true

$$\mathbf{E} \exp\left\{ \frac{s}{\sqrt{2}} \frac{X(t)}{(\mathbf{Var}\ X(t))^{1/2}} \right\} \leq R(s), \qquad [3.28]$$

where $R(s), -A^- < s < A^+$ is a monotonically increasing for $s > 0$ and monotonically decreasing for $s < 0$ continuous function such that $R(0) = 1$.

REMARK 3.6.– Inequality [3.28] makes sense only if $\mathbf{Var}\ X(t) > 0$. But since $\mathbf{Var}\ X(t) = 0$ implies $X(t) = 0$, we will assume that in this case

$$\frac{X(t)}{(\mathbf{Var}\ X(t))^{1/2}} = 0.$$

REMARK 3.7.– It follows from lemma 3.4 (see inequality [3.25]) that at least one function that satisfies inequality [3.28] exists. This function is

$$R(s) = (1 - |s|)^{-1/2} \exp\left\{-\frac{|s|}{2}\right\}.$$

We will use the following notations:

$$\varepsilon_0 = \inf_{t\in\mathbf{T}} \sup_{s\in\mathbf{T}} \rho(t,s), \qquad t_0 = \sigma(\varepsilon_0), \qquad [3.29]$$

$$\gamma_0 = \sup_{t\in\mathbf{T}} (\mathbf{D}\, X(t))^{1/2}, \qquad [3.30]$$

Under $N(u)$ we denote a metric massiveness of the space $\mathbf{T}$ with respect to the metric ρ, it means that $N(u)$ is the least number of closed balls of radius u covering $\mathbf{T}$. $\sigma^{(-1)}(h)$ is the inverse function to the function $\sigma(h)$.

LEMMA 3.5.– Let $X(t) = \{X(t), t \in T\}$ be a separable square-Gaussian stochastic process and the condition [3.28] holds true. Let $r(u) \geq 1, u \geq 1$ be an increasing function such that $r(u) \to \infty$ as $u \to \infty$ and let the function $r(\exp\{t\})$ be convex. If the condition

$$\int_0^{t_0} r(N(\sigma^{(-1)}(u))du < \infty, \tag{3.31}$$

is satisfied, then for all $M = 1, 2, \ldots$ and p such that $0 < p < 1$, and all u such that

$$0 < u < \frac{(1-p)}{\sqrt{2}} \min\left\{\frac{A^+}{\gamma_0}, \frac{1}{t_0 p^{M-1}}\right\} \tag{3.32}$$

the following inequality holds true:

$$\mathbf{E}\exp\left\{u \sup_{t\in T} X(t)\right\}$$

$$\leq \left(R\left(\frac{u\sqrt{2}\gamma_0}{1-p}\right)\right)^{1-p}\left[\left(1 - \frac{p^{M-1}u\sqrt{2}t_0}{1-p}\right)^{-1/2}\exp\left\{-\frac{p^{M-1}u\sqrt{2}t_0}{2(1-p)}\right\}\right]^p$$

$$\times r^{(-1)}\left(\frac{1}{t_0 p^M}\int_0^{t_0 p^M} r(N(\sigma^{(-1)}(v))dv\right). \tag{3.33}$$

Moreover, for all u such that

$$0 < u < \frac{(1-p)}{\sqrt{2}} \min\left\{\frac{A^-}{\gamma_0}, \frac{1}{t_0 p^{M-1}}\right\} \tag{3.34}$$

the following inequality holds true:

$$\mathbf{E}\exp\left\{-u \inf_{t\in T} X(t)\right\}$$

$$\leq \left(R\left(-\frac{u\sqrt{2}\gamma_0}{1-p}\right)\right)^{1-p}\left[\left(1 - \frac{p^{M-1}u\sqrt{2}t_0}{1-p}\right)^{-1/2}\exp\left\{-\frac{p^{M-1}u\sqrt{2}t_0}{2(1-p)}\right\}\right]^p$$

$$\times r^{(-1)}\left(\frac{1}{t_0p^M}\int_0^{t_0p^M} r(N(\sigma^{(-1)}(v))dv\right). \qquad [3.35]$$

PROOF.– Let $\varepsilon_k = \sigma^{(-1)}(t_0p^k), k = 0, 1, \ldots$. Denote by V_{ε_k} the set of the centers of closed balls of radii ε_k that forms a minimal covering of the space (T, ρ). It means that V_{ε_k} is ε_k-net of the set T with respect to the metric ρ. The number of points in V_{ε_k} is equal to $N(\varepsilon_k)$. The set $V = \bigcup_{k=1}^{\infty} V_{\varepsilon_k}$ is a countable everywhere dense set in (T, ρ). It follows from [3.27] and properties of the function $\sigma(u)$ that the process $X(t)$ is continuous in probability. For this reason, any countable everywhere dense in (T, ρ) set (the set V as well) may be a set of separability of the process. That is why with probability 1,

$$\sup_{t\in T} |X(t)| = \sup_{t\in V} |X(t)|. \qquad [3.36]$$

Consider the mapping $\alpha_n(t), n = 0, 1, 2, \ldots$ of the set V into V_{ε_n}: if $t \in V$, then $\alpha_n(t)$ is a point from the set V_{ε_n} such that $\rho(t, \alpha_n(t)) < \varepsilon_n$; if $t \in V_{\varepsilon_n}$, then $\alpha_n(t) = t$. If there exist many points from V_{ε_n} such that $\rho(t, \alpha_n(t)) < \varepsilon_n$, we choose one of them and denote it by $\alpha_n(t)$. The following inequality holds true:

$$\mathbf{P}\left\{|X(t) - X(\alpha_n(t))| > p^{n/2}\right\} \leq$$

$$\frac{\mathbf{Var}\ (X(t) - X(\alpha_n(t)))}{p^n} \leq \frac{\sigma^2(\varepsilon_n)}{p^n} = \frac{t_0^2 p^{2n}}{p^n} = t_0 p^n.$$

This inequality implies that

$$\sum_{n=1}^{\infty} \mathbf{P}\left\{|X(t) - X(\alpha_n(t))| > p^{n/2}\right\} \leq t_0^2 \sum_{n=1}^{\infty} p^n < \infty.$$

It follows from the Borel-Cantelli lemma that for sufficiently large n with probability 1 holds true the inequality $|X(t) - X(\alpha_n(t))| < p^{n/2}$. This implies that $X(t) - X(\alpha_n(t)) \to 0$ with probability 1 as $n \to \infty$. Since V is a countable set, then $X(t) - X(\alpha_n(t)) \to 0$ with probability 1 as $n \to \infty$ for all $t \in V$ simultaneously. Let t be an arbitrary point from V. For any $m \geq 1$, denote $t_m = \alpha_m(t), t_{m-1} = \alpha_{m-1}(t_m), t_{m-2} = \alpha_{m-2}(t_{m-1}), \ldots, t_1 = \alpha_1(t_2)$. Then, for any $M \geq 1, m > M$ the following relations hold true

$$X(t) = (X(t) - X(\alpha_m(t)) + \sum_{k=M+1}^{m} (X(t_k) - X(t_{k-1})) + X(t_M)$$

$$\leq \max_{t\in V_{\varepsilon_M}} X(t) + \sum_{k=M+1}^{m} \max_{t\in V_{\varepsilon_k}} (X(t) - X(\alpha_{k-1}(t)) + (X(t) - X(\alpha_m(t))).$$

Since the last inequality holds true for all $m > M$, then the inequality holds

$$X(t) \leq \liminf_{m\to\infty}\Bigg(\max_{t\in V_{\varepsilon_M}} X(t) + \sum_{k=M+1}^{m} \max_{t\in V_{\varepsilon_k}} (X(t) - X(\alpha_{k-1}(t)))$$

$$+(X(t) - X(\alpha_m(t)))\Bigg)$$

$$= \liminf_{m\to\infty}\left(\max_{t\in V_{\varepsilon_M}} X(t) + \sum_{k=M+1}^{m} \max_{t\in V_{\varepsilon_k}} (X(t) - X(\alpha_{k-1}(t)))\right).$$

Since the right-hand side of the last inequality does not depend on t, then with probability 1 the next inequality holds

$$\sup_{t\in V} X(t) \leq \liminf_{m\to\infty}\left(\max_{t\in V_{\varepsilon_M}} X(t) + \sum_{k=M+1}^{m} \max_{t\in V_{\varepsilon_k}} (X(t) - X(\alpha_{k-1}(t)))\right). \qquad [3.37]$$

It follows from [3.36] and [3.37] and the Fatou lemma that for any $u \geq 0$ the inequality holds

$$\mathbf{E}\exp\left\{u\sup_{t\in T} X(t)\right\} = \mathbf{E}\exp\left\{u\sup_{t\in V} X(t)\right\}$$

$$\leq \mathbf{E}\exp\left\{u\liminf_{m\to\infty}\left(\max_{t\in V_{\varepsilon_M}} X(t) + \sum_{k=M+1}^{m} \max_{t\in V_{\varepsilon_k}} (X(t) - X(\alpha_{k-1}(t)))\right)\right\}$$

$$\leq \liminf_{m\to\infty}\mathbf{E}\exp\left\{u\left(\max_{t\in V_{\varepsilon_M}} X(t) + \sum_{k=M+1}^{m} \max_{t\in V_{\varepsilon_k}} (X(t) - X(\alpha_{k-1}(t)))\right)\right\}. \qquad [3.38]$$

Let $q_k, k = M, M+1, \ldots$ be a sequence such that $q_k > 1$ and $\sum_{k=M}^{\infty} q_k^{-1} = 1$. Then, from the Hölder's inequality we will get the inequality

$$\mathbf{E}\exp\left\{u\max_{t\in V_{\varepsilon_M}} X(t) + u\sum_{k=M+1}^{m} \max_{t\in V_{\varepsilon_k}} (X(t) - X(\alpha_{k-1}(t)))\right\}$$

$$\leq \left(\mathbf{E}\exp\left\{q_M u \max_{t\in V_{\varepsilon_M}} X(t)\right\}\right)^{1/q_M}$$

$$\times \prod_{k=M+1}^{m}\left(\mathbf{E}\exp\left\{q_k u \max_{t\in V_{\varepsilon_k}} (X(t) - X(\alpha_{k-1}(t)))\right\}\right)^{1/q_k}$$

$$\leq (N(\varepsilon_M))^{1/q_M} \left(\max_{t \in V_{\varepsilon_M}} \left(\mathbf{E} \exp\{q_M u X(t)\}\right)^{1/q_M} \right)$$

$$\times \prod_{k=M+1}^{m} \left(N(\varepsilon_k)\right)^{\frac{1}{q_k}} \left(\max_{t \in V_{\varepsilon_k}} \left(\mathbf{E} \exp\left\{ q_k u (X(t) - X(\alpha_{k-1}(t))) \right\} \right)^{\frac{1}{q_k}} \right). \quad [3.39]$$

Let

$$Q(s) = (1 - |s|)^{-1/2} \exp\left\{ -\frac{|s|}{2} \right\}.$$

It follows from [3.26] that for

$$0 < q_k u \sqrt{2} \left(\mathbf{Var}\,(X(t) - X(\alpha_{k-1}(t)))\right)^{1/2} < 1$$

the following inequality holds:

$$\mathbf{E} \exp\{q_k u (X(t) - X(\alpha_{k-1}(t)))\} \leq$$

$$Q\left(q_k u \sqrt{2} \left(\mathbf{Var}\,(X(t) - X(\alpha_{k-1}(t)))\right)^{1/2} \right). \quad [3.40]$$

It follows from [3.27] that

$$\left(\mathbf{Var}\,(X(t) - X(\alpha_{k-1}(t)))\right)^{1/2} \leq \sigma(\varepsilon_{k-1}) = t_0 p^{k-1}. \quad [3.41]$$

For this reason, from [3.41] follows that under the condition

$$q_k u \sqrt{2} t_0 p^{k-1} < 1 \quad [3.42]$$

the inequality holds true:

$$\mathbf{E} \exp\{q_k u (X(t) - X(\alpha_{k-1}(t)))\} \leq Q(q_k u \sqrt{2} t_0 p^{k-1}). \quad [3.43]$$

It follows from [3.28] that under the condition

$$q_M u \sqrt{2} \left(\mathbf{Var}\, X(t)\right)^{1/2} < A^+ \quad [3.44]$$

the inequality

$$\mathbf{E} \exp\{q_M u X(t)\} \leq R\left(q_M u \sqrt{2} \left(\mathbf{Var}\, X(t)\right)^{1/2} \right) \quad [3.45]$$

holds true. Since $(\mathbf{Var}\, X(t))^{1/2} \le \gamma_0$, then from [3.44] and [3.45] follows that under the condition

$$q_M u\sqrt{2}\gamma_0 < A^+ \qquad [3.46]$$

the inequality holds true:

$$\mathbf{E}\exp\{q_M u X(t)\} \le R\left(q_M u\sqrt{2}\gamma_0\right). \qquad [3.47]$$

That is why from [3.38], [3.39], [3.43] and [3.47] it follows that for all u, which satisfy inequalities [3.42] and [3.46] holds true the inequality

$$\mathbf{E}\exp\left\{u\sup_{t\in T} X(t)\right\}$$

$$\le \liminf_{m\to\infty}\left[\prod_{k=M}^{m}(N(\varepsilon_k))^{\frac{1}{q_k}}\left(R\left(q_M u\sqrt{2}\gamma_0\right)\right)^{\frac{1}{q_M}}\prod_{k=M+1}^{m}\left(Q(q_k u\sqrt{2}t_0 p^{k-1})\right)^{\frac{1}{q_k}}\right]$$

$$= \prod_{k=M}^{\infty}(N(\varepsilon_k))^{1/q_k}\left(R\left(q_M u\sqrt{2}\gamma_0\right)\right)^{1/q_M}\prod_{k=M+1}^{\infty}\left(Q(q_k u\sqrt{2}t_0 p^{k-1})\right)^{1/q_k}. \qquad [3.48]$$

Let us take $q_k = p^{M-1}/p^{k-1}(1-p), k = M, M+1, \ldots$ For these q_k, inequality [3.42] holds true when $u < \frac{(1-p)}{\sqrt{2}t_0 p^{M-1}}$ and inequality [3.46] holds true when $u < A^+(1-p)/\sqrt{2}\gamma_0$. If we take $q_k = p^{M-1}/p^{k-1}(1-p), k = M, M+1, \ldots$ in [3.48], then for all u that satisfy [3.32], the following inequality holds

$$\mathbf{E}\exp\left\{u\sup_{t\in T} X(t)\right\}$$

$$\le \left(\prod_{k=M}^{\infty} N(\varepsilon_k)\right)^{1/q_k}\left(R\left(\frac{u\sqrt{2}\gamma_0}{(1-p)}\right)\right)^{1-p}\prod_{k=M+1}^{\infty}\left(Q\left(\frac{u\sqrt{2}t_0 p^{M-1}}{(1-p)}\right)\right)^{\frac{(1-p)p^{k-1}}{p^{M-1}}}$$

$$= \prod_{k=M}^{\infty}(N(\varepsilon_k))^{1/q_k}\left(R\left(\frac{u\sqrt{2}\gamma_0}{(1-p)}\right)\right)^{1-p}\left(Q\left(\frac{u\sqrt{2}t_0 p^{M-1}}{(1-p)}\right)\right)^{p}. \qquad [3.49]$$

Let us estimate

$$\prod_{k=M}^{\infty}(N(\varepsilon_k))^{1/q_k}.$$

Taking into consideration that the function $r(\exp\{t\})$ is convex, we will have the inequality

$$\prod_{k=M}^{\infty} (N(\varepsilon_k))^{1/q_k} = \prod_{k=M}^{\infty} \left(N(\sigma^{(-1)}(t_0 p^k))\right)^{\frac{(1-p)p^{k-1}}{p^{M-1}}}$$

$$= r^{(-1)}\left(r\left(\exp\left\{\sum_{k=M}^{\infty} p^{1-M} p^{k-1}(1-p)\ln N(\sigma^{(-1)}(t_0 p^k))\right\}\right)\right)$$

$$\leq r^{(-1)}\left(\sum_{k=M}^{\infty} p^{k-1}(1-p)p^{1-M}\left(r\left(N(\sigma^{(-1)}(t_0 p^k))\right)\right)\right)$$

$$\leq r^{(-1)}\left(\sum_{k=M}^{\infty} \frac{1}{t_0 p^M}\int_{t_0 p^{k+1}}^{t_0 p^k} r\left(N(\sigma^{(-1)}(u))\right) du\right)$$

$$= r^{(-1)}\left(\frac{1}{t_0 p^M}\int_0^{t_0 p^M} r\left(N(\sigma^{(-1)}(u))\right) du\right). \qquad [3.50]$$

From [3.49] and [3.50] follows inequality [3.33]. Inequality [3.34] may be proved similarly. □

The next theorem follows from lemma 3.5.

THEOREM 3.1.– Let $X(t) = \{X(t), t \in \mathbf{T}\}$ be separable square-Gaussian stochastic process and the conditions of lemma 3.5 are satisfied.

Then, for all integer $M = 1, 2, \ldots, p$, $0 < p < 1$, and u from

$$0 < u < \frac{1-p}{\sqrt{2}} \min\left\{\frac{1}{\gamma_0}, \frac{1}{t_0 p^{M-1}}\right\}, \qquad [3.51]$$

the inequality

$$\mathbf{P}\left\{\sup_{t\in\mathbf{T}} |X(t)| > x\right\} \leq W(p, x), \qquad [3.52]$$

is satisfied, where

$$W(p,x) = 2\left(R\Big(\frac{u\sqrt{2}\gamma_0}{1-p}\Big)\right)^{1-p} \cdot A(p)$$
$$\times \left(1 - \frac{p^{M-1}u\sqrt{2}t_0}{1-p}\right)^{-p/2} \exp\Big\{-\frac{p^M u\sqrt{2}t_0}{2(1-p)} - ux\Big\},$$

the function $R(u) = (1-|s|)^{-1/2}\exp\{-\frac{|s|}{2}\}$, and

$$A(p) = r^{(-1)}\left(\frac{1}{t_0 p^M}\int_0^{t_0 p^M} r(N(\sigma^{(-1)}(v)))\,dv\right).$$

PROOF.– The assertion of the theorem follows from lemma 3.5, remark 3.7 as $A^- = A^+ = 1$. □

COROLLARY 3.5.– Let $X(t) = \{X(t), t \in \mathbf{T}\}$ be a separable square Gaussian random process, $r(u) \geq 1, u \geq 1$ is moronically increasing function such that function $r(\exp\{t\})$ is convex. If the next integral exists

$$\int_0^{t_0} r\left(N\left(\sigma^{(-1)}(u)\right)\right) du,$$

then for all $x > 0$

$$\mathbf{P}\left\{\sup_{t\in\mathbf{T}} |X(t)| > x\right\}$$
$$\leq 2 \inf_{0<p<1}\left\{r^{(-1)}\left(\tfrac{1}{t_0 p}\int_0^{t_0 p} r\left(N\left(\sigma^{(-1)}(\nu)\right)\right) d\nu\right)\right.$$
$$\left.\times \left(1 - \tfrac{\sqrt{2}x(1-p)}{U}\right)^{-\frac{1}{2}} \exp\left\{-\tfrac{x(1-p)}{\sqrt{2}U} - \tfrac{x^2(1-p)^2}{\max(\gamma_0,t_0)U}\right\}\right\},$$

where

$$U = \max(\gamma_0, t_0) + \sqrt{2}x(1-p).$$

PROOF.– Under condition of theorem 3.1, we put $M = A^- = A^+ = 1$,

$$R(s) = (1-|s|)^{-\frac{1}{2}} \exp\left\{-\frac{|s|}{2}\right\}.$$

Let $t_0 > \gamma_0$. Since the function $R(s)$ is increasing as $0 \le s \le 1$, then

$$\inf_{0\le u<\frac{1-p}{\sqrt{2}}\min\left(\frac{A^+}{\gamma_0},\frac{1}{t_0 p^{M-1}}\right)} \left[\left(R\left(\tfrac{u\sqrt{2}\gamma_0}{1-p}\right)\right)^{1-p}\left(1-\tfrac{u\sqrt{2}t_0 p^{M-1}}{1-p}\right)^{-\frac{p}{2}} \times \exp\left\{-\tfrac{up^M\sqrt{2}t_0}{2(1-p)}-ux\right\}\right]$$

$$\le \inf_{0\le u<\frac{1-p}{\sqrt{2}t_0}} \left[\left(R\left(\tfrac{u\sqrt{2}\gamma_0}{1-p}\right)\right)^{1-p}\left(\left(1-\tfrac{u\sqrt{2}t_0}{1-p}\right)^{-\frac{1}{2}}\right)^{p} \times \left(\exp\left\{-\tfrac{u\sqrt{2}t_0}{2(1-p)}\right\}\right)^{p}\exp\{-ux\}\right]$$

$$\le \inf_{0\le u<\frac{1-p}{\sqrt{2}t_0}} \left[\left(R\left(\tfrac{u\sqrt{2}\gamma_0}{1-p}\right)\right)^{1-p}\left(R\left(\tfrac{u\sqrt{2}t_0}{1-p}\right)\right)^{p}\exp\{-ux\}\right]$$

$$\le \inf_{0\le u<\frac{1-p}{\sqrt{2}t_0}} \left[\left(1-\tfrac{u\sqrt{2}t_0}{1-p}\right)^{-\frac{1}{2}}\exp\left\{-\tfrac{u\sqrt{2}t_0}{2(1-p)}\right\}\exp\{-ux\}\right].$$

After finding the minimum of right-hand side, we have

$$\inf_{0\le u<\frac{1-p}{\sqrt{2}}\min\left(\frac{1}{\gamma_0},\frac{1}{t_0 p^{M-1}}\right)} \left[\left(R\left(\frac{u\sqrt{2}\gamma_0}{1-p}\right)\right)^{1-p}\left(1-\frac{u\sqrt{2}t_0 p^{M-1}}{1-p}\right)^{-\frac{p}{2}} \times \exp\left\{-\frac{up^M\sqrt{2}t_0}{2(1-p)}-ux\right\}\right]$$

$$\le \left(1-\frac{\sqrt{2}x(1-p)}{t_0+\sqrt{2}x(1-p)}\right)^{-\frac{1}{2}}\exp\left\{-\frac{x(1-p)}{\sqrt{2}\left(t_0+\sqrt{2}x(1-p)\right)}\right\} \times \exp\left\{-\frac{x^2(1-p)^2}{t_0\left(t_0+\sqrt{2}x(1-p)\right)}\right\}.$$

The same assertion can be obtained in the case $t_0 \le \gamma_0$.

Thereby, the corollary follows from theorem 3.1. □

3.4. The estimations of distribution for supremum of square-Gaussian stochastic processes in the space $[0,T]^d$

Consider now the space $\mathbf{T} = [0,T]^d$, $d \ge 1$, with a metric $\rho(t,s) = \max\limits_{1\le i\le d}|t_i - s_i|$. Let $X = \{X(t),\, t \in \mathbf{T}\}$ be square-Gaussian stochastic process.

Let the function $\sigma(h)$, $h > 0$ be monotonically increasing, continuous and such that inequality [3.27] is satisfied. In the case when $\sigma(h) = C \cdot h^{\alpha}$, $\alpha \in (0, 1]$, where $C > 0$, the constants ε_0 and t_0 are equal to

$$\varepsilon_0 = \inf_{t\in\mathbf{T}} \sup_{s\in\mathbf{T}} \rho(t,s) = \frac{T}{2}, \qquad t_0 = \sigma(\varepsilon_0) = C\Big(\frac{T}{2}\Big)^{\alpha}.$$

The next theorem about distribution of supremum of square-Gaussian stochastic processes follows from theorem 3.1.

THEOREM 3.2.– Let $X(t), t \in [0,T]^d$ be separable square-Gaussian stochastic process and the condition

$$\sup_{\rho(t,s)\le h} (\mathbf{Var}(X(t) - X(s)))^{\frac{1}{2}} \le \sigma(h) = C \cdot h^{\alpha}, \qquad \alpha \in (0,1], C > 0$$

holds true. If for integer $M > 1$ and any $x > 0$

$$x > \frac{\sqrt{2}\gamma_0 M d}{\alpha} \max\{1; \Big((\frac{T}{2})^{\alpha} C \frac{1}{\gamma_0} \Big)^{\frac{1}{M-1}}\}, \qquad [3.53]$$

then the tail of distribution can be estimated as

$$\mathbf{P}\Big\{\sup_{t\in\mathbf{T}} |X(t)| > x\Big\} \le 2^{1+d} e^{\frac{(M+1)d}{\alpha}} \exp\Big\{-\frac{x}{\sqrt{2}\gamma_0}\Big\} \times \Big(\frac{\alpha x}{\sqrt{2}\gamma_0 M d}\Big)^{Md/\alpha} \Big(1 + \frac{2x}{\sqrt{2}\gamma_0}\Big)^{1/2}, \qquad [3.54]$$

where $\gamma_0 = \sup_{t\in\mathbf{T}} (\mathbf{Var} X(t))^{\frac{1}{2}}$.

PROOF.– Since $\sigma(h) = C \cdot h^{\alpha}$, then $\sigma^{(-1)}(h) = (\frac{h}{C})^{1/\alpha}$.

The metric massiveness on $[0,T]^d$ with metric $\rho(t,s) = \max_{1\le i\le d} |t_i - s_i|$ is bounded as

$$N(u) \le (\frac{T}{2u} + 1)^d.$$

Hence,

$$N(\sigma^{(-1)}(u)) \le \Big(\frac{T}{2\sigma^{(-1)}(u)} + 1\Big)^d = \Big(\frac{T}{2}\Big(\frac{C}{u}\Big)^{1/\alpha} + 1\Big)^d.$$

Consider the function $r(u) = u^\beta - 1, \quad \beta \in (0, \frac{\alpha}{d})$ that satisfies the conditions of theorem 3.1. Since $0 < p < 1$ and $t_0 = C\Big(\frac{T}{2}\Big)^\alpha$, then $\frac{T}{2}\Big(\frac{C}{p^M t_0}\Big)^{1/\alpha} > 1$. Therefore, under the condition $0 < u < t_0 p^M$ the inequality

$$N(\sigma^{(-1)}(u)) \le \Big(T\Big(\frac{C}{u}\Big)^{1/\alpha}\Big)^d$$

holds. Since the inverse function of $r(u)$ is equal to $r^{(-1)}(u) = (u+1)^{1/\beta}$, then the value $A(p)$ can be estimated in such a way:

$$\begin{aligned}A(p) &= \Big(\frac{1}{t_0 p^M}\int_0^{t_0 p^M} [(\frac{T}{2}\Big(\frac{C}{u}\Big)^{1/\alpha} + 1)^{d\beta} - 1]du + 1\Big)^{1/\beta}\\ &\le \Big(\frac{1}{t_0 p^M}\int_0^{t_0 p^M} [T\Big(\frac{C}{u}\Big)^{1/\alpha}]^{d\beta} du\Big)^{1/\beta}\\ &= 2^d\big(\frac{\alpha}{\alpha - d\beta}\big)^{1/\beta} p^{-Md/\alpha}.\end{aligned}$$

Find the minimal value of the functional $A(p)$ with respect to β

$$\inf_{\beta\in(0,\frac{\alpha}{d})}\big(\frac{\alpha}{\alpha - d\beta}\big)^{1/\beta} = \lim_{\beta\to 0}\big(\frac{1}{1 - d\beta/\alpha}\big)^{1/\beta} = e^{d/\alpha}.$$

Then, from inequality [3.52] of theorem 3.1, we have

$$\begin{aligned}W(p,x) &\le 2^{1+d} e^{d/\alpha} p^{-Md/\alpha}\Big(R\Big(\frac{u\sqrt{2}\gamma_0}{1-p}\Big)\Big)^{1-p}\\ &\times\Big(1 - \frac{p^{M-1}u\sqrt{2}t_0}{1-p}\Big)^{-\frac{p}{2}}\exp\Big\{-\frac{p^M u\sqrt{2}t_0}{2(1-p)} - ux\Big\}.\end{aligned} \quad [3.55]$$

Let us recall that the function $R(s)$ is monotonically increasing and is equal to

$$R(s) = (1 - |s|)^{-1/2}\exp\{-\frac{|s|}{2}\}.$$

If $t_0 p^{M-1} < \gamma_0$, then from [3.55] it follows that

$$W(p,x) \le 2^{1+d} e^{d/\alpha} p^{-Md/\alpha} R\Big(\frac{u\sqrt{2}\gamma_0}{1-p}\Big)\exp\{-ux\}.$$

The minimum in u on the right-hand side of the last inequality is attained at

$$u_{min} = \frac{1}{z} - \frac{1}{z+2x}, \quad \text{where } z = \frac{\sqrt{2}\gamma_0}{1-p}.$$

Moreover, at the point u_{min} the condition [3.51] holds true.

Substituting u_{min} into the input inequality, we obtain

$$W(p,x) \leq 2^{1+d} e^{d/\alpha} p^{-Md/\alpha} \exp\left\{-\frac{x(1-p)}{\sqrt{2}\gamma_0}\right\}\left(1+\frac{2x(1-p)}{\sqrt{2}\gamma_0}\right)^{1/2}.$$

Since $0 < p < 1$, then

$$W(p,x) \leq 2^{1+d} e^{d/\alpha} p^{-Md/\alpha} \exp\left\{-\frac{x(1-p)}{\sqrt{2}\gamma_0}\right\}\left(1+\frac{2x}{\sqrt{2}\gamma_0}\right)^{1/2}.$$

The minimum of the right-hand side of the inequality with respect to $p \in (0,1)$ is attained in

$$p = \frac{\sqrt{2}\gamma_0 Md}{x\alpha}.$$

From this, it follows that

$$x > \frac{\sqrt{2}\gamma_0 Md}{\alpha},$$

and it is true under the condition of the theorem. Thus,

$$\mathbf{P}\left\{\sup_{t\in\mathbf{T}} |X(t)| > x\right\} \leq 2^{1+d} e^{d/\alpha} \exp\left\{-\frac{x}{\sqrt{2}\gamma_0} + \frac{Md}{\alpha}\right\}$$
$$\times\left(\frac{\alpha x}{\sqrt{2}\gamma_0 Md}\right)^{\frac{Md}{\alpha}}\left(1+\frac{2x}{\sqrt{2}\gamma_0}\right)^{1/2}.$$

From condition [3.53] it follows that $t_0 p^{M-1} < \gamma_0$. The theorem is proved. □

In the case when the space $\mathbf{T}$ is equal to $\mathbf{T} = [0,T]$, the next corollary holds true.

COROLLARY 3.6.– Let $X(t), t \in [0,T]$, be separable square-Gaussian stochastic processes and

$$\sup_{\rho(t,s)\leq h} (\mathbf{Var}(X(t)-X(s)))^{\frac{1}{2}} \leq \sigma(h) = C \cdot h^{\alpha}, \qquad \alpha \in (0,1], C > 0.$$

Suppose that for integer $M > 1$ and $x > 0$

$$x > \frac{\sqrt{2}\gamma_0 M}{\alpha} \max\{1; \left((\frac{T}{2})^{\alpha} C \frac{1}{\gamma_0}\right)^{\frac{1}{M-1}}\},$$

then the following estimator holds true

$$\mathbf{P}\Big\{\sup_{t\in\mathbf{T}}|X(t)|>x\Big\}\leq 4e^{\frac{M+1}{\alpha}}\exp\Big\{-\frac{x}{\sqrt{2}\gamma_0}\Big\}$$
$$\times\Big(\frac{\alpha x}{\sqrt{2}\gamma_0 M}\Big)^{M/\alpha}\Big(1+\frac{2x}{\sqrt{2}\gamma_0}\Big)^{1/2},$$

where $\gamma_0=\sup\limits_{t\in[0,T]}(\mathbf{Var}X(t))^{\frac{1}{2}}$.

Consider now the case when

$$\sigma(h)=\frac{c}{(\ln(e^\alpha+\frac{1}{h}))^\alpha},\quad \alpha>0 \quad\text{and}\quad c>0.$$

Then

$$t_0=\sigma(\varepsilon_0)=\frac{c}{(\ln(e^\alpha+\frac{2}{T}))^\alpha}.$$

THEOREM 3.3.– Let $X(t),t\in[0,T]^d$ be separable square-Gaussian stochastic processes and

$$\sup_{\rho(t,s)\leq h}(\mathbf{Var}(X(t)-X(s)))^{\frac{1}{2}}\leq\sigma(h)=\frac{c}{(\ln(e^\alpha+\frac{1}{h}))^\alpha},\qquad \alpha>1,c>0.$$

If for integer $M>1$ and $x>0$

$$x>\frac{\sqrt{2}\gamma_0 Md\ln(e^\alpha+\frac{2}{T})}{\alpha-1}\max\Big\{1;\Big(\frac{c}{(\gamma_0\ln(e^\alpha+\frac{2}{T}))^\alpha}\Big)^{\frac{M+\alpha}{\alpha(M-1)}}\Big\},\qquad [3.56]$$

then

$$\mathbf{P}\Big\{\sup_{t\in\mathbf{T}}|X(t)|>x\Big\}\leq K_{\alpha,d}\exp\left\{-\frac{x}{\sqrt{2}\gamma_0}+K_{\alpha,d}^M\cdot x^{\frac{M}{M+\alpha}}\right\}\Big(1+\frac{2x}{\sqrt{2}\gamma_0}\Big)^{1/2},[3.57]$$

where

$$K_{\alpha,d}=2D^{\frac{d\alpha}{\alpha-1}},$$

$$K_{\alpha,d}^M=(M+\alpha)\left(\frac{d\ln(e^\alpha+\frac{2}{T})}{\alpha-1}\right)^{\frac{\alpha}{M+\alpha}}(\sqrt{2}\gamma_0 M)^{-\frac{M}{M+\alpha}},$$

$$D = \max\{\frac{T}{2}, e^{-\alpha}\}, \qquad \gamma_0 = \sup_{t \in \mathbf{T}} (\mathbf{Var} X(t))^{\frac{1}{2}}.$$

PROOF.– To prove this assertion, we will use the results of theorem 3.1. By definition of $\sigma(u)$, it follows that

$$\sigma^{(-1)}(u) = \left(\exp\{(c/u)^{1/\alpha}\} - e^{\alpha}\right)^{-1}, \quad 0 < u < \frac{c}{\alpha^{\alpha}}.$$

Hence,

$$N(\sigma^{(-1)}(u)) \le \left(\frac{T}{2\sigma^{(-1)}(u)} + 1\right)^d = \left(\frac{T}{2}(\exp\{(c/u)^{1/\alpha}\} - e^{\alpha}) + 1\right)^d.$$

Remind that $\frac{c}{\alpha^{\alpha}} > t_0 p^M$, since $p \in (0,1)$ and $\frac{(\ln(e^{\alpha} + \frac{2}{T}))^{\alpha}}{\alpha^{\alpha}} \ge 1$ as $\alpha > 0$.

Consider the function

$$r(u) = (\ln u)^{\beta},\ \beta \in [1, \alpha),\ u \ge 1,$$

for which all conditions of theorem 3.1 hold. Then

$$r^{(-1)}(u) = \exp\{x^{1/\beta}\}.$$

For $u \in (0, t_0 p^M)$, the relations are satisfied

$$\begin{aligned}
r(N(\sigma^{(-1)}(u))) &\le r\left(\left(\frac{T}{2}(\exp\{(c/u)^{1/\alpha}\} - e^{\alpha}) + 1\right)^d\right) \\
&\le r\left(D^d \cdot \exp\{d(c/u)^{1/\alpha}\}\right) \\
&\le \left(d \ln D + d(c/u)^{1/\alpha}\right)^{\beta} \\
&\le \left(\frac{c}{u}\right)^{\beta/\alpha}\left(d \ln D\left(\frac{t_0 p^M}{c}\right)^{1/\alpha} + d\right)^{\beta} \\
&= \left(\frac{c}{u}\right)^{\beta/\alpha} Z_M(t_0) \cdot d^{\beta},
\end{aligned} \qquad [3.58]$$

where

$$D = \max(e^{-\alpha}, \frac{T}{2}), \qquad Z_M(t_0) = \left(\ln D \cdot \left(\frac{t_0 p^M}{c}\right)^{1/\alpha} + 1\right)^{\beta}.$$

From [3.58] it follows that

$$\frac{1}{t_0 p^M} \int_0^{t_0 p^M} r(N(\sigma^{(-1)}(u)))du \le \frac{\alpha c^{\beta/\alpha}}{\alpha - \beta}(t_0 p^M)^{-\beta/\alpha} Z_M(t_0) d^{\beta}$$

$$\text{for } \alpha > \beta \ge 1.$$

Hence, $1 \leq \beta < \alpha$

$$A(p) \leq \exp\left\{ \left(\frac{\alpha}{\alpha-\beta}\right)^{1/\beta} (t_0 p^M)^{-1/\alpha} c^{1/\alpha} (Z_M(t_0))^{1/\beta} d \right\}$$

$$= \exp\left\{ \left(\frac{\alpha}{\alpha-\beta}\right)^{1/\beta} \left(d \ln D + d\left(\frac{c}{t_0 p^M}\right)^{1/\alpha}\right) \right\}. \qquad [3.59]$$

Find the minimum of $A(p)$ over β. Since the function $f(\beta) = \left(\frac{\alpha}{\alpha-\beta}\right)^{1/\beta}$ is increasing , when $\beta \in [1, \alpha)$, then $\min f(\beta) = f(1) = \frac{\alpha}{\alpha-1}$. Hence,

$$A(p) \leq \exp\left\{ \frac{\alpha d}{\alpha-1} \left(\ln D + \left(\frac{c}{t_0 p^M}\right)^{1/\alpha} \right) \right\}.$$

We assume that $t_o p^{M-1} < \gamma_0$. Similarly to theorem 3.2, we obtain the inequality

$$W(p,x) \leq 2\exp\left\{ \frac{\alpha d}{\alpha-1} \left(\ln D + \left(\frac{c}{t_0 p^M}\right)^{1/\alpha} \right) \right\} R\left(\frac{u\sqrt{2}\gamma_0}{1-p}\right) \exp\{-ux\}.$$

To minimize right-hand side over u, we get

$$W(p,x) \leq 2\exp\left\{ \frac{\alpha d}{\alpha-1} \left(\ln D + \left(\frac{c}{t_0 p^M}\right)^{1/\alpha} \right) - \frac{x(1-p)}{\sqrt{2}\gamma_0} \right\}$$
$$\times \left(1 + \frac{2x}{\sqrt{2}\gamma_0}\right)^{1/2}. \qquad [3.60]$$

Let us find the minimum of right-hand side [3.60] over $p \in (0,1)$. The minimum is attained at the point

$$p = \left(\frac{\sqrt{2}\gamma_0 M d c^{1/\alpha}}{(\alpha-1) x t_0^{1/\alpha}} \right)^{\frac{\alpha}{M+\alpha}} = \left(\frac{\sqrt{2}\gamma_0 M d \ln(e^\alpha + \frac{2}{T})}{(\alpha-1)x} \right)^{\frac{\alpha}{M+\alpha}}. \qquad [3.61]$$

Therefore, for x from [3.56] the relationship

$$\mathbf{P}\left\{ \sup_{t \in \mathbf{T}} |X(t)| > x \right\}$$
$$\leq 2\exp\left\{ \frac{\alpha d}{\alpha-1} \left(\ln D + \left(\ln(e^\alpha + \frac{2}{T}) \right)^{\frac{\alpha}{M+\alpha}} \left(\frac{(\alpha-1)x}{M d \sqrt{2}\gamma_0} \right)^{\frac{M}{M+\alpha}} \right) \right\}$$
$$\times \exp\left\{ -\frac{x}{\sqrt{2}\gamma_0} + \frac{x}{\sqrt{2}\gamma_0} \left(\frac{M d \sqrt{2}\gamma_0 \ln(e^\alpha + \frac{2}{T})}{(\alpha-1)x} \right)^{\frac{\alpha}{M+\alpha}} \right\} \left(1 + \frac{2x}{\sqrt{2}\gamma_0}\right)^{1/2}$$

holds true, when $\alpha > 1$, and for such integer $M > 1$, that $t_0 p^{M-1} < \gamma_0$, p is from [3.61] and $D = \max(e^{-\alpha}, \frac{T}{2})$.

We have supposed that $t_0 p^{M-1} < \gamma_0$. And it follows from [3.56] and [3.61]. The proof of the theorem is complete. □

In the particular case when $\mathbf{T} = [0, T]$, the next corollary is carried out.

COROLLARY 3.7.– Let $X(t), t \in [0, T]$, be separable square-Gaussian stochastic processes for which

$$\sup_{\rho(t,s) \leq h} (\mathbf{Var}(X(t) - X(s)))^{\frac{1}{2}} \leq \sigma(h) = \frac{c}{(\ln(e^{\alpha} + \frac{1}{u}))^{\alpha}}, \qquad \alpha > 0, c > 0.$$

If for $x > 0$ and integer $M > 1$

$$x > \frac{\sqrt{2}\gamma_0 M \ln(e^{\alpha} + \frac{2}{T})}{\alpha - 1} \max\{1; \left(\frac{c}{(\gamma_0 \ln(e^{\alpha} + \frac{2}{T}))^{\alpha}}\right)^{\frac{M+\alpha}{\alpha(M-1)}}\}, \quad \alpha > 1,$$

then

$$\mathbf{P}\Big\{\sup_{t \in \mathbf{T}} |X(t)| > x\Big\} \leq K_{\alpha,1} \exp\left\{-\frac{x}{\sqrt{2}\gamma_0} + K_{\alpha,1}^{M} \cdot x^{\frac{M}{M+\alpha}}\right\} \left(1 + \frac{2x}{\sqrt{2}\gamma_0}\right)^{1/2},$$

where

$$K_{\alpha,1} = 2D^{\frac{\alpha}{\alpha-1}},$$

$$K_{\alpha,1}^{M} = (M + \alpha) \left(\frac{\ln(e^{\alpha} + \frac{2}{T})}{\alpha - 1}\right)^{\frac{\alpha}{M+\alpha}} (\sqrt{2}\gamma_0 M)^{-\frac{M}{M+\alpha}},$$

$$D = \max\{\frac{T}{2}, e^{-\alpha}\}, \qquad \gamma_0 = \sup_{t \in \mathbf{T}} (\mathbf{Var} X(t))^{\frac{1}{2}}.$$

3.5. Accuracy and reliability of simulation of Gaussian stochastic processes with respect to the output process of some system

Consider the space $\mathbf{T} = [0, T]^d$, $d > 1$, with metric $\rho(t, s) = \max_{1 \leq i \leq d} |t_i - s_i|$, where t, s are vectors from $\mathbf{T}$. Let $\xi = \{\xi(t), t \in \mathbf{T}\}$ be centered Gaussian stochastic process and

$$\xi(t) = \sum_{n=0}^{\infty} \xi_n f_n(t), \qquad [3.62]$$

where the functions $f_n(t)$, $n \geq 0$ are continuous and such that for all $t \in \mathbf{T}$

$$\sum_{n=0}^{\infty} f_n^2(t) < \infty,$$

ξ_n, $n = 0, 1, 2, \ldots$, are independent Gaussian random variables, $\mathbf{E}\xi_n = 0$, $E\xi_n^2 = 1$. Since

$$\mathbf{E}\xi^2(t) = \sum_{n=0}^{\infty} f_n^2(t) < \infty,$$

then the series $\sum_{n=0}^{\infty} \xi_n f_n(t)$ converges in probability (see, for example, [LOE 60]).

Consider the following situation: Let Σ be some system (filter, device) which is intended for transformation of signals (functions) $f_n(t)$. The function that has to be transformed is called the input function on system; the transformed function is called the output function or reaction on input function. Under $g_n(t)$ we will define output function. More information about filters can be found in [GIK 04].

REMARK 3.8.– In particular case, $g_n(t) = z_n \cdot f_n(t)$. It means that transformation does not change the shape of signal.

Another important situation is when $g_n(t) = f_n'(t)$.

If input process on the system Σ is $\xi(t) = \sum_{n=0}^{\infty} \xi_n f_n(t)$, then output process is $\eta(t) = \sum_{n=0}^{\infty} \xi_n g_n(t)$. Suppose that for all $t \in \mathbf{T}$, the series $\sum_{n=0}^{\infty} g_n^2(t)$ converges. It is sufficient condition for convergence in probability of the series $\eta(t) = \sum_{n=0}^{\infty} \xi_n g_n(t)$.

DEFINITION 3.3.– *The process* $\tilde{\xi}_N(t)$ *is called the model of the process* $\xi(t)$, $t \in \mathbf{T}$ *if*

$$\tilde{\xi}_N(t) = \sum_{k=0}^{N} \xi_k f_k(t), \quad t \in \mathbf{T}.$$

Let us define the difference between the process and the model under

$$\xi_N(t) = \xi(t) - \tilde{\xi}_N(t) = \sum_{k=N+1}^{\infty} \xi_k f_k(t), \quad t \in \mathbf{T}.$$

In the same way, $\eta_N(t)$ can be defined:

$$\eta_N(t) = \sum_{k=N+1}^{\infty} \xi_k g_k(t), \quad t \in \mathbf{T}.$$

We will investigate conditions under which the model $\tilde{\xi}_N(t)$ approximates $\xi(t)$ with given accuracy and reliability in Banach space $C([0,T]^d)$ taking into account the process $\eta(t)$. For this purpose, the relationship $\xi^2(t) + \eta^2(t)$ can be analyzed. If this case is generalized, we can consider a semipositive quadratic form

$$X(x,y) = a \cdot x^2 + 2c \cdot x \cdot y + b \cdot y^2,$$

where a, b, c are constants such that $a > 0$, $ab - c^2 > 0$.

For convenience, under $X_N(t)$ we will define a quadratic form that is defined on the processes $\xi_N(t)$, $\eta_N(t)$:

$$X_N(t) = X(\xi_N(t), \eta_N(t)) = a \cdot (\xi_N(t))^2 + 2c \cdot \xi_N(t) \cdot \eta_N(t) + b \cdot (\eta_N(t))^2.$$

Stochastic process $X_N(t)$ is equal to

$$X_N(t) = \sum_{k=N+1}^{\infty} \sum_{n=N+1}^{\infty} \xi_k \xi_n \phi_{kn}(t), \qquad [3.63]$$

where

$$\phi_{kn}(t) = a f_k(t) f_n(t) + c(f_k(t) g_n(t) + g_k(t) f_n(t)) + b g_k(t) g_n(t). \qquad [3.64]$$

Evidently, the function $\phi_{kn}(t)$ is symmetric with respect to k and n. Hence, $\phi_{kn}(t) = \phi_{nk}(t)$.

Denote

$$\bar{X}_N(t) = X_N(t) - \mathbf{E} X_N(t).$$

DEFINITION 3.4.– *The model $\tilde{\xi}_N(t)$ approximates stochastic process $\xi(t)$ on input of the system, taking into account output process, with given reliability $1-\nu$, $\nu \in (0,1)$ and accuracy $\delta > 0$ in Banach space $C(\mathbf{T})$, if*

$$\mathbf{P}\left\{ \sup_{t \in \mathbf{T}} |\bar{X}_N(t)| > \delta \right\} \le \nu.$$

Note that $X_N(t) - \mathbf{E}X_N(t) = \bar{X}_N(t)$, $t \in [0,T]^d$, is a square-Gaussian stochastic process.

REMARK 3.9.– In definition 3.4, the probability

$$\mathbf{P}\left\{\sup_{t\in\mathbf{T}} |X_N(t)| > \delta\right\} \le \nu$$

can be considered. But since

$$\mathbf{P}\left\{\sup_{t\in\mathbf{T}} |X_N(t)| > \delta\right\} \le \mathbf{P}\left\{\sup_{t\in\mathbf{T}} |\bar{X}_N(t)| > \delta - \sup_{t\in\mathbf{T}} |\mathbf{E}X_N(t)|\right\},$$

then all assertions can be easily transformed in this case.

The next additional lemma is proved.

LEMMA 3.6.– Let the series $\sum_{k,n=N+1}^{\infty} \phi_{kn}^2(t)$ be convergent for any $t \in \mathbf{T}$. Define

$$\Delta_{kn}(t,s) = \phi_{kn}(t) - \phi_{kn}(s).$$

Then, for the processes $\bar{X}_N(t)$, $X_N(t)$ the following relationships hold true:

$$\mathbf{E}X_N(t) = \sum_{k=N+1}^{\infty} \phi_{kk}(t),$$

$$\mathbf{Var}\bar{X}_N(t) = 2\sum_{k=N+1}^{\infty}\sum_{n=N+1}^{\infty} \phi_{kn}^2(t), \qquad [3.65]$$

$$\mathbf{Var}(X_N(t) - X_N(s)) = 2\sum_{k=N+1}^{\infty}\sum_{n=N+1}^{\infty} \Delta_{kn}^2(t,s), \qquad [3.66]$$

where the functions $\phi_{kn}(t)$ are from [3.64].

PROOF.– From [3.63], it follows that

$$X_N(t) = \sum_{k=N+1}^{\infty}\sum_{n=N+1}^{\infty} \phi_{kn}(t)\xi_k\xi_n,$$

ξ_k, $k \ge 0$, are zero-mean independent Gaussian random variables with variance 1. Then

$$\mathbf{E}X_N(t) = \sum_{k=N+1}^{\infty} \phi_{kk}(t).$$

Now find $\mathbf{E}(X_N(t))^2$.

$$\mathbf{E}(X_N(t))^2 = \sum_{k=N+1}^{\infty} \sum_{n=N+1}^{\infty} \sum_{k'=N+1}^{\infty} \sum_{n'=N+1}^{\infty} \phi_{kn}(t)\phi_{k'n'}(t)\mathbf{E}\xi_k\xi_n\xi_{k'}\xi_{n'}$$

Use the Isserlis' formula for Gaussian random variables

$$\mathbf{E}\xi_k\xi_n\xi_{k'}\xi_{n'} = \mathbf{E}\xi_k\xi_n\mathbf{E}\xi_{k'}\xi_{n'} + \mathbf{E}\xi_k\xi_{k'}\mathbf{E}\xi_n\xi_{n'} + \mathbf{E}\xi_k\xi_{n'}\mathbf{E}\xi_{k'}\xi_n.$$

From the equality above and the symmetry $\phi_{kn}(t)$ follows that

$$\mathbf{E}(X_N(t))^2 = \sum_{k=N+1}^{\infty} \sum_{n=N+1}^{\infty} \left(\phi_{kk}(t)\phi_{nn}(t) + 2\phi_{kn}^2(t)\right).$$

The relationship [3.65] is obtained from $\mathbf{Var}X_N(t) = \mathbf{Var}(\bar{X}_N(t))$.

Let us find $\mathbf{E}X_N(t)X_N(s)$:

$$\mathbf{E}X_N(t)X_N(s) = \sum_{k=N+1}^{\infty} \sum_{n=N+1}^{\infty} \left(\phi_{kk}(t)\phi_{nn}(s) + 2\phi_{kn}(t)\phi_{kn}(s)\right).$$

Then

$$\begin{aligned}\mathbf{Var}(\bar{X}_N(t) - \bar{X}_N(s)) &= \mathbf{Var}\bar{X}_N(t) + \mathbf{Var}\bar{X}_N(s) \\ &\quad - 2\mathbf{E}X_N(t)X_N(s) + 2\mathbf{E}X_N(t)\mathbf{E}X_N(s),\end{aligned}$$

that yields [3.66]. The lemma is proved. □

Denote $s_{kn} = \sup_{t\in\mathbf{T}} |\phi_{kn}(t)|$, then

$$\sup_{t\in\mathbf{T}}(\mathbf{Var}\bar{X}_N(t))^{\frac{1}{2}} \le \left(2\sum_{k=N+1}^{\infty}\sum_{n=N+1}^{\infty} s_{kn}^2\right)^{\frac{1}{2}}. \qquad [3.67]$$

The following theorem holds true.

THEOREM 3.4.– Let $\xi(t), t \in [0,T]^d$, be separable Gaussian stochastic process from [3.62] such that

$$\sup_{\rho(t,s)\le h} |\phi_{kn}(t) - \phi_{kn}(s)| \le d_{kn}h^{\alpha}, \qquad \alpha \in (0,1], \qquad [3.68]$$

and

$$2\sum_{k=N+1}^{\infty}\sum_{n=N+1}^{\infty} d_{kn}^2 = C^2(N) < \infty,$$

where $\phi_{kn}(t)$ are from [3.64].

The model $\tilde{\xi}_N(t) = \sum_{k=0}^{N} \xi_k f_k(t)$ approximates a separable Gaussian process $\xi(t)$, taking into account the output process, with given accuracy $\delta > 0$ and reliability $1-\nu$, $\nu \in (0,1)$, if for $N \geq 1$ the conditions are fulfilled

$$\delta > \frac{\sqrt{2}\gamma_0(N)Md}{\alpha}\max\{1; \left(\left(\frac{T}{2}\right)^{\alpha}\frac{C(N)}{\gamma_0(N)}\right)^{\frac{1}{M-1}}\},$$

$$2^{1+d}e^{\frac{(M+1)d}{\alpha}}\exp\left\{-\frac{\delta}{\sqrt{2}\gamma_0(N)}\right\}\left(\frac{\alpha\delta}{\sqrt{2}\gamma_0(N)Md}\right)^{Md/\alpha}\left(1+\frac{2\delta}{\sqrt{2}\gamma_0(N)}\right)^{1/2} < \nu,$$

where $M > 1$ is an arbitrary integer number, $\gamma_0(N) = \left(2\sum_{k=N+1}^{\infty}\sum_{n=N+1}^{\infty} s_{kn}^2\right)^{\frac{1}{2}}$.

PROOF.– Since the process $\bar{X}_N(t)$ is square-Gaussian, then the results of theorem 3.2 can be used. From [3.68] and [3.66] follows that

$$\begin{aligned}\sup_{\rho(t,s)\leq h}(\mathbf{Var}(X(t)-X(s)))^{\frac{1}{2}} &= \sup_{\rho(t,s)\leq h}\left(2\sum_{k=N+1}^{\infty}\sum_{n=N+1}^{\infty}(\phi_{kn}(t)-\phi_{kn}(s))^2\right)^{\frac{1}{2}}\\ &\leq \left(2\sum_{k=N+1}^{\infty}\sum_{n=N+1}^{\infty} d_{kn}^2\right)^{\frac{1}{2}} h^{\alpha}\\ &= C(N)h^{\alpha} = \sigma_N(h), \quad \alpha \in (0,1].\end{aligned}$$

Equation [3.67] implies that

$$\sup_{t\in\mathbf{T}}(\mathbf{Var}\bar{X}_N(t))^{\frac{1}{2}} \leq \left(2\sum_{k=N+1}^{\infty}\sum_{n=N+1}^{\infty} d_{kn}^2\right)^{\frac{1}{2}} = \gamma_0(N).$$

If we substitute the obtained relationships in inequalities of theorem 3.2, the theorem will be proved. □

In particular case $\mathbf{T} = [0, T]$, the following corollary holds true.

COROLLARY 3.8.– Let $\xi(t), t \in [0, T]$ be a separable Gaussian stochastic process and

$$\sup_{|t-s|\leq h} |\phi_{kn}(t) - \phi_{kn}(s)| \leq d_{kn} h^{\alpha}, \qquad \alpha \in (0, 1],$$

holds true. Suppose that

$$2 \sum_{k=N+1}^{\infty} \sum_{n=N+1}^{\infty} d_{kn}^2 = C^2(N) < \infty,$$

where $\phi_{kn}(t)$ are from [3.64].

Then, the model $\tilde{\xi}_N(t)$ approximates separable Gaussian process $\xi(t)$, taking into account the output process, with given accuracy $\delta > 0$ and reliability $1-\nu$, $\nu \in (0, 1)$, if for N the next inequalities are satisfied

$$\delta > \frac{\sqrt{2}\gamma_0(N)M}{\alpha} \max\{1; \left((\frac{T}{2})^{\alpha} \frac{C(N)}{\gamma_0(N)} \right)^{\frac{1}{M-1}} \},$$

$$4e^{\frac{(M+1)}{\alpha}} \exp\{-\frac{\delta}{\sqrt{2}\gamma_0(N)}\} \left(\frac{\alpha\delta}{\sqrt{2}\gamma_0(N)M}\right)^{M/\alpha} \left(1 + \frac{2\delta}{\sqrt{2}\gamma_0(N)}\right)^{1/2} < \nu,$$

where $M > 1$ is an arbitrary integer number, $\gamma_0(N) = \left(2 \sum_{k=N+1}^{\infty} \sum_{n=N+1}^{\infty} s_{kn}^2\right)^{\frac{1}{2}}$.

THEOREM 3.5.– Let $\xi(t), t \in [0, T]^d$ be such separable Gaussian random process that

$$\sup_{\rho(t,s)\leq h} |\phi_{kn}(t) - \phi_{kn}(s)| \leq \frac{d_{kn}}{(\ln(e^{\alpha} + \frac{1}{h}))^{\alpha}}, \qquad \alpha > 1, \qquad [3.69]$$

and

$$2 \sum_{k=N+1}^{\infty} \sum_{n=N+1}^{\infty} d_{kn}^2 = c^2(N) < \infty,$$

where $\phi_{kn}(t)$ is defined in [3.64].

The model $\tilde{\xi}_N(t)$ approximates separable Gaussian process $\xi(t)$, taking into account the output process, with given accuracy $\delta > 0$ and reliability $1 - \nu$, $\nu \in (0, 1)$, if for N the next inequalities are satisfied

$$\delta > \frac{\sqrt{2}\gamma_0(N)Md\ln(e^{\alpha} + \frac{2}{T})}{\alpha - 1} \max\{1; \left(\frac{c(N)}{(\gamma_0(N)\ln(e^{\alpha} + \frac{2}{T}))^{\alpha}}\right)^{\frac{M+\alpha}{\alpha(M-1)}}\}, \quad \alpha > 1,$$

$$K_{\alpha,d} \exp\left\{-\frac{\delta}{\sqrt{2}\gamma_0(N)} + K^M_{\alpha,d} \cdot \delta^{\frac{M}{M+\alpha}}\right\} \left(1 + \frac{2\delta}{\sqrt{2}\gamma_0(N)}\right)^{1/2} < \nu,$$

where

$$K_{\alpha,d} = 2D^{\frac{d\alpha}{\alpha-1}},$$

$$K^M_{\alpha,d} = (M+\alpha)\left(\frac{d\ln(e^\alpha + \frac{2}{T})}{\alpha - 1}\right)^{\frac{\alpha}{M+\alpha}} (\sqrt{2}\gamma_0(N)M)^{-\frac{M}{M+\alpha}},$$

$D = \max\{\frac{T}{2}, e^{-\alpha}\}$, $M > 1$ is an arbitrary number,

$$\gamma_0(N) = \left(2\sum_{k=N+1}^{\infty}\sum_{n=N+1}^{\infty} s^2_{kn}\right)^{\frac{1}{2}}.$$

PROOF.– Since the process $\bar{X}_N(t)$ is square-Gaussian, then for $\bar{X}_N(t)$ we can use the result of theorem 3.3. From conditions [3.69] and [3.66] it follows that

$$\begin{aligned}\sup_{\rho(t,s)\le h} (\mathbf{Var}(X(t) - X(s)))^{\frac{1}{2}} &= \sup_{\rho(t,s)\le h} \left(2\sum_{k=N+1}^{\infty}\sum_{n=N+1}^{\infty} (\phi_{kn}(t) - \phi_{kn}(s))^2\right)^{\frac{1}{2}} \\ &\le \left(2\sum_{k=N+1}^{\infty}\sum_{n=N+1}^{\infty} d^2_{kn}\right)^{\frac{1}{2}} \frac{1}{(\ln(e^\alpha + \frac{1}{h}))^\alpha} \\ &= \frac{c(N)}{(\ln(e^\alpha + \frac{1}{h}))^\alpha} = \sigma_N(h), \quad \alpha > 1.\end{aligned}$$

The assertion [3.67] yields

$$\sup_{t\in\mathbf{T}}(\mathbf{Var}\bar{X}_N(t))^{\frac{1}{2}} \le \left(2\sum_{k=N+1}^{\infty}\sum_{n=N+1}^{\infty} s^2_{kn}\right)^{\frac{1}{2}} = \gamma_0(N).$$

The theorem will be proved if the founded values will be substituted in the inequalities of theorem 3.3. □

In the case when $d = 1$, the following result is obtained.

COROLLARY 3.9.– Let $\xi(t), t \in [0, T]$ be a separable Gaussian random process for which

$$\sup_{\rho(t,s)\le h} |\phi_{kn}(t) - \phi_{kn}(s)| \le \frac{d_{kn}}{(\ln(e^\alpha + \frac{1}{h}))^\alpha}, \qquad \alpha > 1,$$

and

$$2\sum_{k=N+1}^{\infty}\sum_{n=N+1}^{\infty} d_{kn}^2 = c^2(N) < \infty,$$

where $\phi_{kn}(t)$ is defined in [3.64].

The model $\tilde{\xi}_N(t)$ approximates separable Gaussian process $\xi(t)$, taking into account the output process, with given accuracy $\delta > 0$ and reliability $1-\nu$, $\nu \in (0,1)$, if for N the next inequalities are satisfied

$$\delta > \frac{\sqrt{2}\gamma_0(N)M\ln(e^\alpha+\frac{2}{T})}{\alpha-1}\max\{1;\left(\frac{c(N)}{(\gamma_0(N)\ln(e^\alpha+\frac{2}{T}))^\alpha}\right)^{\frac{M+\alpha}{\alpha(M-1)}}\},\quad \alpha>1,$$

$$K_{\alpha,1}\exp\left\{-\frac{\delta}{\sqrt{2}\gamma_0(N)}+K_{\alpha,1}^M\cdot\delta^{\frac{M}{M+\alpha}}\right\}\left(1+\frac{2\delta}{\sqrt{2}\gamma_0(N)}\right)^{1/2}<\nu,$$

where $D=\max\{\frac{T}{2},e^{-\alpha}\}$, $M>1$ is an arbitrary integer,

$$K_{\alpha,1}=2D^{\frac{\alpha}{\alpha-1}},$$

$$K_{\alpha,1}^M=(M+\alpha)\left(\frac{\ln(e^\alpha+\frac{2}{T})}{\alpha-1}\right)^{\frac{\alpha}{M+\alpha}}(\sqrt{2}\gamma_0(N)M)^{-\frac{M}{M+\alpha}},$$

$$\gamma_0(N)=\left(2\sum_{k=N+1}^{\infty}\sum_{n=N+1}^{\infty}s_{kn}^2\right)^{\frac{1}{2}}.$$

EXAMPLE 3.2.– Let $\xi=\{\xi(t),t\in[0,T]\}$ be centered Gaussian process that can be represented in the form [3.62], where the functions $f_n(t)$, $n\geq 0$ are continuously differentiable and for all $t\in[0,T]$ $\sum_{n=0}^{\infty}(f_n'(t))^2<\infty$ and $\sum_{n=0}^{\infty}|f_n'(t)|<\infty$.

Consider the case where the output process is equal to $\eta(t)=\xi'(t),t\in[0,T]$. There exists the derivative of stochastic process $\xi'(t)=\sum_{n=0}^{\infty}f_n'(t)\xi_n$ in mean square.

The difference between the process and the model is

$$\xi_N(t)=\xi(t)-\tilde{\xi}_N(t)=\sum_{k=N+1}^{\infty}\xi_k f_k(t),\quad t\in\mathbf{T}$$

and the process $\eta_N(t)$ is equal to

$$\eta_N(t) = \sum_{k=N+1}^{\infty} \xi_k f'_k(t), \quad t \in \mathbf{T}.$$

Let us construct a semipositive quadratic form $X_N(t)$, which is defined on the processes $\xi_N(t)$, $\eta_N(t)$

$$X_N(t) = X(\xi_N(t), \eta_N(t)) = a \cdot (\xi_N(t))^2 + 2c \cdot \xi_N(t) \cdot \eta_N(t) + b \cdot (\eta_N(t))^2.$$

The process $X_N(t)$ can be represented in the form

$$X_N(t) = \sum_{k=N+1}^{\infty} \sum_{n=N+1}^{\infty} \xi_k \xi_n \phi_{kn}(t), \qquad [3.70]$$

where

$$\phi_{kn}(t) = af_k(t)f_n(t) + c(f_k(t)f'_n(t) + f'_k(t)f_n(t)) + bf'_k(t)f'_n(t). \qquad [3.71]$$

Then, corollaries 3.8 and 3.9 can be used for stochastic process [3.70], which gives the conditions under which the model approximates separable Gaussian process, taking into account its derivative, with given accuracy and reliability. It is shown in the following theorem.

THEOREM 3.6.– Let $\xi(t), t \in [0, T]$ be such separable Gaussian stochastic process that

$$\sup_{|t-s| \le h} |\phi_{kn}(t) - \phi_{kn}(s)| \le d_{kn} h^{\alpha}, \qquad \alpha \in (0, 1], \qquad [3.72]$$

and

$$2 \sum_{k=N+1}^{\infty} \sum_{n=N+1}^{\infty} d_{kn}^2 = C^2(N) < \infty,$$

where $\phi_{kn}(t)$ are from [3.71].

The model $\tilde{\xi}_N(t)$ approximates separable Gaussian process $\xi(t)$, taking into account the output process, with given accuracy $\delta > 0$ and reliability $1 - \nu$, $\nu \in (0, 1)$, if for N the next inequalities are satisfied

$$\delta > \frac{\sqrt{2}\gamma_0(N)M}{\alpha} \max\{1; \Big(\big(\frac{T}{2}\big)^{\alpha} \frac{C(N)}{\gamma_0(N)} \Big)^{\frac{1}{M-1}} \},$$

$$4e^{\frac{(M+1)}{\alpha}}\exp\Big\{-\frac{\delta}{\sqrt{2}\gamma_0(N)}\Big\}\Big(\frac{\alpha\delta}{\sqrt{2}\gamma_0(N)M}\Big)^{M/\alpha}\Big(1+\frac{2\delta}{\sqrt{2}\gamma_0(N)}\Big)^{1/2}<\nu,$$

where $M > 1$ is an arbitrary integer number, $\gamma_0(N) = \left(2\sum\limits_{k=N+1}^{\infty}\sum\limits_{n=N+1}^{\infty} s_{kn}^2\right)^{\frac{1}{2}}$.

THEOREM 3.7.– Let $\xi(t), t \in [0,T]$ be separable Gaussian random process such that

$$\sup_{\rho(t,s)\le h}|\phi_{kn}(t)-\phi_{kn}(s)|\le\frac{d_{kn}}{(\ln(e^\alpha+\frac{1}{h}))^\alpha},\qquad \alpha>1, \qquad [3.73]$$

and

$$2\sum_{k=N+1}^{\infty}\sum_{n=N+1}^{\infty}d_{kn}^2=c^2(N)<\infty,$$

where $\phi_{kn}(t)$ is defined in [3.71].

The model $\tilde{\xi}_N(t)$ approximates separable Gaussian process $\xi(t)$, taking into account output process, with given accuracy $\delta > 0$ and reliability $1-\nu$, $\nu \in (0,1)$, if for N the next inequalities are satisfied

$$\delta>\frac{\sqrt{2}\gamma_0(N)M\ln(e^\alpha+\frac{2}{T})}{\alpha-1}\max\Big\{1;\Big(\frac{c(N)}{(\gamma_0(N)\ln(e^\alpha+\frac{2}{T}))^\alpha}\Big)^{\frac{M+\alpha}{\alpha(M-1)}}\Big\},\quad \alpha>1,$$

$$K_{\alpha,1}\exp\left\{-\frac{\delta}{\sqrt{2}\gamma_0(N)}+K_{\alpha,1}^M\cdot\delta^{\frac{M}{M+\alpha}}\right\}\left(1+\frac{2\delta}{\sqrt{2}\gamma_0(N)}\right)^{1/2}<\nu,$$

where $D=\max\{\frac{T}{2},e^{-\alpha}\}$, $M > 1$ is an arbitrary integer number,

$$K_{\alpha,1}=2D^{\frac{\alpha}{\alpha-1}},$$

$$K_{\alpha,1}^M=(M+\alpha)\left(\frac{\ln(e^\alpha+\frac{2}{T})}{\alpha-1}\right)^{\frac{\alpha}{M+\alpha}}(\sqrt{2}\gamma_0(N)M)^{-\frac{M}{M+\alpha}},$$

$$\gamma_0(N)=\left(2\sum_{k=N+1}^{\infty}\sum_{n=N+1}^{\infty}s_{kn}^2\right)^{\frac{1}{2}}.$$

3.6. Model construction of stationary Gaussian stochastic process with discrete spectrum with respect to output process

Consider the space $\mathbf{T} = [0, T]$ with a metric $\rho(t, s) = |t - s|$. Let $\xi(t)$, $t \in \mathbf{T}$ be a stationary Gaussian stochastic process with discrete spectrum, meaning that the process can be represented in the form

$$\xi(t) = \sum_{k=0}^{\infty} (\xi_k b_k \cos \lambda_k t + \eta_k b_k \sin \lambda_k t), \qquad [3.74]$$

where ξ_k, η_k are jointly independent Gaussian random variables,

$$\mathbf{E}\xi_k = E\eta_k = \mathbf{E}\xi_k \eta_l = 0, \quad k = 0, 1, 2 \ldots, \ l = 0, 1, 2 \ldots,$$

and

$$\mathbf{E}\xi_k \xi_l = \mathbf{E}\eta_k \eta_l = 0 \quad \text{as} \quad k \neq l, \quad \mathbf{E}\xi_k^2 = \mathbf{E}\eta_l^2 = 1.$$

Suggest that the coefficients $b_k > 0$, $\sum\limits_{k=0}^{\infty} b_k^2 < \infty$, λ_k are such numbers that $0 \leq \lambda_k \leq \lambda_{k+1}$, and $\lambda_k \to \infty$ as $k \to \infty$. Since

$$\mathbf{E}\xi^2(t) = \sum_{n=0}^{\infty} b_n^2 < \infty,$$

then the series [3.74] converges in probability (see e.g. [LOE 60]).

The random process in the form [3.74] is a particular case of the processes that were investigated in section 3.5.

Assume that on some system Σ the process $\xi(t)$ is entered and the output process of this system is obtained as

$$\eta(t) = \sum_{k=0}^{\infty} z_k \cdot (\xi_k b_k \cos \lambda_k t + \eta_k b_k \sin \lambda_k t).$$

Suppose that the series $\sum\limits_{k=0}^{\infty} z_k^2 b_k^2$, $\sum\limits_{k=0}^{\infty} z_k b_k^2$ are convergent.

DEFINITION 3.5.– *The process $\tilde{\xi}_N(t)$ is called the model of stochastic process $\xi(t)$, $t \in \mathbf{T}$ if*

$$\tilde{\xi}_N(t) = \sum_{k=0}^{N} (\xi_k b_k \cos \lambda_k t + \eta_k b_k \sin \lambda_k t), \quad t \in \mathbf{T}.$$

Under $\xi_N(t)$ denote the difference between the process and the model

$$\xi_N(t) = \xi(t) - \tilde{\xi}_N(t) = \sum_{k=N+1}^{\infty} (\xi_k b_k \cos \lambda_k t + \eta_k b_k \sin \lambda_k t), \quad t \in \mathbf{T}.$$

Similarly, we can define $\eta_N(t)$:

$$\eta_N(t) = \sum_{k=N+1}^{\infty} z_k(\xi_k b_k \cos \lambda_k t + \eta_k b_k \sin \lambda_k t), \quad t \in \mathbf{T}.$$

We will investigate the conditions under which the model $\tilde{\xi}_N(t)$ approximates the process $\xi(t)$ with given accuracy and reliability in Banach space $C([0,T])$ with respect to the output process $\eta(t)$.

As in section 3.5 we consider the semipositive quadratic form $X_N(t)$, defined on the processes $\xi_N(t)$ i $\eta_N(t)$.

$$X_N(t) = a \cdot (\xi_N(t))^2 + 2c \cdot \xi_N(t) \cdot \eta_N(t) + b \cdot (\eta_N(t))^2,$$

where the numbers a, bc are such that $a > 0$, $ab - c^2 > 0$. The process $X_N(t)$ can be represented as

$$X_N(t) = \sum_{k=N+1}^{\infty} \sum_{l=N+1}^{\infty} B_{kl} \big(\xi_k \xi_l \cdot c_{kl}^1(t) + \eta_k \eta_l \cdot c_{kl}^2(t) + \xi_k \eta_l \cdot c_{kl}^3(t)$$
$$+ \eta_k \xi_l \cdot c_{kl}^4(t)\big), \qquad [3.75]$$

where

$$B_{kl} = b_k b_l (a + c(z_k + z_l) + b z_k z_l),$$

$$c_{kl}^1(t) = \cos \lambda_k t \cos \lambda_l t, \qquad c_{kl}^2(t) = \sin \lambda_k t \sin \lambda_l t,$$
$$c_{kl}^3(t) = \cos \lambda_k t \sin \lambda_l t, \qquad c_{kl}^4(t) = \sin \lambda_k t \cos \lambda_l t.$$

REMARK 3.10.– Note that the coefficients B_{kl}, $c_{kl}^i(t)$, $i = 1, 2$, are symmetric with respect to k and l, meaning that $B_{kl} = B_{lk}$, $c_{kl}^i(t) = c_{lk}^i(t)$, $i = 1, 2$, and $c_{kl}^3(t) = c_{lk}^4(t)$.

PROVE AN AUXILIARY LEMMA.–

LEMMA 3.7.– Let $\psi(u), u \geq 0$ be a continuous, monotonically increasing function, $\psi(0) = 0$, such that the function $\frac{u}{\psi(u)}$ is non-decreaasing as $u > u_0$, where the constant $u_0 \geq 0$. Then, for all $u \geq 0$ and $v > 0$, the following inequalities holds true

$$\left|\sin \frac{u}{v}\right| \leq \frac{\psi(u+u_0)}{\psi(v+u_0)}. \qquad [3.76]$$

PROOF.– Inequality [3.76] is obvious when $u \geq v$. Hence, it is enough to prove [3.76] only in the case $u < v$.

Since the function $\frac{z}{\varphi(z)}$ is non-decreasing as $z > u_0$, then

$$|\sin \frac{u}{v}| < \frac{u}{v} \leq \frac{u+u_0}{v+u_0} \leq \frac{\psi(u+u_0)}{\psi(v+u_0)}.$$

Lemma is completely proved. □

LEMMA 3.8.– Let $\psi(u), u \geq 0$ be a function such that all conditions of lemma 3.7 are fulfilled. Assume that $\psi(u) \to \infty$ as $u \to \infty$. Suppose that the series $\sum\limits_{k=0}^{\infty} b_k^2 z_k^i \psi^2(\lambda_k)$, $i = 0, 1, 2$, converge. Then, the following relationships holds for the process $X_N(t)$:

$$\mathbf{E}X_N(t) = \sum_{k=N+1}^{\infty} B_{kk}, \qquad [3.77]$$

$$\mathbf{Var}X_N(t) = 2 \sum_{k=N+1}^{\infty} \sum_{l=N+1}^{\infty} (B_{kl})^2, \qquad [3.78]$$

$$\mathbf{E}(X_N(t) - X_N(s))^2 = \mathbf{Var}(X_N(t) - X_N(s)) \leq \left(\frac{\bar{A}_N^{\psi}}{\psi(\frac{1}{|t-s|} + u_0)} \right)^2, \qquad [3.79]$$

where

$$(\bar{A}_N^{\psi})^2 = 32 \sum_{k=N+1}^{\infty} \sum_{l=N+1}^{\infty} \Big(2B_{kk}B_{ll}\psi(\frac{\lambda_k}{2} + u_0)\psi(\frac{\lambda_l}{2} + u_0)$$
$$+ (B_{kl})^2 (\psi(\frac{\lambda_k}{2} + u_0) + \psi(\frac{\lambda_l}{2} + u_0))^2 \Big).$$

PROOF.– From [3.75] and from relation $(c^1_{kk}(t))^2 + (c^2_{kk}(t))^2 = 1$, follows [3.77].

To prove [3.78] and [3.81], let us use the Isserlis' formula

$$\mathbf{E}\xi_1\xi_2\xi_3\xi_4 = \mathbf{E}\xi_1\xi_2\mathbf{E}\xi_3\xi_4 + \mathbf{E}\xi_1\xi_3\mathbf{E}\xi_2\xi_4 + \mathbf{E}\xi_1\xi_4\mathbf{E}\xi_2\xi_3,$$

where ξ_i, $i = 1, 2, 3, 4$ are Gaussian random variables.

Then, using remark 3.10, we obtain:

$$\begin{aligned}\mathbf{E}\,(X_N(t))^2 = \sum_{k=N+1}^{\infty}\sum_{l=N+1}^{\infty}\Bigg(& B_{kk}B_{ll}\Big(c^1_{kk}(t)\cdot c^1_{ll}(t) + c^2_{kk}(t)\cdot c^2_{ll}(t) \\ & + c^1_{kk}(t)\cdot c^2_{ll}(t) + c^2_{kk}(t)\cdot c^1_{ll}(t)\Big) \\ & +2(B_{kl})^2\Big((c^1_{kl}(t))^2 + (c^2_{kl}(t))^2 + (c^3_{kl}(t))^2 + (c^4_{kl}(t))^2\Big)\Bigg).\end{aligned}$$

It is easy to check

$$\begin{aligned}& c^1_{kk}(t)\cdot c^1_{ll}(t) + c^2_{kk}(t)\cdot c^2_{ll}(t) + c^1_{kk}(t)\cdot c^2_{ll}(t) + c^2_{kk}(t)\cdot c^1_{ll}(t) \\ & = (c^1_{kl}(t))^2 + (c^2_{kl}(t))^2 + (c^3_{kl}(t))^2 + (c^4_{kl}(t))^2 = 1.\end{aligned}$$

Therefore,

$$\mathbf{E}\,(X_N(t))^2 = \sum_{k=N+1}^{\infty}\sum_{l=N+1}^{\infty}\Big(B_{kk}B_{ll} + 2(B_{kl})^2\Big).$$

From the above equality and from [3.77] follows [3.78].

Consider now $\Delta_N(t,s) = X_N(t) - X_N(s)$. Let us define

$$\Delta^i_{kl}(t,s) = c^i_{kl}(t) - c^i_{kl}(s), \qquad i = 1, 2, 3, 4.$$

Then,

$$\begin{aligned}\Delta_N(t,s) = \sum_{k=N+1}^{\infty}\sum_{l=N+1}^{\infty} B_{kl}\,\big(& \xi_k\xi_l\cdot\Delta^1_{kl}(t,s) + \eta_k\eta_l\cdot\Delta^2_{kl}(t,s) \\ & +\xi_k\eta_l\cdot\Delta^3_{kl}(t,s) + \eta_k\xi_l\cdot\Delta^4_{kl}(t,s)\big).\end{aligned}$$

Note that $\mathbf{E}\Delta_N(t,s) = \mathbf{E}(X_N(t) - X_N(s)) = 0$.

Similarly to the process $X_N(t)$, we have

$$\mathbf{E}\left(\Delta_N(t,s)\right)^2 = \sum_{k=N+1}^{\infty}\sum_{l=N+1}^{\infty}\Bigg(B_{kk}B_{ll}\Big(\Delta^1_{kk}(t,s)\Delta^1_{ll}(t,s)+\Delta^2_{kk}(t,s)\Delta^2_{ll}(t,s)$$
$$+\Delta^1_{kk}(t,s)\Delta^2_{ll}(t,s)+\Delta^2_{kk}(t,s)\Delta^1_{ll}(t,s)\Big)$$
$$+2(B_{kl})^2\Big((\Delta^1_{kl}(t,s))^2+(\Delta^2_{kl}(t,s))^2$$
$$+(\Delta^3_{kl}(t,s))^2+(\Delta^4_{kl}(t,s))^2\Big)\Bigg). \qquad [3.80]$$

Estimate now the differences $\Delta^i_{kl}(t,s)$, $i=1,2,3,4$. We will use the inequality from lemma 3.7:

$$\left|\sin\frac{u}{v}\right| \le \frac{\psi(u+u_0)}{\psi(v+u_0)}.$$

We get:

$$|\Delta^1_{kl}(t,s)| = |\cos\lambda_k t\cos\lambda_l t-\cos\lambda_k s\cos\lambda_l s|$$
$$\le|\cos\lambda_k t-\cos\lambda_k s|+|\cos\lambda_k t-\cos\lambda_k s|$$
$$\le 2\left(\left|\sin\frac{\lambda_k(t-s)}{2}\right|+\left|\sin\frac{\lambda_l(t-s)}{2}\right|\right)$$
$$\le\frac{2}{\psi(\frac{1}{|t-s|}+u_0)}\left(\psi(\frac{\lambda_k}{2}+u_0)+\psi(\frac{\lambda_l}{2}+u_0)\right).$$

In the same way, $\Delta^i_{kl}(t,s)$, $i=2,3,4$ can be estimated. Hence,

$$|\Delta^i_{kl}(t,s)|\le\frac{2}{\psi(\frac{1}{|t-s|}+u_0)}\left(\psi(\frac{\lambda_k}{2}+u_0)+\psi(\frac{\lambda_l}{2}+u_0)\right),\quad i=1,2,3,4.$$

If the obtained estimations will be substituted into [3.80], we have [3.81]. □

THEOREM 3.8.– Let $\xi(t)$ be separable Gaussian stochastic process with discrete spectrum [3.74];

$$\sum_{k=0}^{\infty} b_k^2 z_k^i \lambda_k^{2\alpha}<\infty,\quad i=0,1,2,\quad \alpha\in(0,1].$$

if for integer $M>1$ and $x>0$

$$x>\frac{\sqrt{2}\gamma_0(N)M}{\alpha}\max\left\{1;\left(\frac{\bar{A}_N^\alpha}{\gamma_0(N)}\left(\frac{T}{2}\right)^\alpha\right)^{\frac{1}{M-1}}\right\},$$

then for the random process $\bar{X}_N(t) = X_N(t) - \mathbf{E}X_N(t)$ the estimation large deviation probability holds true

$$\mathbf{P}\Big\{\sup_{t\in\mathbf{T}} |\bar{X}_N(t)| > x\Big\} \leq 4e^{\frac{M+1}{\alpha}} \exp\Big\{-\frac{x}{\sqrt{2}\gamma_0(N)}\Big\}$$
$$\times \Big(\frac{\alpha x}{\sqrt{2}\gamma_0 M}\Big)^{M/\alpha}\Big(1+\frac{2x}{\sqrt{2}\gamma_0(N)}\Big)^{1/2},$$

where

$$\cdot\gamma_0(N) = \left(2\sum_{k=N+1}^{\infty}\sum_{l=N+1}^{\infty}(B_{kl})^2\right)^{\frac{1}{2}}, \qquad [3.81]$$

$$\bar{A}_N^\alpha = \left(2^{5-2\alpha}\sum_{k=N+1}^{\infty}\sum_{l=N+1}^{\infty}(2B_{kk}B_{ll}\lambda_k^\alpha\lambda_l^\alpha + (B_{kl})^2(\lambda_k^\alpha+\lambda_l^\alpha)^2)\right)^{\frac{1}{2}}. \qquad [3.82]$$

PROOF.– The theorem follows from corollary 3.6. Note that the function $\psi(u) = u^\alpha$, $\alpha \in (0,1]$ satisfies all conditions of lemma 3.7 as $u_0 = 0$.

A stochastic process $X_N(t) - \mathbf{E}X_N(t)$ is really a Square-Gaussian process, where $X_N(t)$ is defined in [3.75]. If we put $\psi(u) = u^\alpha$, $\alpha \in (0,1]$, $u_0 = 0$, then [3.81] yields the estimation:

$$\sup_{|t-s|\leq h}(\mathbf{Var}(X_N(t) - X_N(s)))^{\frac{1}{2}} \leq \frac{\bar{A}_N^\alpha}{\psi(\frac{1}{h}+u_0)}.$$

Hence,

$$\sigma_N(h) = \bar{A}_N^\alpha h^\alpha, \quad \alpha \in (0,1],$$

where A_N is from [3.82].

From corollary 3.6 $\gamma_0 = \gamma_0(N) = \sup_{t\in[0,T]}(\mathbf{Var}X_N(t))^{\frac{1}{2}}$. It follows from relation [3.78] that:

$$\mathbf{Var}\bar{X}_N(t) = \mathbf{Var}X_N(t) = 2\sum_{k=N+1}^{\infty}\sum_{l=N+1}^{\infty}(B_{kl})^2 = (\gamma_0(N))^2.$$

□

The next corollary follows from theorem 3.8 and definition 3.4.

COROLLARY 3.10.– Let

$$\sum_{k=0}^{\infty} b_k^2 z_k^i \lambda_k^{2\alpha} < \infty, \quad i = 0,1,2, \quad \alpha \in (0,1].$$

The model $\tilde{\xi}_N(t)$ approximates separable Gaussian process $\xi(t)$, taking into account the output process, with given accuracy $\delta > 0$ and reliability $1-\nu$, $\nu \in (0,1)$, if for $N \geq 1$ the conditions are fulfilled:

$$\delta > \frac{\sqrt{2}\gamma_0(N)M}{\alpha} \max\Big\{1; \Big(\frac{\bar{A}_N^{\alpha}}{\gamma_0(N)}\Big(\frac{T}{2}\Big)^{\alpha}\Big)^{\frac{1}{M-1}}\Big\},$$

$$4e^{\frac{(M+1)}{\alpha}} \exp\Big\{-\frac{\delta}{\sqrt{2}\gamma_0(N)}\Big\}\Big(\frac{\alpha\delta}{\sqrt{2}\gamma_0(N)M}\Big)^{M/\alpha}\Big(1+\frac{2\delta}{\sqrt{2}\gamma_0(N)}\Big)^{1/2} < \nu,$$

where $M > 1$ is an arbitrary integer number, $\gamma_0(N)$ is from [3.81].

The following theorem holds true.

THEOREM 3.9.– Let $\xi(t)$ be separable Gaussian stochastic process with discrete spectrum [3.74];

$$\sum_{k=0}^{\infty} b_k^2 z_k^i \ln^{2\alpha}(\lambda_k) < \infty, \quad i = 0,1,2, \quad \alpha > 1.$$

If for integer $M > 1$ and $x > 0$, $\alpha > 1$,

$$x > \frac{\sqrt{2}\gamma_0(N)M\ln(e^{\alpha}+\frac{2}{T})}{\alpha-1} \max\Big\{1; \Big(\frac{\bar{A}_N^{\ln}}{\gamma_0(N)(\ln(e^{\alpha}+\frac{2}{T}))^{\alpha}}\Big)^{\frac{M+\alpha}{\alpha(M-1)}}\Big\},$$

then for stochastic process $\bar{X}_N(t) = X_N(t) - \mathbf{E}X_N(t)$ the inequality comes true

$$\mathbf{P}\Big\{\sup_{t\in\mathbf{T}} |\bar{X}_N(t)| > x\Big\} \leq K_{\alpha,1} \exp\Big\{-\frac{x}{\sqrt{2}\gamma_0(N)} + K_{\alpha,1}^M \cdot x^{\frac{M}{M+\alpha}}\Big\} \times \Big(1+\frac{2x}{\sqrt{2}\gamma_0(N)}\Big)^{1/2},$$

where $D = \max\{\frac{T}{2}, e^{-\alpha}\}$, $\gamma_0(N)$ is from [3.81], $K_{\alpha,1} = 2D^{\frac{\alpha}{\alpha-1}}$,

$$K_{\alpha,1}^M = (M+\alpha)\left(\frac{\ln(e^{\alpha}+\frac{2}{T})}{\alpha-1}\right)^{\frac{\alpha}{M+\alpha}} (\sqrt{2}\gamma_0(N)M)^{-\frac{M}{M+\alpha}},$$

$$(\bar{A}_N^{\ln})^2 = 32 \sum_{k=N+1}^{\infty} \sum_{l=N+1}^{\infty} \Big(2B_{kk}B_{ll} \ln^\alpha(\frac{\lambda_k}{2} + e^\alpha) \ln(\frac{\lambda_l}{2} + e^\alpha)$$

$$+(B_{kl})^2 (\ln^\alpha(\frac{\lambda_k}{2} + e^\alpha) + \ln(\frac{\lambda_l}{2} + e^\alpha))^2 \Big).$$

PROOF.– The theorem follows from corollary 3.6. Note that the function $\psi(u) = \ln^\alpha(u+1)$, $\alpha > 0$ satisfies all conditions of lemma 3.7 as $u_0 = e^\alpha - 1$.

It follows from [3.81] that:

$$\sup_{|t-s|\le h} (\mathbf{Var}(X_N(t) - X_N(s)))^{\frac{1}{2}} \le \sigma_N(h) = \frac{\bar{A}_N^{\ln}}{\psi(\frac{1}{h} + u_0)}.$$

As a function $\psi(u)$ can be obtained, the function $\psi(u) = \ln^\alpha(u+1)$, $\alpha > 1$, with constant $u_0 = e^\alpha - 1$. This function satisfies all conditions of lemma 3.7. Hence,

$$\sigma_N(h) = \frac{\bar{A}_N^{\ln}}{\ln^\alpha(\frac{1}{h} + e^\alpha)}, \quad \alpha > 1.$$

From [3.78] it follows that:

$$\mathbf{Var}\bar{X}_N(t) = \mathbf{Var}X_N(t) = 2 \sum_{k=N+1}^{\infty} \sum_{l=N+1}^{\infty} (B_{kl})^2 = (\gamma_0(N))^2.$$

The theorem will be complete if the obtained values will be substituted in corollary 3.7. □

From theorem 3.9 and definition 3.4 follows the following corollary.

COROLLARY 3.11.– Let

$$\sum_{k=0}^{\infty} b_k^2 z_k^i \ln^{2\alpha}(\lambda_k) < \infty, \quad i = 0, 1, 2, \quad \alpha \in (0, 1].$$

The model $\tilde{\xi}_N(t)$ approximates separable Gaussian process with discrete spectrum $\xi(t)$, taking into account the output process, with given accuracy $\delta > 0$ and reliability $1 - \nu$, $\nu \in (0, 1)$, if for $N \ge 1$ the conditions are fulfilled:

$$\delta > \frac{\sqrt{2}\gamma_0(N) M \ln(e^\alpha + \frac{2}{T})}{\alpha - 1} \max\Big\{1; \Big(\frac{\bar{A}_N^{\ln}}{\gamma_0(N)(\ln(e^\alpha + \frac{2}{T}))^\alpha} \Big)^{\frac{M+\alpha}{\alpha(M-1)}} \Big\},$$

$$K_{\alpha,1} \exp\Big\{ -\frac{\delta}{\sqrt{2}\gamma_0(N)} + K_{\alpha,1}^M \cdot \delta^{\frac{M}{M+\alpha}} \Big\} \Big(1 + \frac{2\delta}{\sqrt{2}\gamma_0(N)}\Big)^{1/2} < \nu,$$

where $\alpha > 1$, $K_{\alpha,1}$, D, A_N and $K_{\alpha,1}^{M}$ are defined in theorem 3.9, $M > 1$ is an arbitrary integer number, $\gamma_0(N)$ is from [3.81].

EXAMPLE 3.3.– Let $\xi(t)$, $t \in \mathbf{T} = [0, T]$ be a stationary Gaussian stochastic process with discrete spectrum [3.74]. Suppose that the series $\sum\limits_{k=0}^{\infty} b_k^2 \lambda_k^4$ converges.

Consider the case where the output process is an derivative of the input process $\eta(t) = \xi'(t), t \in [0, T]$. The derivative of stochastic process $\xi'(t) = \sum\limits_{k=0}^{\infty} b_k \lambda_k(-\xi_k \sin \lambda_k t + \eta_k \cos \lambda_k t)$ exists in mean square.

Let us write the difference of the input process and the model from definition 3.5:

$$\xi_N(t) = \xi(t) - \tilde{\xi}_N(t) = \sum_{k=N+1}^{\infty} b_k(\xi_k \cos \lambda_k t + \eta_k \sin \lambda_k t), \quad t \in \mathbf{T}$$

Similarly, we can define $\eta_N(t)$ as

$$\eta_N(t) = \sum_{k=N+1}^{\infty} b_k \lambda_k(-\xi_k \sin \lambda_k t + \eta_k \cos \lambda_k t), \quad t \in \mathbf{T}.$$

Construct a semiadditive quadratic form $X_N(t)$ with respect to:

$$X_N(t) = X(\xi_N(t), \eta_N(t)) = a \cdot (\xi_N(t))^2 + 2c \cdot \xi_N(t) \cdot \eta_N(t) + b \cdot (\eta_N(t))^2.$$

Stochastic process $X_N(t)$ can be given as

$$X_N(t) = \sum_{k=N+1}^{\infty} \sum_{l=N+1}^{\infty} b_k b_l \left(\xi_k \xi_l c_{kl}^1(t) + \xi_k \eta_l c_{kl}^2(t) + \eta_k \xi_l c_{kl}^3(t) + \eta_k \eta_l c_{kl}^4(t)\right),$$

where

$$\begin{aligned} c_{kl}^1(t) &= a \cos \lambda_k t \cos \lambda_l t - c(\lambda_l \cos \lambda_k t \sin \lambda_l t + \lambda_k \cos \lambda_l t \sin \lambda_k t) \\ &\quad + b \lambda_k \lambda_l \sin \lambda_k t \sin \lambda_l t, \\ c_{kl}^2(t) &= a \cos \lambda_k t \sin \lambda_l t + c(\lambda_l \cos \lambda_k t \cos \lambda_l t - \lambda_k \sin \lambda_k t \sin \lambda_l t) \\ &\quad - b \lambda_k \lambda_l \sin \lambda_k t \cos \lambda_l t, \end{aligned}$$

$$c_{kl}^3(t) = a\sin\lambda_k t\cos\lambda_l t - c(\lambda_l\sin\lambda_k t\sin\lambda_l t - \lambda_k\cos\lambda_k t\cos\lambda_l t) \\ - b\lambda_k\lambda_l\cos\lambda_k t\sin\lambda_l t,$$

$$c_{kl}^4(t) = a\sin\lambda_k t\sin\lambda_l t + c(\lambda_l\sin\lambda_k t\cos\lambda_l t + \lambda_k\cos\lambda_k t\sin\lambda_l t) \\ + b\lambda_k\lambda_l\cos\lambda_k t\cos\lambda_l t.$$

REMARK 3.11.– Note that the function $c_{kl}^i(t)$, $i = 1, 4$ is symmetric with respect to k and l, $c_{kl}^i(t) = c_{lk}^i(t)$, $i = 1, 4$ and $c_{kl}^3(t) = c_{lk}^2(t)$.

Denote $B_{kl} = b_k b_l(a + c(\lambda_k + \lambda_l) + b\lambda_k\lambda_l)$.

LEMMA 3.9.– Let the function $\psi(u)$, $u \geq 0$ satisfies the conditions of lemma 3.7, and $\psi(u) \to \infty$ as $u \to \infty$.

Suggest that the series $\sum\limits_{k=0}^{\infty} b_k^2\lambda_k^2\psi^2(\lambda_k)$ converges.

Then, the following relationships holds for stochastic process $X_N(t)$:

$$\mathbf{E}X_N(t) = \sum_{k=N+1}^{\infty} b_k^2(a + b\lambda_k^2), \qquad [3.83]$$

$$\mathbf{Var}X_N(t) \leq \sum_{k=N+1}^{\infty}\sum_{l=N+1}^{\infty} \Big(4B_{kk}B_{ll} + 8(B_{kl})^2 \\ - b_k^2 b_l^2(a + b\lambda_k^2)(a + b\lambda_l^2)\Big) = (\gamma_0(N))^2, \qquad [3.84]$$

$$\mathbf{E}(X_N(t) - X_N(s))^2 = \mathbf{Var}(X_N(t) - X_N(s)) \leq \left(\frac{\bar{A}_N^{\psi}}{\psi(\frac{1}{|t-s|} + u_0)}\right)^2, \qquad [3.85]$$

where

$$(\bar{A}_N^{\psi})^2 = 32\sum_{k=N+1}^{\infty}\sum_{l=N+1}^{\infty}\left(2B_{kk}B_{ll}\psi(\frac{\lambda_k}{2} + u_0)\psi(\frac{\lambda_l}{2} + u_0) \\ + (B_{kl})^2(\psi(\frac{\lambda_k}{2} + u_0) + \psi(\frac{\lambda_l}{2} + u_0))^2\right).$$

PROOF.– The proof of the lemma is the same as for lemma 3.8. □

Note that the values $\bar{A}_N^{\psi}$ in lemmas 3.9 and 3.8 coincide.

The approximation theorems for Gaussian stochastic process with discrete spectrum also hold true.

THEOREM 3.10.– Let $\sum_{k=0}^{\infty} b_k^2 \lambda_k^{2+2\alpha} < \infty, \quad \alpha \in (0,1]$.

The model $\tilde{\xi}_N(t)$ approximates separable Gaussian process with discrete spectrum $\xi(t)$, taking into account the derivative of the process, with given accuracy $\delta > 0$ and reliability $1 - \nu$, $\nu \in (0,1)$, if for $N \geq 1$ the conditions are fulfilled:

$$\delta > \frac{\sqrt{2}\gamma_0(N)M}{\alpha} \max\Big\{1; \Big(\frac{\bar{A}_N^{\alpha}}{\gamma_0(N)}\Big(\frac{T}{2}\Big)^{\alpha}\Big)^{\frac{1}{M-1}}\Big\},$$

$$4e^{\frac{(M+1)}{\alpha}} \exp\Big\{-\frac{\delta}{\sqrt{2}\gamma_0(N)}\Big\}\Big(\frac{\alpha\delta}{\sqrt{2}\gamma_0(N)M}\Big)^{M/\alpha}\Big(1 + \frac{2\delta}{\sqrt{2}\gamma_0(N)}\Big)^{1/2} < \nu,$$

where $\bar{A}_N^{\alpha}$ is from [3.82], $M > 1$ is an arbitrary integer number, $\gamma_0(N)$ is defined in [3.84].

EXAMPLE 3.4.– Consider a Gaussian stationary process with discrete spectrum from [3.74]. Suppose that

$$\lambda_k = k, \quad b_k = \frac{1}{k^s},$$

where $s > 2,5$. Consider a particular case when $\alpha = 1$. Evidently, the series in theorem 3.10 will be convergent if $2s - 4 > 1$. We consider now semi-positive quadratic form $X_N(t)$ defined on the processes $\xi_N(t)$ and $\eta_N(t)$ as

$$X_N(t) = (\xi_N(t))^2 + (\xi'_N(t))^2.$$

In this case, the value B_{kl} is equal to

$$B_{kl} = \frac{1}{(kl)^s}(1 + kl).$$

Let us estimate now the series $\sum_{k=N+1}^{\infty}\sum_{l=N+1}^{\infty}\frac{1}{k^m l^p}$, where $m > 1$ and $p > 1$ are some numbers:

$$\begin{aligned}\sum_{k=N+1}^{\infty}\sum_{l=N+1}^{\infty}\frac{1}{k^m l^p} &= \sum_{k=N+1}^{\infty}\sum_{l=N+1}^{\infty}\int_{k-1}^{k}\int_{l-1}^{l}\frac{1}{k^m l^p}dudv \\ &\leq \int_{N}^{\infty}\int_{N}^{\infty}\frac{1}{v^m u^p}dudv \\ &= \frac{1}{(m-1)(p-1)}\frac{1}{N^{m+p-2}}. \end{aligned} \quad [3.86]$$

By [3.86], we obtain that

$$\sum_{k=N+1}^{\infty}(B_{kl})^2 \leq \frac{1}{(2s-1)^2 N^{(2s-1)2}} + \frac{2}{(2s-2)^2 N^{(2s-2)2}} + \frac{1}{(2s-3)^2 N^{(2s-3)2}}$$

and

$$\sum_{k=N+1}^{\infty} B_{kk} \leq \frac{1}{(2s-1)N^{2s-1}} + \frac{1}{(2s-3)N^{2s-3}}.$$

Then, the quantity $\gamma_0(N)$ from [3.84] can be estimated as:

$$\begin{aligned}(\gamma_0(N))^2 &= 3\left(\sum_{k=N+1}^{\infty} B_{kk}\right)^2 + 8\sum_{k=N+1}^{\infty}\sum_{l=N+1}^{\infty}(B_{kl})^2 \\ &\leq \frac{1}{N^{4s-2}}\left(3\left(\frac{1}{2s-1}+\frac{N^2}{2s-3}\right)^2\right. \\ &\quad \left. +8\left(\frac{1}{(2s-1)^2}+\frac{2N^2}{2s-2}+\frac{N^4}{(2s-3)^2}\right)\right).\end{aligned}$$

It is easy to show that $\bar{A}_N^\alpha$ from [3.82] in the case $\alpha = 1$ is transformed in

$$(\bar{A}_N^\alpha)^2 = \frac{4}{N^{4s-2}}\Bigg(\Big(\frac{N}{2s-2}+\frac{N^3}{2s-4}\Big)^2+\frac{N^2}{(2s-1)(2s-3)}$$
$$+\frac{2N^4}{(2s-2)(2s-4)}+\frac{N^6}{(2s-3)(2s-5)}+\frac{1}{(2s-2)^2}$$
$$+\frac{2N^4}{(2s-3)^2}+\frac{N^6}{(2s-4)^2}\Bigg).$$

Consider the case when the values from theorem 3.10 are $T = 1$ and $M = 2$. Then, the inequalities of theorem 3.10 will be rewritten as

$$\delta > 2\sqrt{2}\gamma_0(N)\max\Big\{1; \frac{\bar{A}_N^\alpha}{2\gamma_0(N)}\Big\}$$

and

$$4e^3\exp\Big\{-\frac{\delta}{\sqrt{2}\gamma_0(N)}\Big\}\Big(\frac{\delta}{2\sqrt{2}\gamma_0(N)}\Big)^2\Big(1+\frac{2\delta}{\sqrt{2}\gamma_0(N)}\Big)^{1/2} < \nu.$$

Assume that $s = 3$. The values of N dependent on accuracy δ and reliability $1-\nu$ are found in environment for statistical computing R and are shown in Table 3.1.

	$\nu = 0.1$	$\nu = 0.05$	$\nu = 0.01$
$\delta = 0.1$	6	7	7
$\delta = 0.06$	7	8	8
$\delta = 0.01$	13	13	14

Table 3.1. *The result of the simulation of stationary Gaussian process with discrete spectrum taking into account the derivative of the process*

THEOREM 3.11.– Let $\sum\limits_{k=0}^{\infty} b_k^2\lambda_k^2\ln^{2\alpha}(\lambda_k) < \infty, \quad \alpha > 1.$

The model $\tilde{\xi}_N(t)$ approximates separable Gaussian process with discrete spectrum $\xi(t)$, taking into account the derivative of the process, with given accuracy $\delta > 0$ and reliability $1-\nu$, $\nu \in (0,1)$, if for $N \geq 1$ the conditions are fulfilled

$$\delta > \frac{\sqrt{2}\gamma_0(N)M\ln(e^\alpha+\frac{2}{T})}{\alpha-1}\max\Big\{1; \Big(\frac{\bar{A}_N^{\ln}}{\gamma_0(N)(\ln(e^\alpha+\frac{2}{T}))^\alpha}\Big)^{\frac{M+\alpha}{\alpha(M-1)}}\Big\},$$

$$K_{\alpha,1}\exp\left\{-\frac{\delta}{\sqrt{2}\gamma_0(N)}+K_{\alpha,1}^{M}\cdot\delta^{\frac{M}{M+\alpha}}\right\}\left(1+\frac{2\delta}{\sqrt{2}\gamma_0(N)}\right)^{1/2}<\nu,$$

where $\alpha > 1$, $K_{\alpha,1}$, D, $\bar{A}_N^{\ln}$ and $K_{\alpha,1}^{M}$ are defined in theorem 3.9, $M > 1$ is an arbitrary integer number, $\gamma_0(N)$ is from [3.84].

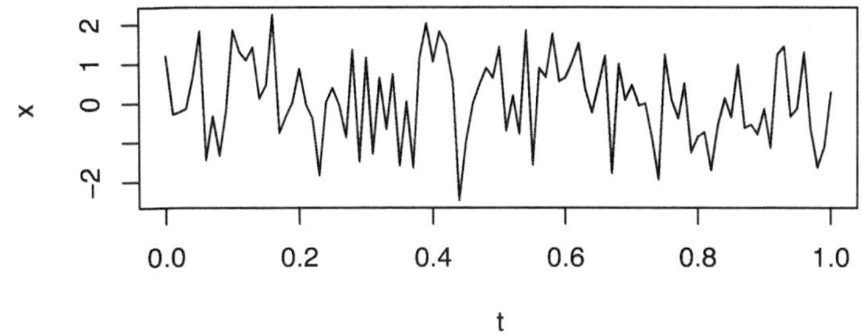

Figure 3.1. *The sample path of the model of Gaussian stationary process with discrete spectrum taking into account the derivative of the process with accuracy 0.01 and reliability 0.99 in space* $C([0,1])$

3.7. Simulation of Gaussian stochastic fields

Let $(\mathbf{T}, \rho)$ be a compact metric space with the metric ρ, where $\mathbf{T} = [0,T]^d$ and $\rho(t,s) = \max\limits_{1\le i\le d}|t_i - s_i|$, where t, s are vectors from $\mathbf{T}$. Let $\xi = \{\xi(t), t \in \mathbf{T}\}$ be a centered Gaussian random field that can be represented in the form

$$\xi(t) = \sum_{n=1}^{\infty}\xi_n f_n(t), \qquad [3.87]$$

where $f_n(t)$, $n \ge 1$ are continuous functions and ξ_n, $n = 1,2,\ldots$ are independent Gaussian random variables such that $\mathbf{E}\xi_k = 0$, $E\xi_k^2 = 1$. We also suppose that the following series converges for all $t \in \mathbf{T}$:

$$\sum_{n=1}^{\infty}E(\xi_n f_n(t))^2 = \sum_{n=1}^{\infty}f_n^2(t) < \infty. \qquad [3.88]$$

From [3.88] it follows that the series [3.87] converges with probability one for every $t \in \mathbf{T}$ (see, for example, [LOE 60]).

Suppose the continuous derivatives $\frac{\partial f_n(t)}{\partial t_i}$, $i = \overline{1,d}$, $t \in \mathbf{T}$, $n = 1, 2, \ldots$ exist.

For convenience, define $\xi_N(t) = \sum\limits_{n=N}^{\infty} \xi_n f_n(t)$, $N = 1, 2, \ldots$, a vector $t \in [0, T]^d$.

Consider a positive semidefinite quadratic form

$$X_N(t) = \xi_N^2(t) + \sum_{i=1}^{d} \left(\frac{\partial \xi_N(t)}{\partial t_i}\right)^2, N \geq 1. \qquad [3.89]$$

Consider

$$\sum_{k=N}^{\infty} \sum_{l=N}^{\infty} \xi_k \xi_l \left(f_k(t) f_l(t) + \sum_{i=1}^{d} \frac{\partial f_k(t)}{\partial t_i} \frac{\partial f_l(t)}{\partial t_i} \right) = \sum_{k=N}^{\infty} \sum_{l=N}^{\infty} \xi_k \xi_l \phi_{kl}(t), \qquad [3.90]$$

where

$$\phi_{kl} = f_k(t) f_l(t) + \sum_{i=1}^{d} \frac{\partial f_k(t)}{\partial t_i} \frac{\partial f_l(t)}{\partial t_i}. \qquad [3.91]$$

Suppose $\sum\limits_{k=N}^{\infty} \sum\limits_{l=N}^{\infty} \phi_{kl}^2(t) < \infty$. Show now that under some conditions [3.89] is equal to [3.90] and this series converges uniformly for $t \in \mathbf{T}$ with probability 1. The theorem follows from [BUL 00].

THEOREM 3.12.– Let there exist the continuous derivatives $\frac{\partial f_n(t)}{\partial t_i}$, $1 \leq i \leq d$ and the series $\sum\limits_{k=N}^{\infty} \sum\limits_{l=N}^{\infty} \phi_{kl}^2(t)$ converges for all $t \in \mathbf{T}$. Suppose,

$$\sup_{|t-s| \leq h} |\phi_{kl}(t) - \phi_{kl}(s)| \leq c_{kl} \varrho(h), \quad \sum_{k,l=1}^{\infty} c_{kl} < \infty,$$

where $\varrho(h)$, $h > 0$ is a continuous non-decreasing function such that $\varrho(h) \to 0$ as $h \to 0$, for which

$$\int_{0+} \left| \ln \varrho^{(-1)}(\varepsilon) \right|^{1/2} d\varepsilon < \infty,$$

holds true. Then, the series $\sum_{k=N}^{\infty}\sum_{l=N}^{\infty}\phi_{kl}(t)\xi_k\xi_l$ converges uniformly as $t\in\mathbf{T}$ with probability 1 and

$$X_N(t)=\xi_N^2(t)+\sum_{i=1}^{d}\left(\frac{\partial\xi_N(t)}{\partial t_i}\right)^2=\sum_{k=N}^{\infty}\sum_{l=N}^{\infty}\xi_k\xi_l\phi_{kl}(t),$$

where $\phi_{kl}(t)$ are from [3.91].

Let us give some definitions.

DEFINITION 3.6.– *A field* $\tilde{\xi}_N(t)=\sum_{n=1}^{N}\xi_n f_n(t)$, $t\in[0,T]^d$ *is called a model of the field* $\xi(t)$ *from [3.87].*

DEFINITION 3.7.– *The model* $\tilde{\xi}_N(t)$ *approximates* $\xi(t)$ *with given reliability* $1-\nu$, $\nu\in(0,1)$ *and accuracy* $\delta>0$ *in Banach space* $C^1_{[o,T]}$ *if the next inequality holds*

$$\mathbf{P}\left\{\sup_{t\in\boldsymbol{T}}|X_{N+1}(t)-\mathbf{E}X_{N+1}(t)|>\delta\right\}\leq\nu,$$

where

$$\tilde{X}(t)=\tilde{\xi}_N^2(t)+\sum_{i=1}^{d}\left(\frac{\partial\tilde{\xi}_N(t)}{\partial t_i}\right)^2,$$

$$X_{N+1}(t)=X(t)-\tilde{X}(t).$$

We want to construct a model $\tilde{\xi}_N(t)$ of the Gaussian field $\xi(t)$ such that $\tilde{\xi}_N(t)$ approximates $\xi(t)$ with given reliability and accuracy.

Then, the next lemma is similar to lemma 3.6.

LEMMA 3.10.– For $X_N(t)$, $N=1,\ldots,$ $t\in[0,T]^d$, the following relationships hold true

$$\mathbf{E}X_N(t)=\sum_{n=N}^{\infty}\phi_{nn}(t),$$

$$\mathbf{Var}X_N(t)=2\sum_{k=N}^{\infty}\sum_{l=N}^{\infty}\phi_{kl}^2(t),$$

$$\mathbf{Var}(X_N(t)-X_N(s))=2\sum_{k=N}^{\infty}\sum_{l=N}^{\infty}(\phi_{kl}(t)-\phi_{kl}(s))^2, \qquad [3.92]$$

where $\phi_{kl}(t)$ is from [3.91].

THEOREM 3.13.– Let $\xi(t) = \{\xi(t), t \in \mathbf{T}\}$ be a centered Gaussian separable stochastic process and there exist continuous derivatives $\frac{\partial \xi(t)}{\partial t_i}$, $i = \overline{1,d}$.

Assume that

$$\sup_{\rho(t,s)\le h} |\phi_{kl}(t) - \phi_{kl}(s)| \le d_{kl} \cdot |h|^{\alpha}, \quad 2\sum_{k=N+1}^{\infty}\sum_{l=N+1}^{\infty} d_{kl}^2 = C^2 < \infty, \qquad [3.93]$$

where $\alpha \in (0;1]$, a constant $C > 0$ and $\phi_{kl}(t)$ are from [3.91]. The model $\tilde{\xi}_N(t)$ approximates separable Gaussian field $\xi(t)$ with given reliability $1-\nu$, $\nu \in (0,1)$ and accuracy $\delta > 0$ if for $\alpha \in (0,1]$ and integer $M \ge 1$ the following inequalities hold true:

$$\delta > \frac{\sqrt{2}\gamma_0 M d}{\alpha} \max\{1; \left(\left(\frac{T}{2}\right)^{\alpha}\frac{C}{\gamma_0}\right)^{\frac{1}{M-1}}\},$$

$$2^{1+d} d^{d/2} e^{\frac{(M+1)d}{\alpha}} \exp\left\{-\frac{\delta}{\sqrt{2}\gamma_0}\right\}\left(\frac{\alpha\delta}{\sqrt{2}\gamma_0 M d}\right)^{Md/\alpha}\left(1 + \frac{2\delta}{\sqrt{2}\gamma_0}\right)^{1/2} < \nu,$$

where $\gamma_0 = \sup\limits_{t\in\mathbf{T}}(\mathbf{Var} X_{N+1})^{1/2}$.

PROOF.– From condition [3.93] it follows that theorem 3.2 holds true. Hence, $X_{N+1}(t) = \xi_{N+1}^2(t) + \sum_{i=1}^{d}\left(\frac{\partial \xi_{N+1}(t)}{\partial t_i}\right)^2 = \sum_{k=N+1}^{\infty}\sum_{l=N+1}^{\infty} \xi_k \xi_l \phi_{kl}^2(t)$ and this series converge uniformly as $t \in \mathbf{T}$. From [3.92] and [3.93] it follows:

$$\sup_{\rho(t,s)\le h} \left(\mathbf{Var}(X_{N+1}(t) - X_{N+1}(s))\right)^{1/2} = \left(2 \sup_{\rho(t,s)\le h} \sum_{k,l=N+1}^{\infty} (\phi_{kl}(t) - \phi_{kl}(s))^2\right)^{1/2}$$

$$\le \left(2\sum_{k=N+1}^{\infty}\sum_{l=N+1}^{\infty} d_{kl}^2\right)^{1/2} |h|^{\alpha} = C|h|^{\alpha} = \sigma(h).$$

Now the proof of the theorem follows from theorem 3.2. □

Using theorems 3.3 and 3.12, the following theorem is obtained.

THEOREM 3.14.– Let $\xi(t) = \{\xi(t), t \in \mathbf{T}\}$ be a centered Gaussian separable stochastic process and there exist continuous derivatives $\frac{\partial \xi(t)}{\partial t_i}$, $i = \overline{1,d}$.

Assume that

$$\sup_{\rho(t,s)\le h} |\phi_{kl}(t) - \phi_{kl}(s)| \le \frac{n_{kl}}{(\ln(e^{\alpha} + \frac{1}{h}))^{\alpha}}, \quad 2\sum_{k=N+1}^{\infty}\sum_{l=N+1}^{\infty} n_{kl}^2 = c^2 < \infty,$$

where $\alpha > 1$, a constant $c > 0$ and $\phi_{kl}(t)$ is from [3.91]. The model $\tilde{\xi}_N(t)$ approximates separable Gaussian field $\xi(t)$ with given reliability $1-\nu$, $\nu \in (0,1)$ and accuracy $\delta > 0$ if for $\alpha > 1$ and integer $M \geq 1$, the following inequalities hold true:

$$\delta > \frac{\sqrt{2}\gamma_0 M d \ln(e^{\alpha} + \frac{2}{T\sqrt{d}})}{\alpha - 1} \max\{1; \Big(\frac{c}{(\gamma_0 \ln(e^{\alpha} + \frac{2}{T}))^{\alpha}}\Big)^{\frac{M+\alpha}{\alpha(M-1)}}\}, \qquad \alpha > 1,$$

$$2\exp\left\{\frac{d\alpha}{\alpha-1}\Big(\ln D + \Big(\ln(e^{\alpha} + \frac{2}{T})\Big)^{\frac{\alpha}{M+\alpha}}\Big(\frac{(\alpha-1)\delta}{Md\sqrt{2}\gamma_0}\Big)^{\frac{M}{M+\alpha}}\Big)\right\}$$

$$\times \exp\left\{-\frac{\delta}{\sqrt{2}\gamma_0} + \frac{\delta}{\sqrt{2}\gamma_0}\Big(\frac{Md\sqrt{2}\gamma_0 \ln(e^{\alpha} + \frac{2}{T})}{(\alpha-1)\delta}\Big)^{\frac{\alpha}{M+\alpha}}\right\}\Big(1 + \frac{2\delta}{\sqrt{2}\gamma_0}\Big)^{1/2} < \nu,$$

where $D = \max\{\frac{T\sqrt{d}}{2}, e^{-\alpha}\}$, $\gamma_0 = \sup_{t\in\mathbf{T}}(\mathbf{Var}X_{N+1}(t))^{1/2}$.

3.7.1. *Simulation of Gaussian fields on spheres*

Consider now the Gaussian field on unit sphere $S_d \in \mathbb{R}^d$, $d \geq 3$. Let us give some well-known designations; $(r, \theta_1, \ldots, \theta_{d-2}, \mathbf{P})$ are spherical coordinates of the point x, where $0 \leq \theta_i \leq \pi$, $0 \leq \mathbf{P} \leq 2\pi$ and in our case $r = 1$. Then, the coordinates of the point x can be rewritten as

$$\begin{aligned} x_i &= \sin\theta_1 \sin\theta_2 \cdots \sin\theta_{i-1}\cos\theta_i, \quad i = 1, \ldots, d-2, \\ x_{d-1} &= \sin\theta_1 \sin\theta_2 \cdots \sin\theta_{d-3}\sin\theta_{d-2}\cos\mathbf{P}, \\ x_d &= \sin\theta_1 \sin\theta_2 \cdots \sin\theta_{d-3}\sin\theta_{d-2}\sin\mathbf{P}. \end{aligned}$$

In this section, we will use spherical coordinates. In this case, the sphere S_d is transformed in $[0,\pi]^{d-2} \times [0, 2\pi]$ and the metric $\rho(x, x') = \max_{1\leq i\leq d-1} |\theta_i - \theta_i'|$ is considered, where for convenience we denote $\theta_{d-1} = \mathbf{P}$. Consider the full system of orthogonal spherical harmonics. Then, for them the following theorem is fulfilled.

THEOREM 3.15.– [BAT 53] Let $m_0, \ldots, m_{d-2}$ be such integer numbers that

$$m = m_0 \geq m_1 \geq \ldots \geq m_{d-2} \geq 0,$$

then the spherical harmonic polynomials:

$$\begin{aligned} Y(m_k, \theta_k, \pm\mathbf{P}) &= Y(m, m_1, \ldots, m_{d-2}, \theta_1, \ldots, \theta_{d-2}, \mathbf{P}) = \\ &= e^{\pm i m_{d-2}\mathbf{P}} \prod_{k=0}^{d-3} (\sin\theta_{k+1})^{m_{k+1}} C_{m_k - m_{k+1}}^{m_{k+1} + \frac{d-2}{2} - \frac{k}{2}}(\cos\theta_{k+1}) \end{aligned} \quad [3.94]$$

make up the full system of linearly independent harmonic polynomials of the power of m on S_d.

Let us number the harmonics from theorem 3.15 and denote them by $S^l_m(\theta_1, \ldots, \theta_{d-2}, \mathbf{P}), l = 1, \ldots, h(m, d))$, where m is a power of polynomials and

$$h(m, d) = (2m + d - 2)\frac{(m + d - 3)!}{(d - 2)!m!}$$

is the number of linearly independent harmonics of the power of m.

Let $C^\nu_m(x)$, $\nu \neq 0$ be the Gegenbauer polynomials that are defined by such productive function

$$(1 - 2xt + t^2)^{-\nu} = \sum_{m=0}^{\infty} C^\nu_m(x)t^m.$$

DEFINITION 3.8.– *A stochastic field $\xi(x)$ on sphere S_d is called isotropic in wide sense if*

$$\mathbf{E}\xi(x_1)\xi(x_2) = B(\cos\theta)$$

depends only on the angular distance θ between x_1 and x_2.

Then, the correlation function of isotropic stochastic field can be represented in the form [YAD 83]

$$B(\cos\theta) = \frac{1}{\omega_d}\sum_{m=0}^{\infty} b_m \frac{C_m^{\frac{d-2}{2}}(\cos\theta)}{C_m^{\frac{d-2}{2}}(1)} h(m, d), \qquad [3.95]$$

where $b_m \geq 0$ $\sum_{m=0}^{\infty} b_m h(m, d) < \infty$, ω_d is a surface area of the sphere S_d,

$$\omega_d = \frac{2\pi^{\frac{d}{2}}}{\Gamma(\frac{d}{2})},$$

and the process can be represented in the form of the series

$$\xi(x) = \sum_{m=0}^{\infty}\sum_{l=1}^{h(m,d)} \xi^l_m \sqrt{b_m} S^l_m(x), \qquad [3.96]$$

where the sequence of random variables ξ_m^l is such that $\mathbf{E}\xi_m^l = 0$, $\mathbf{E}\xi_m^l \xi_{m_1}^{l_1} = \delta_m^{m_1}\delta_l^{l_1}$ $(m = 0, 1 \ldots; l = 1, \ldots, h(m,d))$.

For Gegenbauer polynomials, we will use explicit representation [KOZ 76]:

$$C_l^n(\cos\theta) = \sum_{k=0}^{l} \frac{(n)_k (n)_{l-k}}{k!(l-k)!} \cos[(l-2k)\theta], \qquad [3.97]$$

where $(\lambda)_k = \lambda(\lambda+1)\cdots(\lambda+k-1)$.

Consider a quadratic semipositive form:

$$X(x) = \xi(x)^2 + \sum_{i=1}^{d-2} \left(\frac{\partial \xi(x)}{\partial \theta_i}\right)^2 + \left(\frac{\partial \xi(x)}{\partial \mathbf{P}}\right)^2.$$

Suggest that all conditions of theorem 3.12 for $\xi(x)$ from [3.96] and for respective $X(x)$ are satisfied, then

$$X(x) = \sum_{m=0}^{\infty} \sum_{m'=0}^{\infty} \sum_{l=1}^{h(m,d)} \sum_{l'=1}^{h(m',d)} \phi_{mm'}(x) \xi_m^l \xi_{m'}^{l'},$$

where

$$\phi_{mm'}(x) = \sqrt{b_m b_{m'}} \times \left(S_m^l(x) S_{m'}^{l'} + \sum_{i=1}^{d-2} \frac{\partial S_m^l(x)}{\partial \theta_i} \frac{\partial S_{m'}^{l'}(x)}{\partial \theta_i'} + \frac{\partial S_m^l(x)}{\partial \mathbf{P}} \frac{\partial S_{m'}^{l'}(x)}{\partial \mathbf{P}'} \right). \qquad [3.98]$$

We assume that $\sum_{m=0}^{\infty} \sum_{m'=0}^{\infty} \sum_{l=1}^{h(m,d)} \sum_{l'=1}^{h(m',d)} \phi_{mm'}^2(x)$ converges uniformly as $x \in S_d$.

Let us denote $\xi_N(x) = \sum_{m=N}^{\infty} \sum_{l=1}^{h(m,d)} \xi_m^l \sqrt{b_m} S_m^l(x)$ and $X_N(x) = \xi_N(x)^2 + \sum_{i=1}^{d-2} \left(\frac{\partial \xi_N(x)}{\partial \theta_i}\right)^2 + \left(\frac{\partial \xi_N(x)}{\partial \mathbf{P}}\right)^2$, $N = 0, 1, \ldots$ Similarly to lemma 3.10, the following relationships can be obtained for $X_N(x)$:

$$\mathbf{E}X_N(x) = \sum_{m=N}^{\infty} \sum_{l=1}^{h(m,d)} \phi_{mm}(x),$$

$$\mathbf{Var}X_N(t) = 2\sum_{m=N}^{\infty}\sum_{m'=N}^{\infty}\sum_{l=1}^{h(m,d)}\sum_{l'=1}^{h(m',d)} \phi_{mm'}^2(x),$$

$$\mathbf{Var}(X_N(x) - X_N(x')) = 2\sum_{m=N}^{\infty}\sum_{m'=N}^{\infty}\sum_{l=1}^{h(m,d)}\sum_{l'=1}^{h(m',d)} \times(\phi_{mm'}(x) - \phi_{mm'}(x'))^2. \qquad [3.99]$$

Estimate now the difference:

$$\begin{aligned}|\phi_{mm'}(x) - \phi_{mm'}(x')| &\le \sqrt{b_m b_{m'}} \\ &\times\Big[|S_m^l(x)||S_{m'}^{l'}(x) - S_{m'}^{l'}(x')| + |S_{m'}^{l'}(x')||S_m^l(x) - S_m^l(x')| \\ &+ \sum_{i=1}^{d-2}\Big|\frac{\partial S_m^l(x)}{\partial\theta_i}\Big|\Big|\frac{\partial S_{m'}^{l'}(x)}{\partial\theta_i} - \frac{\partial S_{m'}^{l'}(x')}{\partial\theta_i}\Big| + \Big|\frac{\partial S_{m'}^{l'}(x')}{\partial\theta_i}\Big|\Big|\frac{\partial S_m^l(x)}{\partial\theta_i} - \frac{\partial S_m^l(x')}{\partial\theta_i}\Big| \\ &+ \Big|\frac{\partial S_m^l(x)}{\partial\mathbf{P}}\Big|\Big|\frac{\partial S_{m'}^{l'}(x)}{\partial\mathbf{P}} - \frac{\partial S_{m'}^{l'}(x')}{\partial\mathbf{P}}\Big| + \Big|\frac{\partial S_{m'}^{l'}(x')}{\partial\mathbf{P}}\Big|\Big|\frac{\partial S_m^l(x)}{\partial\mathbf{P}} - \frac{\partial S_m^l(x')}{\partial\mathbf{P}}\Big|\Big].\end{aligned} \qquad [3.100]$$

Theorem 3.15 allows us to use the exact representation of $S_m^l(x)$ for estimation of [3.100]. At first, we estimate the Gegenbauer polynomial in [3.97]. We have:

$$\begin{aligned}&\Big|C_{m_k-m_{k+1}}^{m_{k+1}+\frac{d-2}{2}-\frac{k}{2}}(\cos\theta_{k+1})\Big| \\ &\le \sum_{i=0}^{m_k-m_{k+1}}\left|\frac{(m_{k+1}+\frac{d-2}{2}-\frac{k}{2})_i(m_{k+1}+\frac{d-2}{2}-\frac{k}{2})_{m_k-m_{k+1}-i}}{i!(m_k-m_{k+1}-i)!}\right| \\ &\le \sum_{i=0}^{m}\left|\frac{(m_k+\frac{d-2}{2}-\frac{k}{2}-1)^i(m_k+\frac{d-2}{2}-\frac{k}{2}-1)^{m_k-m_{k+1}-i}}{2^i\,2^{m_k-m_{k+1}-i}}\right| \\ &\le (m+1)\Big(\frac{2m+d-4}{4}\Big)^{m_k-m_{k+1}}.\end{aligned}$$

Then, from [3.94] it follows that

$$|S_m^l(x)| \le \prod_{k=0}^{d-3}(m+1)\left(\frac{2m+d-4}{4}\right)^{m_k-m_{k+1}} \le (m+1)^{d-2}\left(\frac{2m+d-4}{4}\right)^m,$$

for all $l=\overline{1,h(m,d)}$. Denote $L(m)=(m+1)^{d-2}\left(\frac{2m+d-4}{4}\right)^m$

Estimate now:

$$\sup_{\rho(x,x')<h}|S_m^l(x)-S_m^l(x')| \le (m+1)^{d-2}\left(\frac{2m+d-4}{4}\right)^m$$

$$\times \left|e^{\pm i m_{d-2}\mathbf{P}}\prod_{k=0}^{d-3}\left(\sin\theta_{k+1}\right)^{m_{k+1}} - e^{\pm i m_{d-2}\mathbf{P}'}\prod_{k=0}^{d-3}\left(\sin\theta'_{k+1}\right)^{m_{k+1}}\right| \qquad [3.101]$$

We use the following obvious inequality:

$$|\prod_{i=0}^{n}a_i-\prod_{i=0}^{n}b_i| \le \sum_{i=0}^{n}|a_i-b_i|\prod_{j\ne i}\max\{|a_j|,|b_j|\} \le \sum_{i=0}^{n}|a_i-b_i|, \qquad [3.102]$$

if $|a_i|\le 1$ and $|b_i|\le 1$.

For $m_{i+1}\ge 1$, we get

$$|(\sin\theta_{i+1})^{m_{i+1}}-(\sin\theta'_{i+1})^{m_{i+1}}| = m_{i+1}|\int_{\theta'_{i+1}}^{\theta_{i+1}}(\sin t)^{m_{i+1}-1}d\sin t|$$

$$\le m_{i+1}|\int_{\theta'_{i+1}}^{\theta_{i+1}}d\sin t| \le m|\sin\theta_{i+1}-\sin\theta'_{i+1}| \le 2^{1-\alpha}mh^{\alpha}, \qquad [3.103]$$

where $\rho(x,x')=\max_{1\le i\le d-1}|\theta_i-\theta'_i|<h$ and $\alpha\in(0,1]$

Substituting [3.102] and [3.103] into [3.101], we obtain

$$\sup_{\rho(x,x')<h}|S_m^l(x)-S_m^l(x')| \le L(m)\sum_{i=0}^{d-3}|(\sin\theta_{i+1})^{m_{i+1}}-(\sin\theta'_{i+1})^{m_{i+1}}|$$

$$\times \prod_{k\neq i} \max\{(\sin\theta_{i+1})^{m_{i+1}}, (\sin\theta'_{i+1})^{m_{i+1}}\} \le 2^{1-\alpha} m(d-2)L(m)h^{\alpha} := K(m)h^{\alpha},$$

where $K(m) = 2^{1-\alpha}m(d-2)L(m) = 2^{1-\alpha}m(d-2)(m+1)^{d-2}\left(\frac{2m+d-4}{4}\right)^m$.

Find now the estimations for derivatives of $S^l_m(x)$. Since Gegenbauer polynomials $C^{m_{k+1}+\frac{d-2}{2}-\frac{k}{2}}_{m_k-m_{k+1}}(\cos\theta_{k+1})$ are trigonometric polynomials of the power of $m_k - m_{k+1}$, then for them Bernstein inequality holds true;

$$\sup_{0\le\theta_{k+1}\le 2\pi} \left|\left(C^{m_{k+1}+\frac{d-2}{2}-\frac{k}{2}}_{m_k-m_{k+1}}(\cos\theta_{k+1})\right)'\right|$$

$$\le (m_k - m_{k+1}) \sup_{0\le\theta_{k+1}\le 2\pi} \left|C^{m_{k+1}+\frac{d-2}{2}-\frac{k}{2}}_{m_k-m_{k+1}}(\cos\theta_{k+1})\right|$$

$$\le mC^{m_{k+1}+\frac{d-2}{2}-\frac{k}{2}}_{m_k-m_{k+1}}(\cos\theta_{k+1})m.$$

Then, from [3.94] the following inequalities can be obtained:

$$\left|\frac{\partial S^l_m(x)}{\partial\theta_i}\right| \le 2mL(m) = 2m(m+1)^{d-2}\left(\frac{2m+d-4}{4}\right)^m,$$

$$\left|\frac{\partial S^l_m(x)}{\partial\theta_i} - \frac{\partial S^l_m(x')}{\partial\theta'_i}\right| \le 2mK(m)h^{\alpha}.$$

Note that $\left|\frac{\partial S^l_m(x)}{\partial \mathbf{P}}\right| = |m_{d-2}||S^l_m(x)| \le mL(m)$ and

$$\left|\frac{\partial S^l_m(x)}{\partial \mathbf{P}} - \frac{\partial S^l_m(x')}{\partial \mathbf{P}'}\right| \le 2mK(m)h^{\alpha}, \qquad \alpha\in(0,1].$$

The estimation for [3.100] follows from inequalities above. Hence,

$$|\phi_{mm'}(x) - \phi_{mm'}(x')| \le \sqrt{b_m b_{m'}}(1+(4d-6)mm')$$
$$\times (L(m)K(m') + L(m')K(m))h^{\alpha}$$

$$\le 2^{1-\alpha}(d-2)\sqrt{b_m b_{m'}}(m'+m)(1+(4d-6)mm')L(m)L(m')h^{\alpha},$$

where $\alpha\in(0,1]$, $L(m) = (m+1)^{d-2}\left(\frac{2m+d-4}{4}\right)^m$.

Using the representation [3.99], we get the following inequality for $\alpha \in (0,1]$,

$$\sup_{\rho(x,x')<h} \left(\mathbf{Var}(X_N(x) - X_N(x'))\right)^{1/2} \le A_N h^\alpha = \sigma(h), \quad N = 0, 1, 2 \ldots,$$

where

$$A_N^2 = 2^{3-2\alpha}(d-2)^2 \sum_{m=N}^{\infty} \sum_{m'=N}^{\infty} b_m b_{m'} h(m,d) h(m',d)(m+m')^2$$

$$\times (1 + (4d-6)mm')^2 L^2(m) L^2(m'). \qquad [3.104]$$

We suggest that the series $\sum_{m=0}^{\infty} b_m h(m,d) m^4 L^2(m) < \infty$.

So, the next theorems are proved.

THEOREM 3.16.– Let $\xi(x)$ be Gaussian separable isotropic field on sphere S_d. If for all integers $M \ge 1$ and $y > 0$

$$\pi^\alpha A_N \Big(\frac{\sqrt{2}\gamma_0 M(d-1)}{y\alpha}\Big)^{M-1} < \gamma_0,$$

and

$$y > \frac{\sqrt{2}\gamma_0 M(d-1)}{\alpha} \max\{1, (\frac{2}{\sqrt{d-1}})^{\alpha/M}\}$$

then the following inequality holds true:

$$\mathbf{P}\Big\{\sup_{x \in \mathbf{T}} |X_N(x) - \mathbf{E}X_N(x)| > y\Big\} \le 4(\sqrt{d-1}e^{1/\alpha})^{d-1}$$

$$\times \exp\Big\{-\frac{y}{\sqrt{2}\gamma_0} + \frac{M(d-1)}{\alpha}\Big\}\Big(\frac{\alpha y}{\sqrt{2}\gamma_0 M(d-1)}\Big)^{M(d-1)/\alpha}\Big(1 + \frac{2y}{\sqrt{2}\gamma_0}\Big)^{1/2},$$

where A_N is defined in [3.104] and $\gamma_0 = \sup_{x \in S_d} (\mathbf{Var} X_N(x))^{1/2}$.

PROOF.– Recall that for spherical coordinates the sphere S_d is transformed to $[0,\pi]^{d-2}[0,2\pi] = \mathbf{T} \in \mathbb{R}^{d-1}$ and we consider the metric $\rho(x,x') = \max_{1 \le i \le d-1} |\theta_i - \theta_i'|$. Then, the following inequality for the number of closed balls of radius u, which covers S_d, holds true:

$$N(u) \le 2\Big(\frac{\pi\sqrt{d-1}}{u} + 1\Big)^{d-1}.$$

Note that $\varepsilon_0 = \inf_{t\in\mathbf{T}} \sup_{s\in\mathbf{T}} \rho(t,s) = \pi$ and $t_0 = \sigma(\varepsilon_0)$. We proved $\sigma(h) = A_N h^\alpha, \quad \alpha \in (0,1], \quad A_N$ is from [3.104]. Hence, the result of theorem 3.13 can be used as for $d-1$ dimensional space, which proves the theorem. □

From theorem 3.16 follows the next theorem.

THEOREM 3.17.– Let all conditions of the previous theorem be satisfied. The model $\tilde{\xi}_N(x)$ approximates the Gaussian separable isotropic field $\xi(x)$ on S_d with given reliability $1-\nu$, $\nu \in (0,1)$ and accuracy $\delta > 0$ if for $\alpha \in (0,1]$ and integer $M \geq 1$ the following inequalities hold true:

$$\pi^\alpha A_{N+1}\Big(\frac{\sqrt{2}\gamma_0 M(d-1)}{\delta\alpha}\Big)^{M-1} < \gamma_0,$$

$$\delta > \frac{\sqrt{2}\gamma_0 M(d-1)}{\alpha}\max\{1, (\frac{2}{\sqrt{d-1}})^{\alpha/M}\},$$

$$4\ \left(\sqrt{d-1}e^{1/\alpha}\right)^{d-1}\exp\Big\{-\tfrac{\delta}{\sqrt{2}\gamma_0} + \tfrac{M(d-1)}{\alpha}\Big\}$$

$$\times\left(\tfrac{\alpha\delta}{\sqrt{2}\gamma_0 M(d-1)}\right)^{M(d-1)/\alpha}\left(1+\tfrac{2\delta}{\sqrt{2}\gamma_0}\right)^{1/2} < \nu.$$

4

The Construction of the Model of Gaussian Stationary Processes

This chapter offers two approaches to construct the models of Gaussian stationary stochastic processes. These results can be found in the works of [ANT 02a] and [KOZ 12]. The methods of model construction are generalized in the case of random fields. Similar statements are discussed in Tegza's studies [TEG 07, TEG 08, TEG 11].

The proposed methods of modeling can be applied in different areas of science and technology, particularly in radio, physics and meteorology. The models can be interpreted as a set of the signals with limited energy, harmonic signals and signals with limited duration.

Let $\{\Omega, \mathcal{B}, \mathbf{P}\}$ be a probabilistic space.

Let $X = \{X(t),\ t \in \mathbf{R}\}$ be a real-valued Gaussian stationary centered second-order stochastic process with covariance function:

$$B(\tau) = \mathbf{E}X(t+\tau) \cdot X(t) = \int\limits_0^\infty \cos \lambda\tau dF(\lambda),$$

where $F(\lambda)$ is a spectral function. Assume that the function $F(\lambda)$ is continuous.

We assume that the process $X(t)$ is separable and almost sure sample continuous on any interval $[0, T]$. All the necessary and sufficient conditions for sample continuous separable stationary Gaussian processes on a compact can be found in [FER 75].

Let us consider the close to the necessary sufficient conditions of sample continuity of Gaussian stationary separable random processes.

THEOREM 4.1.– [HUN 51] Let $X = \{X(t), 0 \leq t \leq T\}$ be a separable Gaussian stationary real-valued stochastic process. It will be sample continuous if the following relationship holds true

$$\int_0^\infty (\ln(1+\lambda))^{1+\varepsilon}\, dF(\lambda) < \infty, \qquad [4.1]$$

where $F(\lambda)$ is a spectral function, $\varepsilon > 0$.

Note that the statement and the proof of this theorem in a weaker form (under condition $\varepsilon > 2$) is contained in [CRA 67].

The process X can be represented as:

$$X(t) = \int_0^\infty \cos \lambda t d\eta_1(\lambda) + \int_0^\infty \sin \lambda t d\eta_2(\lambda), \qquad [4.2]$$

where $\eta_1(\lambda)$ and $\eta_2(\lambda)$ are such independent centered Gaussian processes with independent increments that

$$\mathbf{E}(\eta_i(\lambda_2) - \eta_i(\lambda_1))^2 = F(\lambda_2) - F(\lambda_1), \quad \lambda_1 < \lambda_2,\ i = 1, 2.$$

Consider the partition of the set $[0, \infty]$ $\Lambda = \{\lambda_0, \ldots, \lambda_{M+1}\}$ such that $\lambda_0 = 0$, $\lambda_k < \lambda_{k+1}$, $\lambda_{M+1} = \infty$, then

$$X(t) = \sum_{k=0}^{M} \int_{\lambda_k}^{\lambda_{k+1}} \cos \lambda t d\eta_1(\lambda) + \sum_{k=0}^{M} \int_{\lambda_k}^{\lambda_{k+1}} \sin \lambda t d\eta_2(\lambda). \qquad [4.3]$$

Let us consider a process

$$X_\Lambda(t) = \sum_{k=0}^{M} (\eta_{k1} \cos \zeta_k t + \eta_{k2} \sin \zeta_k t)\,,$$

where ζ_k are independent for any k, and are defined on $[\lambda_k, \lambda_{k+1}]$ with cumulative distribution function

$$F_k(\lambda) = P\{\zeta_k < \lambda\} = \frac{F(\lambda) - F(\lambda_k)}{F(\lambda_{k+1}) - F(\lambda_k)}\,.$$

Put $\eta_{k1} = \int\limits_{\lambda_k}^{\lambda_{k+1}} d\eta_1(\lambda)$, $\eta_{k2} = \int\limits_{\lambda_k}^{\lambda_{k+1}} d\eta_2(\lambda)$.

The process $X_\Lambda(t)$ is called a model of the Gaussian random process $X(t)$. And it is clear that a different selection of the number *M* provides different levels of accuracy and reliability for the computer-simulated models of the process.

There are some models that will be investigated in the following sections. Let us consider these models and their properties.

1) As a model for Gaussian random process $X(t)$, a random process

$$X_\Lambda(t) = \sum_{k=0}^{M} (\eta_{k1} \cos \zeta_k t + \eta_{k2} \sin \zeta_k t) \qquad [4.4]$$

is studied, where $\eta_{k1} = \int\limits_{\lambda_k}^{\lambda_{k+1}} d\eta_1(\lambda)$, $\eta_{k2} = \int\limits_{\lambda_k}^{\lambda_{k+1}} d\eta_2(\lambda)$; $\eta_{l1}, \eta_{m2}, \zeta_k$ are independent random variables for any l, m and k, $\Lambda = \{\lambda_0, \lambda_1, \ldots, \lambda_{M+1}\}$ is a division of the set $[0, \infty]$, where $\lambda_0 = 0$, $\lambda_k < \lambda_{k+1}$, $\lambda_{M+1} = \infty$, η_{k1}, η_{k2} are Gaussian random variables such that $\mathbf{E}\eta_{k1} = \mathbf{E}\eta_{k2} = 0$, $\mathbf{E}\eta_{k1}^2 = \mathbf{E}\eta_{k2}^2 = F(\lambda_{k+1}) - F(\lambda_k) = b_k^2$, ζ_k are random variables that take values on the segments $[\lambda_k, \lambda_{k+1}]$, and if $b_k^2 > 0$, then

$$F_k(\lambda) = P\{\zeta_k < \lambda\} = \frac{F(\lambda) - F(\lambda_k)}{F(\lambda_{k+1}) - F(\lambda_k)}.$$

If $b_k^2 = 0$, then $\eta_{k1} = 0, \eta_{k2} = 0, \zeta_k = 0$ with probability of 1.

It is easy to check that the model is zero-mean random process:

$$\mathbf{E}X_\Lambda(t) = \mathbf{E}\sum_{k=0}^{M} (\eta_{k1} \cos \zeta_k t + \eta_{k2} \sin \zeta_k t)$$
$$= \sum_{k=0}^{M} (\mathbf{E}\eta_{k1}\mathbf{E}\cos \zeta_k t + \mathbf{E}\eta_{k2}\mathbf{E}\sin \zeta_k t) = 0.$$

The covariance function of the process $X_\Lambda(t)$ coincides with the covariance function of stochastic process $X(t)$

$$\mathbf{E}X_\Lambda(t+\tau)X_\Lambda(t)$$
$$= \mathbf{E}\left(\sum_{k=0}^{M} (\eta_{k1} \cos \zeta_k (t+\tau) + \eta_{k2} \sin \zeta_k (t+\tau))\right)$$

$$\times\left(\sum_{k=0}^{M}(\eta_{k1}\cos\zeta_k t+\eta_{k2}\sin\zeta_k t)\right)$$

$$=\sum_{k=0}^{M}\left[\mathbf{E}\eta_{k1}^2\mathbf{E}\cos\zeta_k(t+\tau)\cos\zeta_k t+\mathbf{E}\eta_{k2}^2\mathbf{E}\sin\zeta_k(t+\tau)\sin\zeta_k t\right]$$

$$=\sum_{k=0}^{M}b_k^2\mathbf{E}\cos\zeta_k\tau=\sum_{k=0}^{M}b_k^2\int_{\lambda_k}^{\lambda_{k+1}}\cos\lambda\tau dF_k(\lambda)$$

$$=\int_{0}^{\infty}\cos\lambda\tau dF(\lambda)=r(\tau), \qquad [4.5]$$

$(\mathbf{E}\eta_{i1}\cdot\eta_{j1}=0,\ \mathbf{E}\eta_{i2}\cdot\eta_{j2}=0,\ \ i\neq j, \mathbf{E}\eta_{k1}\eta_{k2}=0,\ k=0,\cdots,M)$.

But $X_\Lambda(t)$ is not Gaussian process. Our goal is to identify how the process $X_\Lambda(t)$ is closed to Gaussian process $X(t)$.

Consider the model $X_\Lambda(t)$ and put

$$\eta_{k1}=\int_{\lambda_k}^{\lambda_{k+1}}d\eta_1(\lambda),\quad \eta_{k2}=\int_{\lambda_k}^{\lambda_{k+1}}d\eta_2(\lambda).$$

Let $\eta_\Lambda(t)=X(t)-X_\Lambda(t)$. Then

$$\eta_\Lambda(t)=\sum_{k=0}^{M}\int_{\lambda_k}^{\lambda_{k+1}}\cos\lambda t d\eta_1(\lambda)+\sum_{k=0}^{M}\int_{\lambda_k}^{\lambda_{k+1}}\sin\lambda t d\eta_2(\lambda)$$

$$-\sum_{k=0}^{M}\int_{\lambda_k}^{\lambda_{k+1}}\cos\zeta_k t d\eta_1(\lambda)-\sum_{k=0}^{M}\int_{\lambda_k}^{\lambda_{k+1}}\sin\zeta_k t d\eta_2(\lambda);$$

$$\eta_\Lambda(t)=\sum_{k=0}^{M}\left(\int_{\lambda_k}^{\lambda_{k+1}}(\cos\lambda t-\cos\zeta_k t)d\eta_1(\lambda)\right.$$

$$\left.+\int_{\lambda_k}^{\lambda_{k+1}}(\sin\lambda t-\sin\zeta_k t)d\eta_2(\lambda)\right). \qquad [4.6]$$

LEMMA 4.1.– The following relationships hold true:

$$\mathbf{E}\left(\int_{\lambda_k}^{\lambda_{k+1}}(\cos\lambda t-\cos\zeta_k t)d\eta_1(\lambda)\right)^{2m+1}=0,$$

$$\mathbf{E}\left(\int_{\lambda_k}^{\lambda_{k+1}}(\sin\lambda t-\sin\zeta_k t)d\eta_2(\lambda)\right)^{2m+1}=0,$$

$$\mathbf{E}\left(\int_{\lambda_k}^{\lambda_{k+1}}(\cos\lambda t-\cos\zeta_k t)d\eta_1(\lambda)\right)^{2m}\leq Z_{km},$$

$$\mathbf{E}\left(\int_{\lambda_k}^{\lambda_{k+1}}(\sin\lambda t-\sin\zeta_k t)d\eta_2(\lambda)\right)^{2m}\leq Z_{km},$$

where $Z_{km}=4^m\Delta_{2m}\mathbf{E}\left(\int_{\lambda_k}^{\lambda_{k+1}}\left|\sin\frac{t(\zeta_k-\lambda)}{2}\right|^2dF(\lambda)\right)^m$, $\Delta_{2m}=\frac{(2m)!}{2^m m!}$.

PROOF.– Since for zero-mean Gaussian random variable ξ, we have

$\mathbf{E}\xi=0$, $\mathbf{E}\xi^2=\sigma^2$, $\mathbf{E}\xi^{2k}=\Delta_{2k}\sigma^{2k}$, $k=1,2,\ldots$, $\Delta_{2k}=(2k-1)!!=\frac{(2k)!}{2^k k!}$, and random variable ζ_k does not depend on $\eta_i(\lambda)$, then it follows from Fubini theorem that:

$$\mathbf{E}\left(\int_{\lambda_k}^{\lambda_{k+1}}(\cos\lambda t-\cos\zeta_k t)d\eta_1(\lambda)\right)^{2m}$$

$$=\mathbf{E}\mathbf{E}_{\zeta k}\left(\int_{\lambda_k}^{\lambda_{k+1}}(\cos\lambda t-\cos\zeta_k t)d\eta_1(\lambda)\right)^{2m}$$

$$\leq\Delta_{2m}\mathbf{E}\left(\int_{\lambda_k}^{\lambda_{k+1}}|\cos\lambda t-\cos\zeta_k t|^2dF(\lambda)\right)^m$$

$$\leq\Delta_{2m}\mathbf{E}\left(\int_{\lambda_k}^{\lambda_{k+1}}\left|2\sin\frac{t(\zeta_k-\lambda)}{2}\cdot\sin\frac{t(\zeta_k+\lambda)}{2}\right|^2dF(\lambda)\right)^m$$

$$\leq \Delta_{2m} \mathbf{E} \left(\int\limits_{\lambda_k}^{\lambda_{k+1}} \left| 2 \sin \frac{t(\zeta_k - \lambda)}{2} \right|^2 dF(\lambda) \right)^m = Z_{km},$$

where $\mathbf{E}_{\zeta k}$ denotes conditional mathematical expectation with respect to ζ_k.

Similarly, the second inequality is obtained:

$$\mathbf{E} \left(\int\limits_{\lambda_k}^{\lambda_{k+1}} (\sin \lambda t - \sin \zeta_k t) d\eta_2(\lambda) \right)^{2m}$$

$$\leq \Delta_{2m} \mathbf{E} \left(\int\limits_{\lambda_k}^{\lambda_{k+1}} \left| 2 \sin \frac{t(\lambda - \zeta_k)}{2} \cdot \cos \frac{t(\zeta_k + \lambda)}{2} \right|^2 dF(\lambda) \right)^m$$

$$\leq \Delta_{2m} \mathbf{E} \left(\int\limits_{\lambda_k}^{\lambda_{k+1}} \left| 2 \sin \frac{t(\zeta_k - \lambda)}{2} \right|^2 dF(\lambda) \right)^m = Z_{km}.$$

□

THEOREM 4.2.– Random process $\eta_\Lambda(t)$ is sub-Gaussian.

PROOF.– Show that

$$\chi_{k1} = \int\limits_{\lambda_k}^{\lambda_{k+1}} (\cos \lambda t - \cos \zeta_k t) d\eta_1(\lambda) \text{ òà } \chi_{k2} = \int\limits_{\lambda_k}^{\lambda_{k+1}} (\sin \lambda t - \sin \zeta_k t) d\eta_2(\lambda)$$

are sub-Gaussian random variables. It follows from lemma 4.1 that

$$\left(\mathbf{E} \left(\int\limits_{\lambda_k}^{\lambda_{k+1}} (\cos \lambda t - \cos \zeta_k t) d\eta_1(\lambda) \right)^{2m} \right)^{\frac{1}{2m}}$$

$$\leq \sqrt[2m]{4^m \Delta_{2m}} \left(\int\limits_{\lambda_k}^{\lambda_{k+1}} \left(\int\limits_{\lambda_k}^{\lambda_{k+1}} \left(\sin \frac{t(u - \lambda)}{2} \right)^2 dF(u) \right)^m dF_k(\lambda) \right)^{\frac{1}{2m}}$$

$$\leq \sqrt[2m]{4^m \Delta_{2m}} \left(\int\limits_{\lambda_k}^{\lambda_{k+1}} (F(\lambda_{k+1}) - F(\lambda_k))^m \frac{dF(\lambda)}{F(\lambda_{k+1}) - F(\lambda_k)} \right)^{\frac{1}{2m}}$$

$$\leq \sqrt[2m]{4^m \Delta_{2m}} \, (F(\lambda_{k+1}) - F(\lambda_k))^{\frac{1}{2}} ,$$

and

$$\left(\mathbf{E} \left(\int\limits_{\lambda_k}^{\lambda_{k+1}} (\sin \lambda t - \sin \zeta_k t) d\eta_2(\lambda) \right)^{2m} \right)^{1/2m}$$

$$\leq \sqrt[2m]{4^m \Delta_{2m}} \left(\int\limits_{\lambda_k}^{\lambda_{k+1}} \left(\int\limits_{\lambda_k}^{\lambda_{k+1}} \left(\sin \frac{t(u-\lambda)}{2} \right)^2 dF(u) \right)^m \right.$$

$$\left. \times \frac{dF(\lambda)}{F(\lambda_{k+1}) - F(\lambda_k)} \right)^{\frac{1}{2m}}$$

$$\leq \sqrt[2m]{4^m \Delta_{2m}} \, (F(\lambda_{k+1}) - F(\lambda_k))^{\frac{1}{2}} .$$

Then,

$$\Theta_1(\chi_{ki}) = \sup_{m \geq 1} \left[\frac{1}{\Delta_{2m}} \mathbf{E} \chi_{ki}^{2m} \right]^{\frac{1}{2m}}$$

$$= \sup_{m \geq 1} \left[\frac{2^m m!}{(2m)!} \cdot 4^m \cdot \frac{(2m)!}{2^m m!} (F(\lambda_{k+1}) - F(\lambda_k))^m \right]^{\frac{1}{2m}}$$

$$= 2 \, (F(\lambda_{k+1}) - F(\lambda_k))^{\frac{1}{2}} < \infty , \quad i = 1, 2.$$

From the well-known theorem about the necessary and sufficient conditions of a sub-Gaussian random variable [BUL 00] follows that χ_{k1} and χ_{k2} are sub-Gaussian centered random variables. That is, for any $t \in \mathbf{T}$, the value $\eta_\Lambda(t)$ is a sub-Gaussian random variable. Since $\eta_\Lambda(t)$ is a boundary sum of sub-Gaussian random variables, then this process is sub-Gaussian. □

THEOREM 4.3.– For sub-Gaussian process $\eta_\Lambda(t)$, the next inequality is satisfied

$$\tau(\eta_\Lambda(t)) \leq 4 \left(\sum_{k=0}^{M} b_k^2 \sup_{m \geq 1} \left(\mathbf{E} \left| \sin \frac{t(\zeta_k - \zeta_k^*)}{2} \right|^{2m} \right)^{\frac{1}{m}} \right)^{\frac{1}{2}} = (B_\Lambda(t))^{\frac{1}{2}} , \qquad [4.7]$$

where $b_k^2 = F(\lambda_{k+1}) - F(\lambda_k)$, ζ_k^* are random variables that are not dependent on ζ_k and have the same distributions as ζ_k.

PROOF.– It follows from lemma 4.1 that

$$\tau^2 \left(\int_{\lambda_k}^{\lambda_{k+1}} (\cos \lambda t - \cos \zeta_k t) d\eta_1(\lambda) \right) \leq \Theta_1^2 \left(\int_{\lambda_k}^{\lambda_{k+1}} (\cos \lambda t - \cos \zeta_k t) d\eta_1(\lambda) \right)$$

$$\leq \sup_{m \geq 1} b_k^2 \left(\mathbf{E} \left(\int_{\lambda_k}^{\lambda_{k+1}} \left| 2 \sin \frac{t(\zeta_k - \lambda)}{2} \right|^2 dF_k(\lambda) \right)^m \right)^{\frac{1}{m}}$$

$$= \sup_{m \geq 1} 4b_k^2 \left(\int_{\lambda_k}^{\lambda_{k+1}} \left(\int_{\lambda_k}^{\lambda_{k+1}} \left| \sin \frac{t(u - \lambda)}{2} \right|^2 dF_k(\lambda) \right)^m dF_k(u) \right)^{\frac{1}{m}} = I_k.$$

Similarly

$$\tau^2 \left(\int_{\lambda_k}^{\lambda_{k+1}} (\sin \lambda t - \sin \zeta_k t) d\eta_2(\lambda) \right) \leq \Theta_1^2 \left(\int_{\lambda_k}^{\lambda_{k+1}} (\sin \lambda t - \sin \zeta_k t) d\eta_2(\lambda) \right)$$

$$\leq \sup_{m \geq 1} 4b_k^2 \left(\int_{\lambda_k}^{\lambda_{k+1}} \left(\int_{\lambda_k}^{\lambda_{k+1}} \left| \sin \frac{t(u - \lambda)}{2} \right|^2 dF_k(\lambda) \right)^m dF_k(u) \right)^{\frac{1}{m}} = I_k.$$

Then

$$\tau^2 \left(\int_{\lambda_k}^{\lambda_{k+1}} (\cos \lambda t - \cos \zeta_k t) d\eta_1(\lambda) + \int_{\lambda_k}^{\lambda_{k+1}} (\sin \lambda t - \sin \zeta_k t) d\eta_2(\lambda) \right)$$

$$\leq \left(\tau \left(\int_{\lambda_k}^{\lambda_{k+1}} (\cos \lambda t - \cos \zeta_k t) d\eta_1(\lambda) \right) + \tau \left(\int_{\lambda_k}^{\lambda_{k+1}} (\sin \lambda t - \sin \zeta_k t) d\eta_2(\lambda) \right) \right)^2$$

$$\leq 4 I_k.$$

Since the items of the series in [4.6] for different k are independent, then last inequality and lemma 1.7 yield

$$\tau^2\left(\eta_\Lambda(t)\right) \le 4\sum_{k=0}^{M} I_k\,,$$

$$\tau\left(\eta_\Lambda(t)\right)$$

$$\le 4\left[\sum_{k=0}^{M}\sup_{m\ge 1}\frac{1}{b_k^{2/m}}\left(\int_{\lambda_k}^{\lambda_{k+1}}\left(\int_{\lambda_k}^{\lambda_{k+1}}\left|\sin\frac{t(u-\lambda)}{2}\right|^2 dF(\lambda)\right)^m dF(u)\right)^{\frac{1}{m}}\right]^{\frac{1}{2}}$$

$$\le 4\left[\sum_{k=0}^{M}\sup_{m\ge 1} b_k^2\left(\mathbf{E}\left|\sin\frac{t(\zeta_k-\zeta_k^*)}{2}\right|^{2m}\right)^{\frac{1}{m}}\right]^{\frac{1}{2}}.$$

□

2) A stochastic process

$$X_\Lambda(t) = \sum_{k=0}^{M}(\eta_{k1}\cos\lambda_k t + \eta_{k2}\sin\lambda_k t) \qquad [4.8]$$

will be used as a model of the Gaussian stationary random process, where $\eta_{l1}, \eta_{m2}, \zeta_k$ are independent random variables for any l, m and k, $D_\Lambda = \{\lambda_0, \lambda_1, \ldots, \lambda_{M+1}\}$ is a division of the set $[0, \Lambda]$, $\Lambda \in \mathbf{R}_+$, where $\lambda_0 = 0,\ \lambda_k < \lambda_{k+1},\ \lambda_{M+1} = \Lambda$ (Λ can be equal to ∞), η_{k1}, η_{k2} are Gaussian random variables such that $\mathbf{E}\eta_{k1} = \mathbf{E}\eta_{k2} = 0$, $\mathbf{E}\eta_{k1}^2 = \mathbf{E}\eta_{k2}^2 = F(\lambda_{k+1}) - F(\lambda_k) = b_k^2$, ζ_k is any point of the segment $[\lambda_k, \lambda_{k+1}]$.

3) A similar approach will be used to build the models of homogeneous random field.

Let $\left\{Y\left(\vec{t}\right), \vec{t} \in \mathbf{T}\right\}$ be a centered, homogeneous, continuous in mean square random field, $\left\{\mathbf{R}_+^n, \mathfrak{U}, \Phi\right\}$ is a measurable space and $\Phi(\cdot)$ is a finite measure. For the covariance function $B(\vec{\tau})$ of random field, the next representation holds true

$$B(\vec{\tau}) = \int_{\mathbf{R}_+^n}\cos\left(\vec{\lambda}, \vec{\tau}\right) d\Phi\left(\vec{\lambda}\right),$$

where $\Phi\left(\vec{\lambda}\right)$, $\vec{\lambda} \in \mathbf{R}_{+}^{n}$, is the measure that $\Phi\left(\mathbf{R}_{+}^{n}\right) = B(\vec{0})$. By Karhunen theorem homogeneous, centered field $Y\left(\vec{t}\right)$ can be represented as

$$Y\left(\vec{t}\right) = \int\limits_{\mathbf{R}_{+}^{n}} \cos\left(\vec{\lambda}, \vec{t}\right) dZ_1\left(\vec{\lambda}\right) + \int\limits_{\mathbf{R}_{+}^{n}} \sin\left(\vec{\lambda}, \vec{t}\right) dZ_2\left(\vec{\lambda}\right), \qquad [4.9]$$

where $Z_1(S)$ and $Z_2(S)$, $S \in \mathfrak{U}$ are uncorrelated random measures that are subordinated to Φ. It means that $\mathbf{E}Z_i(S_1) Z_i(S_2) = \Phi(S_1 \cap S_2)$, $S_1, S_2 \in \mathfrak{U}$, $i = 1, 2$, $(\cdot, \cdot)$ is a scalar product.

For the model of this random field, the sum $\widetilde{Y}\left(\vec{t}\right)$

$$\widetilde{Y}\left(\vec{t}\right) = \sum_{i_1, \ldots, i_n = 0}^{N-1} \cos\left(\vec{t}, \vec{\lambda}\left(\lambda_1^{i_1}, \ldots, \lambda_n^{i_n}\right)\right) Z_1\left(\Delta\left(i_1, \ldots, i_n\right)\right)$$
$$+ \sum_{i_1, \ldots, i_n = 0}^{N-1} \sin\left(\vec{t}, \vec{\lambda}\left(\lambda_1^{i_1}, \ldots, \lambda_n^{i_n}\right)\right) Z_2\left(\Delta\left(i_1, \ldots, i_n\right)\right) \qquad [4.10]$$

will be taken, where $\vec{\lambda}\left(\lambda_1^{i_1}, \ldots, \lambda_n^{i_n}\right)$ are the points of some partition D_{Λ^n}:

$$\Delta\left(i_1, \ldots, i_n\right) = \left\{\left[\lambda_1^{i_1}, \lambda_1^{i_1+1}\right) \times \ldots \times \left[\lambda_n^{i_n}, \lambda_n^{i_n+1}\right) \mid \lambda_m^{i_m} < \lambda_m^{i_m+1}, \right.$$
$$\left. \lambda_m^{i_m+1} - \lambda_m^{i_m} = \frac{\Lambda}{N}, \ \Lambda \in \mathbf{R}_{+}, \ N \in \mathbf{N}, \ m = \overline{1, n}, \ i_m = \overline{1, N-1}\right\}$$

of domain $[0, \Lambda]^n$, $\Lambda \in \mathbf{R}$.

4) The model construction of inhomogeneous Gaussian fields.

Let $\{\Omega, \mathcal{B}, \mathbf{P}\}$ be a standard probability space $\left\{Y\left(\vec{t}\right), \vec{t} \in \mathbf{T}\right\}$, $\mathbf{T} \subset \mathbf{R}^n$centered, Gaussian, continuous in mean square random field. Then, its covariance function $B\left(\vec{t}, \vec{s}\right)$ is continuous on $\mathbf{T} \times \mathbf{T}$.

Consider the Fredholm integral equation

$$\phi\left(\vec{t}\right) = \lambda \int_{\mathbf{T}} B\left(\vec{t}, \vec{s}\right) \phi\left(\vec{s}\right) d\vec{s}. \qquad [4.11]$$

It is known that the set of eigenvalues λ_k of this equation for continuous and non-negative definite kernel is at most countable set, eigen functions $\phi_k\left(\vec{t}\right)$ are continuous and eigenvalues λ_k are non-negative [VLA 67]. Let us consider the eigenvalues

λ_k, $k = 1, 2, \ldots$ in ascending order: $0 < \lambda_1 \leq \lambda_2 \leq \ldots \leq \lambda_n \leq \ldots$. We assume that corresponding eigen functions are orthonormalized, i.e.

$$\int_{\mathbf{T}} \phi_k(\vec{s})\phi_l(\vec{s})ds = \begin{cases} 1, \ k = l, \\ 0, \ k \neq l. \end{cases}$$

The covariance function has the following representation

$$B\left(\vec{t}, \vec{s}\right) = \sum_{k=1}^{\infty} \frac{\phi_k\left(\vec{t}\right) \phi_k\left(\vec{s}\right)}{\lambda_k}, \qquad [4.12]$$

where the series on the right side is uniformly convergent as $s \in \mathbf{T}$, and $\sum\limits_{k=1}^{\infty} \frac{1}{\lambda_k}$ is convergent [VLA 67].

Then, the field $Y\left(\vec{t}\right)$ admits the representation

$$Y\left(\vec{t}\right) = \sum_{k=1}^{\infty} \frac{\xi_k}{\sqrt{\lambda_k}} \phi_k\left(\vec{t}\right), \qquad [4.13]$$

where $\xi_k \sim N(0, 1)$, $\mathbf{E}\xi_k\xi_l = \delta_{kl}$, δ_{kl} is Kronecker symbol, i.e. ξ_k are independent and the series in [4.13] converges in mean square (it is follows from Karhunen theorem).

As a model of this field, a stochastic process

$$\widetilde{Y}\left(\vec{t}\right) = \sum_{k=1}^{N} \frac{\xi_k}{\sqrt{\lambda_k}} \phi_k\left(\vec{t}\right) \qquad [4.14]$$

will be accepted.

5

The Modeling of Gaussian Stationary Random Processes with a Certain Accuracy and Reliability

In this chapter, the accuracy and reliability of the models of stationary Gaussian random processes are studied in spaces $L_p([0,T])$, $p \geq 1$; in Orlicz spaces and in the space of continuous functions $C([0,T])$. The properties of models of stationary Gaussian processes in a uniform metric, applying the theory of $Sub_\varphi(\Omega)$ spaces, are investigated. A generalized model of Gaussian stationary processes is also considered.

5.1. Reliability and accuracy in $L_p(\mathbf{T})$, $p \geq 1$ of the models for Gaussian stationary random processes

In section 5.1.1, the theorems on approximation of a model to the Gaussian random process in space $L_1([0,T])$, $L_p([0,T])$, $1 < p \leq 2$ where a given accuracy and reliability are proved. These issues are discussed in [ANT 02b].

In section 5.1.2, the theorems are considered on estimates of the "tails" of norm distributions of random processes under different conditions in the space $L_p(\mathbf{T})$, where $\mathbf{T}$ is some parametric set, $p \geq 1$. These statements are applied to investigate the partition selection of the set $[0,\infty]$ such that there exists a Gaussian process that is approximated by the model with some accuracy and reliability in the space $L_p([0,T])$ when $p \geq 1$. These results can be found in [ANT 02a, TEG 04a].

Note that obtained estimates in section 5.1.2 for $1 < p \leq 2$ are worse than estimates from section 5.1.1.

In section 5.1.3, a theorem on model approximation of Gaussian random processes with a given accuracy and reliability of Orlicz space $L_U(\Omega)$ is presented. These issues are discussed in [ANT 02a].

5.1.1. *The accuracy of modeling stationary Gaussian processes in* $L_p([0,T])$, $1 \leq p \leq 2$

Let $X = \{X(t), t \in \mathbf{T}\}$ be a Gaussian stationary real centered continuous in a mean square random process, where $\mathbf{T}$ is some parametric set. The definitions of this process and its model $X_\Lambda(t)$ are described in sections 4.2 and 4.4, respectively.

DEFINITION 5.1.– *Random process $X_\Lambda(t)$ approximates the process $X(t)$ with reliability $(1-\beta)$, $0 < \beta < 1$ and accuracy $\delta > 0$ in $L_p([0,T])$, if there exists a partition of Λ such that*

$$P\left\{\left(\int_0^T |\eta_\Lambda(t)|^p dt\right)^{\frac{1}{p}} > \delta\right\} \leq \beta,$$

where $\eta_\Lambda(t) = X(t) - X_\Lambda(t)$

THEOREM 5.1.– The model $X_\Lambda(t)$ approximates Gaussian random process $X(t)$ with reliability $(1-\beta)$, $0 < \beta < 1$ and accuracy $\delta > 0$ in $L_1([0,T])$, if for the partition Λ

$$\int_0^T 4\left(\sum_{k=0}^{M} b_k^2 \sup_{m\geq 1}\left(\mathbf{E}\left(\sin\frac{t(\zeta_k-\zeta_k^*)}{2}\right)^{2m}\right)^{\frac{1}{m}}\right)^{\frac{1}{2}} dt \leq \left(\frac{\delta^2}{2(-\ln\frac{\beta}{2})}\right)^{\frac{1}{2}}. \quad [5.1]$$

PROOF.– From corollary 1.2 and theorem 4.2 follows that for all $U > 0$

$$\mathbf{E}\exp\left\{U\int_0^T |\eta_\Lambda(t)|dt\right\} \leq 2\exp\left\{\frac{U^2}{2}\left(\int_0^T (B_\Lambda(t))^{\frac{1}{2}}\, dt\right)^2\right\}.$$

From [1.20], we have

$$P\left\{\int_0^T |\eta_\Lambda(t)|dt > \delta\right\} \leq 2\exp\left\{-\frac{\delta^2}{2}\left(\int_0^T (B_\Lambda(t))^{\frac{1}{2}}\, dt\right)^{-2}\right\}.$$

Then, by definition 5.1, the next inequality is satisfied

$$2\exp\left\{-\frac{\delta^2}{2}\left(\int_0^T (B_\Lambda(t))^{\frac{1}{2}}\, dt\right)^{-2}\right\} \leq \beta.$$

If in the above inequality we determine $\int_0^T (B_\Lambda(t))^{\frac{1}{2}}\, dt$, where $B_\Lambda(t)$ is described in [4.7], then condition [5.1] is obtained. □

THEOREM 5.2.– The model $X_\Lambda(t)$ approximates Gaussian random process $X(t)$ with reliability $(1-\beta)$, $0<\beta<1$ and accuracy $\delta>0$ in $L_p([0,T])$, $1<p\le 2$ if for the partition of Λ

$$\int_0^T (B_\Lambda(t))^{\frac{p}{2}}\, dt \le \delta^p,$$

and

$$\int_0^T (B_\Lambda(t))^{\frac{p}{2}}\, dt \le Z(p,\delta), \tag{5.2}$$

where $B_\Lambda(t)$ is described in [4.7], $Z(p,\delta)$ is a root of the equation

$$\left(\frac{\delta^p}{Z}\cdot\frac{2}{p}+1-\frac{2}{p}\right)^{1/2}\exp\left\{\frac{1}{p}\right\}\exp\left\{-\frac{\delta^p}{pZ}\right\}=\beta. \tag{5.3}$$

PROOF.– From [1.19] and theorem 4.3 follows that for all $0\le s<1$

$$\mathbf{E}\exp\left\{\frac{s}{p}\int_0^T|\eta_\Lambda(t)|^p dt\cdot\left(\int_0^T|B_\Lambda(t)|^{\frac{p}{2}}dt\right)^{-1}\right\}\le(1-s)^{-\frac{1}{2}}\exp\left\{\frac{(2-p)s}{2p}\right\}. \tag{5.4}$$

From [1.21] and [5.4], we have that for $\int_0^T(B_\Lambda(t))^{\frac{p}{2}}dt\le\delta^p$

$$P\left\{\left(\int_0^T|\eta_\Lambda(t)|^p dt\right)^{\frac{1}{p}}>\delta\right\}$$

$$=P\left\{\int_0^T|\eta_\Lambda(t)|^p dt\cdot\left(\int_0^T|B_\Lambda(t)|^{\frac{p}{2}}dt\right)^{-1}>\delta^p\left(\int_0^T|B_\Lambda(t)|^{\frac{p}{2}}dt\right)^{-1}\right\}$$

$$\le\left(\delta^p\cdot\frac{2}{p}\left(\int_0^T|B_\Lambda(T)|^{\frac{p}{2}}dt\right)^{-1}+1-\frac{2}{p}\right)^{\frac{1}{2}}$$

$$\times \exp\left\{\frac{1}{p}\right\} \cdot \exp\left\{-\frac{\delta^p}{p}\left(\int\limits_0^T |B_\Lambda(t)|^{\frac{p}{2}} dt\right)^{-1}\right\}.$$

The left side of [5.3] is increasing as $Z < \delta^p$ with respect to Z. Then

$$P\left\{\left(\int\limits_0^T |\eta_\Lambda(t)|^p dt\right)^{\frac{1}{p}} > \delta\right\} \leq \beta.$$

It means that [5.2] is satisfied. □

EXAMPLE 5.1.– In theorem 4.3 evaluating the integrand expression, we obtain

$$\tau(\eta_\Lambda(t)) \leq 4\left(\sum_{k=0}^{M-1}\left(\frac{t}{2}\right)^{2\gamma} |\lambda_{k+1} - \lambda_k|^{2\gamma}\left(F(\lambda_{k+1}) - F(\lambda_k)\right) + \right.$$

$$\left. + F(+\infty) - F(\Lambda)\right)^{\frac{1}{2}}.$$

Let $|\lambda_{k+1} - \lambda_k| = \frac{\Lambda}{M}$, then

$$\tau(\eta_\Lambda(t)) \leq 4\left(\left(\frac{\Lambda t}{2M}\right)^{2\gamma} F(\Lambda) + F(+\infty) - F(\Lambda)\right)^{\frac{1}{2}}, \qquad [5.5]$$

Let $\delta = 0.01, \beta = 0.01, p = 2$. Then, it follows from theorem 5.2 that $\frac{\delta}{\sqrt{z}} \exp\{\frac{1}{2} - \frac{\delta^2}{2z}\} = \beta$, whence it appears $z = 8.04 \times 10^{-6}$; [5.2] yields

$$\int\limits_0^T 16\left(\left(\frac{\Lambda t}{2M}\right)^{2\gamma} F(\Lambda) + F(+\infty) - F(\Lambda)\right)$$

$$= 16\left(\frac{T^{2\gamma+1}}{2\gamma+1}\left(\frac{\Lambda}{2M}\right)^{2\gamma} F(\Lambda) + T(F(+\infty) - F(\Lambda))\right)$$

Let $\gamma = 1$, $T = 1$, $F(\lambda) = 1 - e^{-\lambda}$. Then, from [5.2] and the last equality, we have

$$\frac{4}{3}\left(\frac{\Lambda}{M}\right)^2 (1 - e^{-\Lambda}) + 16e^{-\Lambda} \leq 8.04 \times 10^{-6}$$

$$M \geq \frac{\Lambda(1 - e^{-\Lambda})^{\frac{1}{2}}}{(6 \times 10^{-6} - 12e^{-\Lambda})^{\frac{1}{2}}}$$

With the graphical editor SciDaVis, the approximate minimum of this function is found in Λ (see Figure 5.1): $M(16.75) \approx 7,233$.

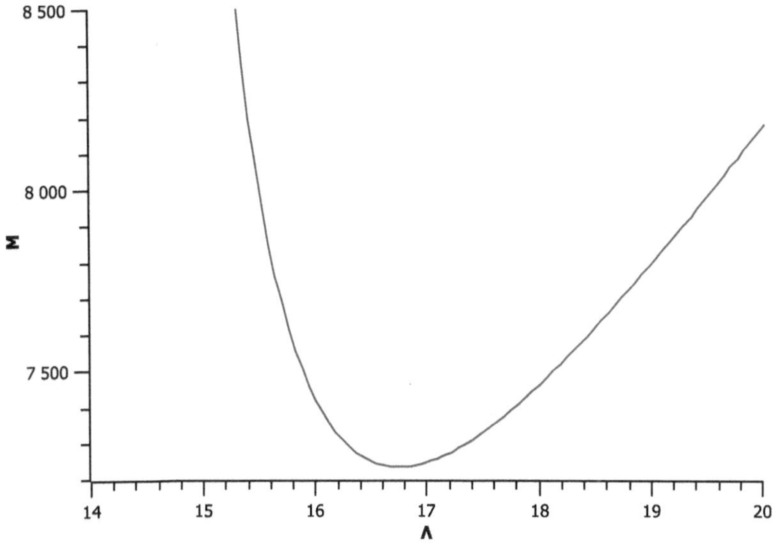

Figure 5.1. *Graph of function* $M(\Lambda) = \frac{\Lambda(1-e^{-\Lambda})^{\frac{1}{2}}}{(6 \cdot 10^{-6} - 12e^{-\Lambda})^{\frac{1}{2}}}$

Using computer-simulated Gaussian random variables η_{k1}, η_{k2} and variables ζ_k (in the process $X(t)$), the model of Gaussian stationary random process is obtained with accuracy 0.01 and reliability 0.99 in space $L_2([0, 1])$ (see Figure 5.2).

EXAMPLE 5.2.– Let $\gamma = 1$, $T = 1$, $F(\Lambda) = \frac{1}{9} - \frac{1}{9(1+\Lambda)^9}$, $F(+\infty) = \frac{1}{9}$.

Then, from [5.2], we obtain

$$M \geq \frac{\Lambda}{3} \left(\frac{1 - (1 + \Lambda)^{-9}}{6 \times 10^{-6} - \frac{4}{3}(1 + \Lambda)^{-9}} \right)^{\frac{1}{2}}$$

With the graphical editor SciDaVis, we shall find the minimum of this function in Λ. Figure 5.3 shows which minimum of function equals $M(3.67) = 562$, building a model $X_\Lambda(t)$ of Gaussian process $X(t)$ as $M = 562$ (see Figure 5.4).

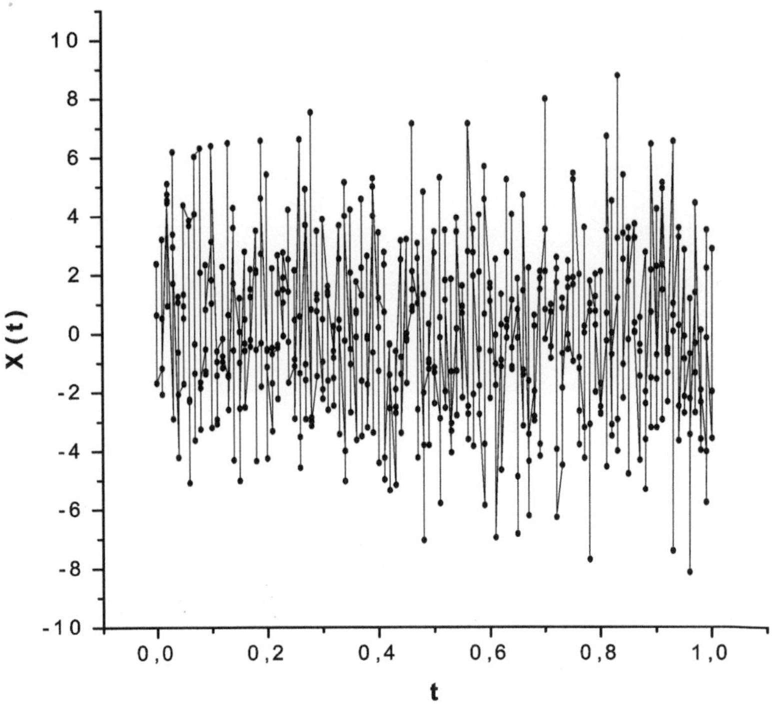

Figure 5.2. *The model of Gaussian random process in space* $L_2([0,1])$, *with spectral density* $f(\lambda) = e^{-\lambda}$

EXAMPLE 5.3.– Let $\gamma = 1$, $T = 1$, $F(\Lambda) = \ln \frac{2e^\Lambda}{1+e^\Lambda}$.

Then, from [5.2], we obtain

$$M \geq \Lambda \left(\frac{\ln \frac{2e^\Lambda}{1+e^\Lambda}}{6 \times 10^{-6} + 12 \ln \frac{e^\Lambda}{1+e^\Lambda}} \right)^{\frac{1}{2}}$$

With the graphical editor SciDaVis, we shall find the minimum of this function in Λ. Figure 5.5 shows which minimum of function equals $M(16.78) = 6,022$, building a model $X_\Lambda(t)$ of Gaussian process $X(t)$ as $M = 6022$ (see Figure 5.6).

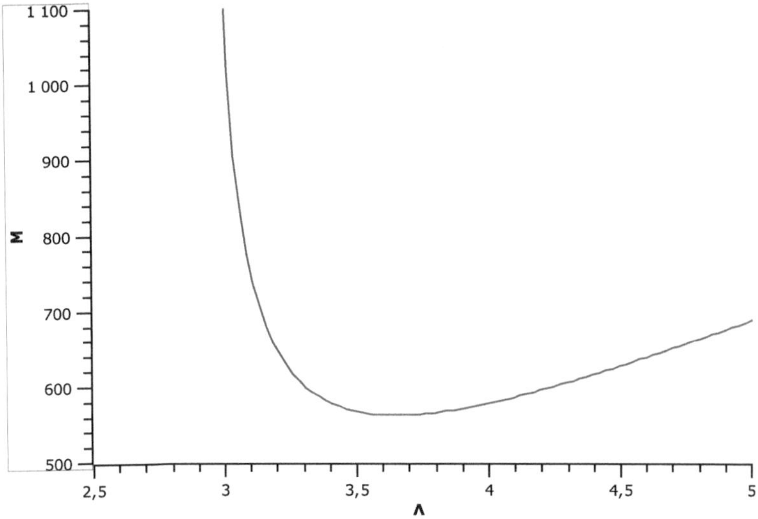

Figure 5.3. *Graph of function* $M(\Lambda) = \frac{\Lambda}{3}\left(\frac{1-(1+\Lambda)^{-9}}{6\cdot 10^{-6}-\frac{4}{3}(1+\Lambda)^{-9}}\right)^{\frac{1}{2}}$

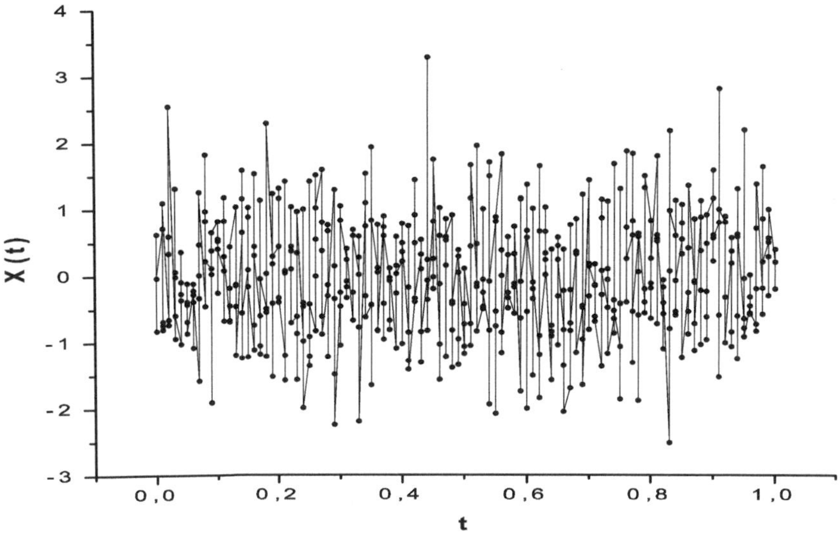

Figure 5.4. *The model of Gaussian random process in space* $L_2([0,1])$ *with spectral density* $f(\lambda) = \frac{1}{(1+\lambda)^{10}}$

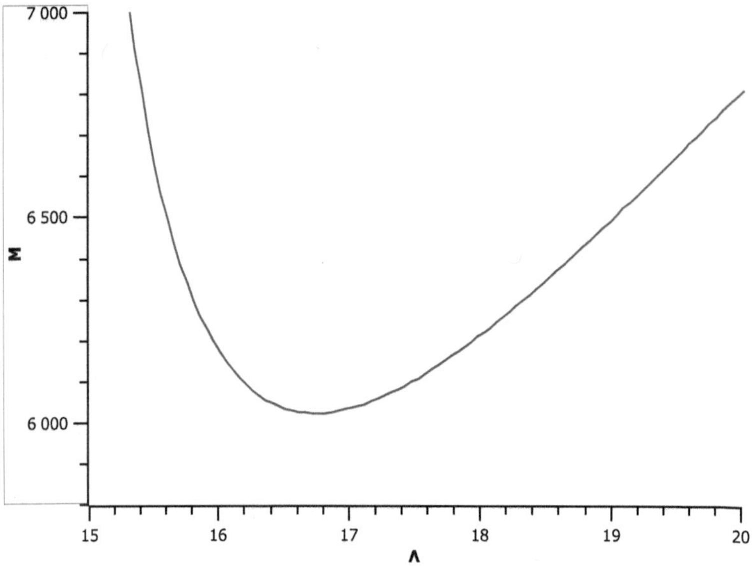

Figure 5.5. *Graph of function* $M(\Lambda) = \Lambda \left(\frac{\ln \frac{2e^{\Lambda}}{1+e^{\Lambda}}}{6 \cdot 10^{-6} + 12 \ln \frac{e^{\Lambda}}{1+e^{\Lambda}}} \right)^1 /2$

5.1.2. *The accuracy of modeling stationary Gaussian processes* $L_p([0,T])$ *at* $p \geq 1$

Let $X = \{X(t), t \in \mathbf{T}\}$ be Gaussian stationary real centered continuous in a mean square random process, where $\mathbf{T}$ is some parametric set. The definition of this process and its model $X_\Lambda(t)$ are described in section 5.4.

Let $\{\mathbf{T}, \mathcal{A}, \mu\}$ be a measurable space, $\mu(\mathbf{T}) < \infty$, $\tau(t) = \tau(X(t))$.

LEMMA 5.1.– Let $\int\limits_{\mathbf{T}} (\tau(t))^p \, d\mu(t) < \infty$, $p \geq 1$. Then, $X \in L_p(\mathbf{T})$ with probability 1.

PROOF.– The proof of the lemma follows from lemma 1.3 because of

$$\mathbf{E} \int\limits_{\mathbf{T}} |X(t)|^p d\mu(t) = \int\limits_{\mathbf{T}} \mathbf{E}|X(t)|^p d\mu(t) \leq 2 \left(\frac{p}{e}\right)^{\frac{p}{2}} \int\limits_{\mathbf{T}} (\tau(X(t)))^p d\mu(t) < \infty.$$

So, $\int\limits_{\mathbf{T}} |X(t)|^p d\mu(t) < \infty$ with probability one. □

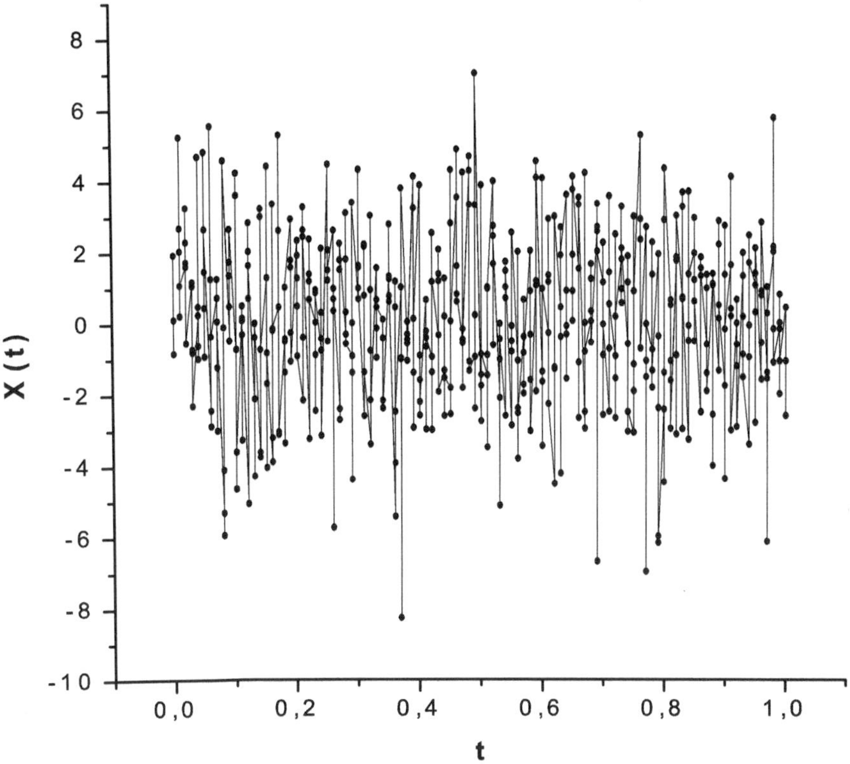

Figure 5.6. *The model of Gaussian random processes in space $L_2([0,1])$ with spectral density $f(\lambda) = \frac{1}{1+e^{\lambda}}$*

LEMMA 5.2.– For all $s \geq p \geq 1$, $\varepsilon > 0$, the relationship

$$P\left\{\|X\|_{L_p} > \varepsilon\right\} \leq 2\varepsilon^{-s}\left(\frac{s}{e}\right)^{\frac{s}{2}} \int_{\mathbf{T}} (\tau(t))^s \, d\mu(t) \cdot \left(\mu(\mathbf{T})\right)^{\frac{s}{p}-1} \qquad [5.6]$$

holds true.

PROOF.– By the Tchebychev's inequality

$$P\left\{\|X\|_{L_p} > \varepsilon\right\} \leq \frac{\mathbf{E}\|X\|_{L_p}^s}{\varepsilon^s},$$

$$\mathbf{E}\|X\|_{L_p}^s = \mathbf{E}\left(\int\limits_{\mathbf{T}} |X(t)|^p \, d\mu(t)\right)^{\frac{s}{p}} \leq \mathbf{E}\left(\int\limits_{\mathbf{T}} |X(t)|^p \, d\left(\frac{\mu(t)}{\mu(\mathbf{T})}\right)\right)^{\frac{s}{p}} (\mu(\mathbf{T}))^{\frac{s}{p}}$$

$$\leq \mathbf{E}\left(\int\limits_{\mathbf{T}} |X(t)|^s \, d\left(\frac{\mu(t)}{\mu(\mathbf{T})}\right)\right) (\mu(\mathbf{T}))^{\frac{s}{p}} = \int\limits_{\mathbf{T}} \mathbf{E}\,|X(t)|^s \, d\mu(t) \cdot (\mu(\mathbf{T}))^{\frac{s}{p}-1}.$$

Then, from lemma 1.3 follows that

$$P\left\{\|X\|_{L_p} > \varepsilon\right\} \leq \varepsilon^{-s} \cdot 2\left(\frac{s}{e}\right)^{\frac{s}{2}} \int\limits_{\mathbf{T}} (\tau(t))^s \, d\mu(t) \cdot (\mu(\mathbf{T}))^{\frac{s}{p}-1}.$$

□

PROPOSITION 5.1.– Let $\tau = \sup_{t\in\mathbf{T}} \tau(t) < \infty$. Then, for all $\varepsilon \geq p^{\frac{1}{2}} (\mu(\mathbf{T}))^{\frac{1}{p}} \cdot \tau$ the following inequality holds

$$P\left\{\|X\|_{L_p} > \varepsilon\right\} \leq 2\exp\left\{-\frac{\varepsilon^2}{2\tau^2 \cdot (\mu(\mathbf{T}))^{\frac{2}{p}}}\right\}.$$

PROOF.– By [5.6] for all $\varepsilon > 0, \ s \geq p$, we have

$$P\left\{\|X\|_{L_p} > \varepsilon\right\} \leq 2 s^{\frac{s}{2}} a^s,$$

where $a = \frac{(\mu(T))^{\frac{1}{p}} \cdot \tau}{\varepsilon\sqrt{e}}$. Consider $s = a^{-2} e^{-1}$. It is a minimum point of the right-hand side of last inequality.

Then, for $s = \frac{1}{a^2 e} \geq p$, that is for $\varepsilon > (\mu(\mathbf{T}))^{\frac{1}{p}} p^{\frac{1}{2}} \tau$, we have

$$P\left\{\|X\|_{L_p} > \varepsilon\right\} \leq 2(a^2 e)^{-\frac{1}{2a^2 e}} a^{\frac{1}{a^2 e}}$$

$$= 2\exp\left\{-\frac{1}{2a^2 e}\right\} = 2\exp\left\{-\frac{\varepsilon^2}{2(\mu(\mathbf{T}))^{\frac{2}{p}} \tau^2}\right\}.$$

□

PROPOSITION 5.2.– Let $\mathbf{T} = [0, T]$, $T > 0$, $\mu(\cdot)$ be Lebesgue measure and $\tau(t) \leq t^\nu b$ for some $\nu > 0$, $b > 0$. Then, for $\varepsilon > p^{\frac{1}{2}} T^{\nu + \frac{1}{p}} b$ the inequality

$$P\left\{\|X\|_{L_p} > \varepsilon\right\} \leq 2\exp\left\{-\frac{\varepsilon^2}{2T^{2\nu+\frac{2}{p}} b^2}\right\} \left(\frac{\nu\varepsilon^2}{T^{2\nu+\frac{2}{p}} b^2} + 1\right)^{-1} \qquad [5.7]$$

holds.

PROOF.– From theorem's conditions $\int_0^T (\tau(t))^s \, dt \leq \frac{b^s T^{\nu s+1}}{\nu s+1}$ and [5.6], we obtain

$$P\left\{\|X\|_{L_p} > \varepsilon\right\} \leq 2\varepsilon^{-s}\left(\frac{s}{e}\right)^{\frac{s}{2}} \frac{b^s T^{\nu s+1}}{\nu s+1} \cdot T^{\frac{s}{p}-1} = 2s^{\frac{s}{2}} \frac{1}{\nu s+1}\left(\frac{T^{\nu+\frac{1}{p}} b}{\sqrt{e}\varepsilon}\right)^s.$$

If we take $s = \frac{\varepsilon^2}{T^{2\nu+\frac{2}{p}} b^2}$, then as $s \geq p$, that is as $\varepsilon \geq p^{\frac{1}{2}} T^{\nu+\frac{1}{p}} b$ we obtain

$$P\left\{\|X\|_{L_p} > \varepsilon\right\} \leq 2\left(\frac{\varepsilon^2}{T^{2\nu+\frac{2}{p}} b^2}\right)^{\frac{\varepsilon^2}{2T^{2\nu+\frac{2}{p}} b^2}} \frac{T^{2\nu+\frac{2}{p}} b^2}{\nu\varepsilon^2 + T^{2\nu+\frac{2}{p}} b^2}\left(\frac{T^{\nu+\frac{1}{p}} b}{\sqrt{e}\varepsilon}\right)^{\frac{\varepsilon^2}{T^{2\nu+\frac{2}{p}} b^2}}$$

$$= 2\left(\nu \frac{\varepsilon^2}{T^{2\nu+\frac{2}{p}} b^2} + 1\right)^{-1} \exp\left\{-\frac{\varepsilon^2}{2T^{2\nu+\frac{2}{p}} b^2}\right\}.$$

□

THEOREM 5.3.– Let in model $X_\Lambda(t)$ the partition of Λ be such that

$\tau\left(\eta_\Lambda(t)\right) \leq \tau(\Lambda, T)$,

where $\tau(\Lambda, T) = (B_\Lambda(t))^{\frac{1}{2}}$ is defined in [4.7],

$$\tau(\Lambda, T) \leq \frac{\delta}{p^{\frac{1}{2}} T^{\frac{1}{p}}}, \qquad [5.8]$$

$$\tau(\Lambda, T) \leq \frac{\delta}{T^{\frac{1}{p}}\left(2\ln\frac{2}{\beta}\right)^{\frac{1}{2}}}, \qquad [5.9]$$

then the model approximates Gaussian process $X(t)$ with reliability $1-\beta$, $0 < \beta < 1$ and accuracy $\delta > 0$ in $L_p([0,T])$, $p \geq 1$.

PROOF.– This assertion follows from proposition 5.1.

Really, if $\delta > p^{\frac{1}{2}} T^{\frac{1}{p}} \tau(\Lambda, T)$ (this is the condition [5.8]), then from proposition 5.1 and definition 5.1 we have

$$P\left\{\|\eta_\Lambda(t)\|_{L_p} > \delta\right\} \leq 2\exp\left\{-\frac{\delta^2}{2\tau^2(\Lambda, T) \cdot T^{\frac{2}{p}}}\right\} \leq \beta.$$

And last inequality is satisfied when the condition [5.9] holds. □

THEOREM 5.4.– Let in the model $X_\Lambda(t)$ the partition of Λ be such that

$$\tau(\eta_\Lambda(t)) \leq t^\nu \tau_\Lambda, \quad \nu > 0;$$

(another form of condition [4.7])

$$\tau_\Lambda < \frac{\delta}{p^{\frac{1}{2}} T^{\nu+\frac{1}{p}}}; \quad \tau_\Lambda < \frac{\delta}{T^{\nu+\frac{1}{p}} (y_\beta)^{\frac{1}{2}}},$$

where y_β is a root of the equation $2\exp\left\{-\frac{y_\beta}{2}\right\}(\nu y_\beta + 1)^{-1} = \beta$. Then, the model approximates Gaussian process $X(t)$ with reliability $(1-\beta), 0 < \beta < 1$ and accuracy $\delta > 0$ in space $L_p([0,T]), p \geq 1$.

PROOF.– This theorem follows from proposition 5.2. Really, let τ_Λ be such that $\delta > p^{\frac{1}{2}} T^{\nu+\frac{1}{p}} \tau_\Lambda$. Then

$$\frac{\delta}{T^{\nu+\frac{1}{p}} \tau_\Lambda^{\frac{1}{2}}} \geq \frac{\delta}{p^{\frac{1}{2}} T^{\nu+\frac{1}{p}} \tau_\Lambda} > 1.$$

A function $f(y_\beta) = 2\exp\left\{-\frac{y_\beta}{2}\right\}(\nu y_\beta + 1)^{-1}$ decreases as $y_\beta > 1$. Thus, statement 5.2 implies the estimate

$$P\left\{\|\eta_\Lambda(t)\|_{L_p} > \delta\right\} \leq 2\exp\left\{-\frac{\delta^2}{2T^{2\nu+\frac{2}{p}}\tau_\Lambda^2}\right\} \cdot \left(\nu \frac{\delta^2}{T^{2\nu+\frac{2}{p}}\tau_\Lambda^2} + 1\right)^{-1},$$

when $\frac{\delta^2}{T^{2\nu+\frac{2}{p}}\tau_\Lambda^2} \geq y_\beta$. □

EXAMPLE 5.4.– Let the spectral function $F(\lambda)$ of the process $X(t)$ be such that $F(+\infty) = 1$, $F(+\infty) - F(\lambda) \leq \frac{1}{\lambda^{2\gamma}}$, $0 \leq \gamma \leq 1$, $\lambda > 0$. Using theorem 4.2, we have

$$\tau^2(\eta_\Lambda(t)) \leq 4\sum_{k=0}^{M} I_k,$$

where

$$I_k = \sup_{m\geq 1} 4b_k^2 \left(\int_{\lambda_k}^{\lambda_{k+1}} \left(\int_{\lambda_k}^{\lambda_{k+1}} \left| \sin\frac{t(u-\lambda)}{2} \right|^2 dF_k(\lambda) \right)^m dF_k(u) \right)^{\frac{1}{m}},$$

$$b_k^2 = F(\lambda_{k+1}) - F(\lambda_k).$$

For $k = 0, \ldots, M-1$:

$$I_k \leq \sup_{m\geq 1} 4b_k^2 \left(\int\limits_{\lambda_k}^{\lambda_{k+1}} \left(\int\limits_{\lambda_k}^{\lambda_{k+1}} \frac{t^{2\gamma}|u-v|^{2\gamma}}{4^\gamma} dF_k(v) \right)^m dF_k(u) \right)^{\frac{1}{m}}$$

$$\leq 4^{1-\gamma} t^{2\gamma} b_k^2 \sup_{m\geq 1} \left(E|\zeta_k - \zeta_k^*|^{2\gamma m} \right)^{\frac{1}{m}} = 4^{1-\gamma} t^{2\gamma} b_k^2 |\lambda_{k+1} - \lambda_k|^{2\gamma}.$$

For $k = M$

$$I_M \leq \sup_{m\geq 1} \frac{4}{b_M^{\frac{2}{m}}} \left(\int\limits_{\lambda_M}^{\infty} \left(\int\limits_{\lambda_M}^{\infty} dF(u) \right)^m dF(v) \right)^{\frac{1}{m}} \leq 4\left(F(+\infty) - F(\lambda_M)\right).$$

Hence,

$$\tau^2(\eta_\Lambda(t))$$

$$\leq 4 \left(\sum_{k=0}^{M-1} 4^{1-\gamma} t^{2\gamma} |\lambda_{k+1} - \lambda_k|^{2\gamma} \left(F(\lambda_{k+1}) - F(\lambda_k)\right) + 4\left(F(+\infty) - F(\lambda_M)\right) \right).$$

Let $|\lambda_{k+1} - \lambda_k| = \frac{\lambda_M}{M}$, then

$$\tau^2(\eta_\Lambda(t)) \leq 4^{2-\gamma} t^{2\gamma} \left(\frac{\lambda_M}{M} \right)^{2\gamma} F(\lambda_M) + 16\left(F(+\infty) - F(\lambda_M)\right)$$

$$\leq 16 \left(\frac{\lambda_M T}{2M} \right)^{2\gamma} + 16\lambda_M^{-2\gamma},$$

$$\tau(\Lambda, T) = 4 \left(\left(\frac{\lambda_M T}{2M} \right)^{2\gamma} + \frac{1}{\lambda_M^{2\gamma}} \right)^{\frac{1}{2}}.$$

Find the minimum of the function $\tau(\Lambda, T)$ with respect to $a = \lambda_M^{2\gamma}$:

$$y = 4 \left(\left(\frac{T}{2M} \right)^{2\gamma} a + \frac{1}{a} \right)^{\frac{1}{2}},$$

$$y' = 2 \left(\left(\frac{T}{2M} \right)^{2\gamma} a + \frac{1}{a} \right)^{-\frac{1}{2}} \left(\left(\frac{T}{2M} \right)^{2\gamma} - \frac{1}{a^2} \right) = 0,$$

$a = \left(\frac{2M}{T} \right)^{\gamma}$ is a point of minimum.

So, $\lambda_M^{2\gamma} = \left(\frac{2M}{T}\right)^\gamma$. Then

$$\tau(\Lambda, T) = 4\left[\left(\frac{2M}{T}\right)^\gamma \left(\frac{T}{2M}\right)^{2\gamma} + \left(\frac{T}{2M}\right)^\gamma\right]^{\frac{1}{2}} = 4\sqrt{2}\left(\frac{T}{2M}\right)^{\frac{\gamma}{2}}.$$

Suppose, for example, in theorem 5.3, $2\ln\frac{2}{\beta} > p$. Then

$$\tau < \frac{\delta}{T^{\frac{1}{p}}\left(2\ln\frac{2}{\beta}\right)^{\frac{1}{2}}} < \frac{\delta}{p^{\frac{1}{2}}T^{\frac{1}{p}}},$$

$$4\sqrt{2}\left(\frac{T}{2M}\right)^{\frac{\gamma}{2}} < \frac{\delta}{T^{\frac{1}{p}}\left(2\ln\frac{2}{\beta}\right)^{\frac{1}{2}}},$$

$$4\sqrt{2}\frac{T^{\frac{\gamma}{2}}}{\delta 2^{\frac{\gamma}{2}}} T^{\frac{1}{p}}\left(2\ln\frac{2}{\beta}\right)^{\frac{1}{2}} < M^{\frac{\gamma}{2}},$$

$$M > 2^{\frac{5}{\gamma}-1}\frac{T^{1+\frac{2}{p\gamma}}}{\delta^{\frac{2}{\gamma}}}\left(2\ln\frac{2}{\beta}\right)^{\frac{1}{\gamma}}.$$

Hence, the model $X_\Lambda(t)$ approximates the process $X(t)$ with reliability $1-\beta$, $0<\beta<1$, and accuracy $\delta > 0$ in space $L_p([0,T])$, if $\lambda_M = \sqrt{2}M^{\frac{1}{2}}T^{-\frac{1}{2}}$ and

$$M > 2^{\frac{5}{\gamma}-1}\frac{T^{1+\frac{2}{p\gamma}}}{\delta^{\frac{2}{\gamma}}}\left(2\ln\frac{2}{\beta}\right)^{\frac{1}{\gamma}}.$$

EXAMPLE 5.5.– Let for the function $F(\lambda)$ the condition $\int_0^\infty \lambda^{2\gamma} dF(\lambda) < \infty$ for some $0 \le \gamma \le 1$ holds and $F(+\infty) = 1$. Then, by theorem 4.3

$$\tau^2(\eta_\lambda(t)) \le 4\sum_{k=0}^{M} I_k,$$

$$I_k = \sup_{m\ge 1} 4b_k^2 \left(\int_{\lambda_k}^{\lambda_{k+1}}\left(\int_{\lambda_k}^{\lambda_{k+1}}\left|\sin\frac{t(u-\lambda)}{2}\right|^2 dF_k(\lambda)\right)^m dF_k(u)\right)^{\frac{1}{m}}.$$

If $k = 0, \ldots, M-1$, then

$$I_k \le 4^{1-\gamma}t^{2\gamma}b_k^2 \sup_{m\ge 1}\left(E|\zeta_k - \zeta_k^*|^{2\gamma m}\right)^{\frac{1}{m}} \le 4^{1-\gamma}t^{2\gamma}b_k^2|\lambda_{k+1}-\lambda_k|^{2\gamma},$$

where ζ_k, ζ_k^* are independent random variables with the same cumulative distribution function (cdf) $F_k(\lambda) = \frac{F(\lambda)-F(\lambda_k)}{F(\lambda_{k+1})-F(\lambda_k)}$.

If $k = M$, then

$$I_M \leq \sup_{m\geq 1} \frac{4}{b_M^{\frac{2}{m}}} \left(\int_{\lambda_M}^{\infty} \left(\int_{\lambda_M}^{\infty} \left| \sin \frac{t(u-\lambda)}{2} \right|^2 dF(u) \right)^m dF(\lambda) \right)^{\frac{1}{m}}$$

$$\leq \sup_{m\geq 1} \frac{4}{b_M^{\frac{2}{m}}} \left(\int_{\lambda_M}^{\infty} \left(\int_{\lambda_M}^{\infty} \frac{t^{2\gamma}}{4^\gamma} |u-\lambda|^{2\gamma} dF(u) \right)^m dF(\lambda) \right)^{\frac{1}{m}}$$

$$\leq \sup_{m\geq 1} \frac{t^{2\gamma} 4^{1-\gamma}}{b_M^{\frac{2}{m}}} \left(\int_{\lambda_M}^{\infty} \left(\int_{\lambda_M}^{\infty} |u-\lambda_M|^{2\gamma} dF(u) \right)^m dF(\lambda) \right)^{\frac{1}{m}}$$

$$= t^{2\gamma} 4^{1-\gamma} \int_{\lambda_M}^{\infty} |u-\lambda_M|^{2\gamma} dF(u),$$

$$\tau^2(\eta_\Lambda(t))$$

$$\leq 4 \left(\sum_{k=0}^{M-1} 4^{1-\gamma} t^{2\gamma} b_k^2 |\lambda_{k+1}-\lambda_k|^{2\gamma} + 4^{1-\gamma} t^{2\gamma} \int_{\lambda_M}^{\infty} |\lambda-\lambda_M|^{2\gamma} dF(\lambda) \right)$$

$$= 4^{2-\gamma} t^{2\gamma} \left(\max_{0\leq k\leq M-1} |\lambda_{k+1}-\lambda_k|^{2\gamma} F(\lambda_M) + \int_{\lambda_M}^{\infty} |\lambda-\lambda_M|^{2\gamma} dF(\lambda) \right)$$

$$\leq \frac{16 t^{2\gamma}}{4^\gamma} \left(\left(\frac{\lambda_M}{M} \right)^{2\gamma} + \int_{\lambda_M}^{\infty} |\lambda-\lambda_M|^{2\gamma} dF(\lambda) \right).$$

Therefore,

$$\tau(\eta_\Lambda(t)) \leq t^\gamma \tau_\Lambda ,$$

where

$$\tau_\Lambda = \frac{4}{2^\gamma} \left(\left(\frac{\lambda_M}{M} \right)^{2\gamma} + \tilde{J}(\lambda_M) \right)^{\frac{1}{2}}; \quad \tilde{J}(\lambda_M) = \int_{\lambda_M}^{\infty} |\lambda-\lambda|^{2\gamma} dF(\lambda).$$

Suppose, for example, $y_\beta > p$ in theorem 5.6. Then, $\tau_\Lambda \leq \frac{\delta}{T^{\gamma+\frac{1}{p}}(y_\beta)^{\frac{1}{2}}}$,

$$\frac{4}{2^\gamma}\left(\left(\frac{\lambda_M}{M}\right)^{2\gamma} + \tilde{J}(\lambda_M)\right)^{\frac{1}{2}} \leq \frac{\delta}{T^{\gamma+\frac{1}{p}}(y_\beta)^{\frac{1}{2}}},$$

$$\left(\frac{\lambda_M}{M}\right)^{2\gamma} + \tilde{J}(\lambda_M) \leq \frac{2^{2\gamma}\delta^2}{16T^{2\gamma+\frac{2}{p}}y_\beta},$$

$$M \geq \lambda_M\left(\frac{2^{2\gamma}\delta^2}{16T^{2\gamma+\frac{2}{p}}y_\beta} - \tilde{J}(\lambda_M)\right)^{-\frac{1}{2\gamma}}$$

$$= \lambda_M 4^{\frac{1}{\gamma}} T^{1+\frac{1}{p\gamma}} y_\beta^{\frac{1}{2\gamma}} \left(2^{2\gamma}\delta^2 - \tilde{J}(\lambda_M)16T^{2\gamma+\frac{2}{p}}y_\beta\right)^{-\frac{1}{2\gamma}}.$$

Thus, the model $X_\Lambda(t)$ approximates the process $X(t)$ with reliability $1-\beta$, $0 < \beta < 1$ and accuracy $\delta > 0$ in space $L_p([0,T])$, if

$$M \geq \lambda_M 2^{\frac{2}{\gamma}} T^{1+\frac{1}{p\gamma}} y_\beta^{\frac{1}{2\gamma}} (4^\gamma\delta^2 - \tilde{J}(\lambda_M)16T^{2\gamma+\frac{2}{p}}y_\beta)^{-\frac{1}{2\gamma}},$$

where y_β is a root of the equation

$$2\exp\left\{-\frac{y_\beta}{2}\right\}(\gamma y_\beta + 1)^{-1} = \beta, \quad 0 \leq \gamma \leq 1.$$

EXAMPLE 5.6.– From [5.5] and theorem 5.3 follows that

$$4\left(\left(\frac{\Lambda T}{2M}\right)^{2\gamma} F(\Lambda) + F(+\infty) - F(\Lambda)\right)^{\frac{1}{2}} \leq \frac{\delta}{T^{\frac{1}{p}}\left(2\ln\frac{2}{\beta}\right)^{\frac{1}{2}}},$$

where

$$M \geq 2\Lambda T^{1+\frac{1}{p}}\left(\frac{2F(\Lambda)\ln\frac{2}{\beta}}{\delta^2 - 32T^{\frac{2}{p}}\ln\frac{2}{\beta}(F(+\infty) - F(\Lambda))}\right)^{\frac{1}{2}}$$

Let the spectral density be equal to $f(\lambda) = \exp(-\lambda)$, it means that $F(\lambda) = 1 - e^{-\lambda}$. Let $T = 1$, $\delta = 0.01$, $\beta = 0.01$. Then, we have

$$M \geq 2\Lambda\left(\frac{2(1-e^{-\Lambda})\ln 200}{0.0001 - 32e^{-\Lambda}\ln 200}\right)^{\frac{1}{2}}$$

A minimum of this function on Λ is approximately equal to $M(16.55) \approx 11,422$ (see Figure 5.7). Thus, choosing the minimum breakdown on the level of $M = 11,422$, we can computer simulate the model $X_\Lambda(t)$ of Gaussian process $X(t)$ with spectral function $F(\lambda) = 1 - e^{-\lambda}$ (see Figure 5.8).

EXAMPLE 5.7.– Let the spectral density be equal to $f(\lambda) = \frac{1}{(1+\lambda)^{10}}$, such that $F(\Lambda) = \frac{1}{9} - \frac{1}{9(1+\lambda)^9}$, $F(+\infty) = \frac{1}{9}$. Let $T = 1$, $\delta = 0.01$, $\beta = 0.01$. Then, we have

$$M \geq \Lambda \left(\frac{\frac{8}{9}\left(1 - (1+\Lambda)^{-9}\right) \ln 200}{0.0001 - \frac{32}{9}(1+\Lambda)^{-9} \ln 200} \right)^{\frac{1}{2}}$$

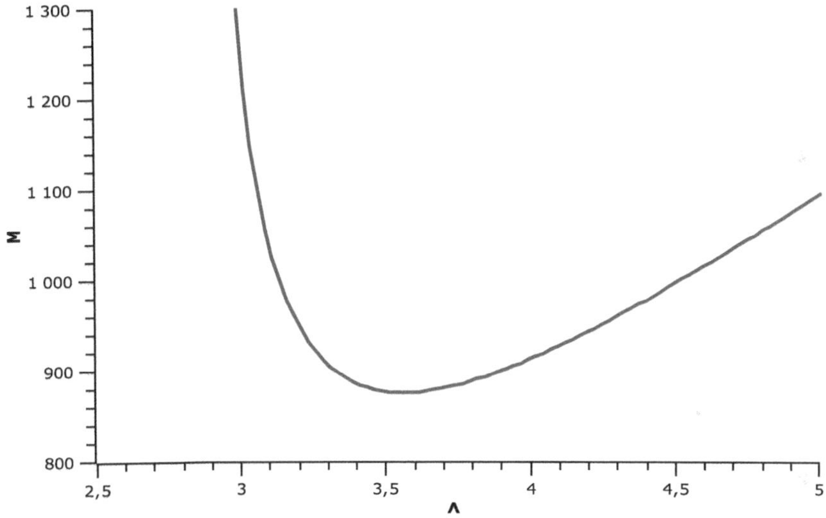

Figure 5.7. *Graph of function* $M(\Lambda) = 2\Lambda \left(\frac{2(1-e^{-\Lambda}) \ln 200}{0.0001 - 32e^{-\Lambda} \ln 200} \right)^{\frac{1}{2}}$

A minimum of this function on Λ is approximately equal to $M(3.56) = 876$ (see Figure 5.9).

Thus, choosing the minimum breakdown on the level of $M = 876$, we can computer simulate the model $X_\Lambda(t)$ of Gaussian process $X(t)$ with spectral function $F(\Lambda) = \frac{1}{9} - \frac{1}{9(1+\lambda)^9}$, $F(+\infty) = \frac{1}{9}$ (see Figure 5.10).

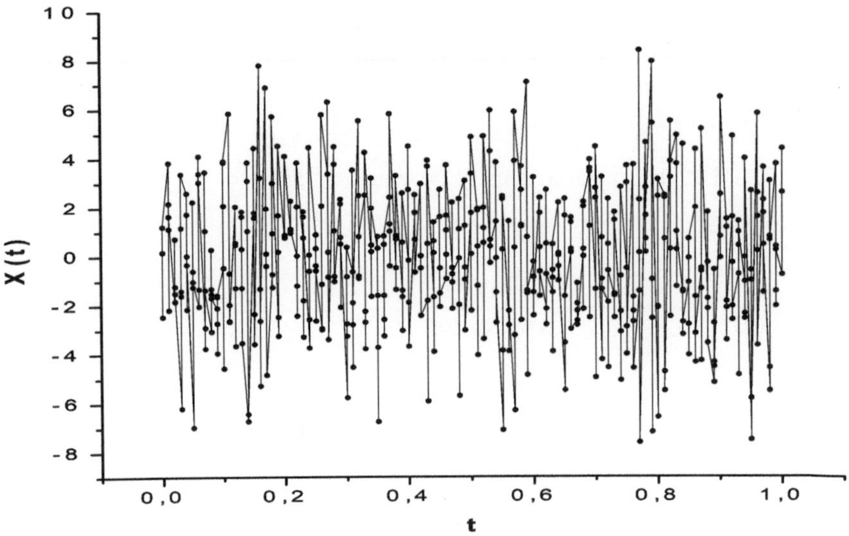

Figure 5.8. *The model of Gaussian random processes in space* $L_p([0,1])$, $p \geq 1$ *with spectral density* $f(\lambda) = e^{-\lambda}$

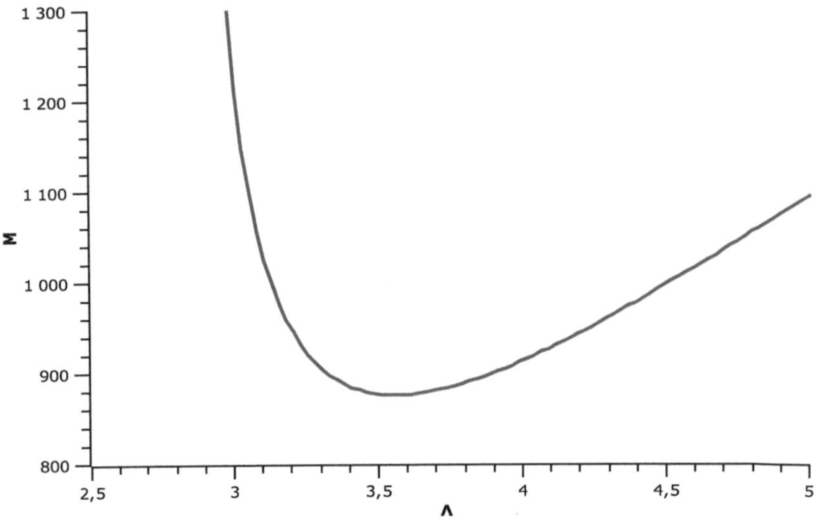

Figure 5.9. *Graph of function* $M(\Lambda) = \Lambda \left(\frac{\frac{1}{9}\left(1-(1+\Lambda)^{-9}\right) \ln 200}{0.0001 - \frac{32}{9}(1+\Lambda)^{-9} \ln 200} \right)^{\frac{1}{2}}$

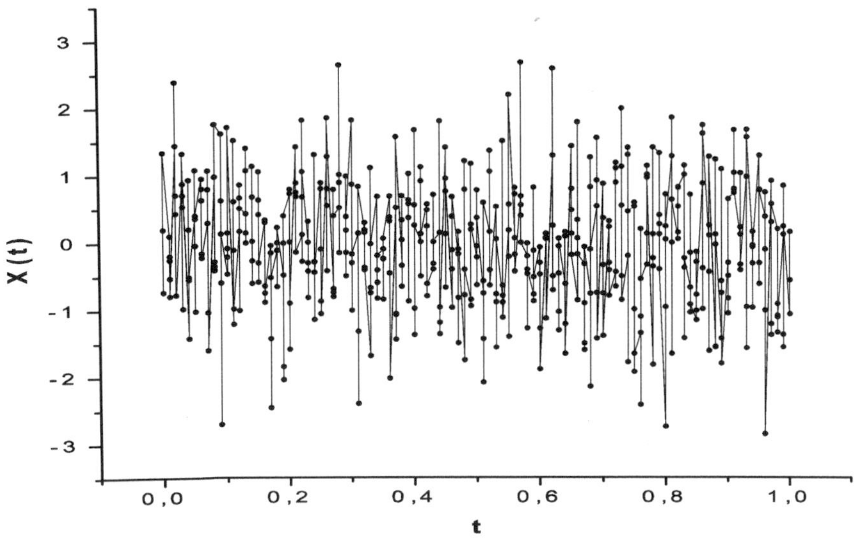

Figure 5.10. *The model of Gaussian random processes in space $L_p([0,1])$, $p \geq 1$ with spectral density $f(\lambda) = \frac{1}{(1+\lambda)^{10}}$*

EXAMPLE 5.8.– Let the spectral density be equal to $f(\lambda) = \frac{1}{1+e^{\lambda}}$, such that $F(\Lambda) = \ln \frac{2e^{\Lambda}}{1+e^{\Lambda}}$, $F(+\infty) = \ln 2$. Let $T = 1$, $\delta = 0.01$, $\beta = 0.01$. Then, we have

$$M \geq \Lambda \left(\frac{8 \ln \frac{2e^{\Lambda}}{1+e^{\Lambda}} \ln 200}{0.0001 + 32 \ln 200 \ln \frac{e^{\Lambda}}{1+e^{\Lambda}}} \right)^{\frac{1}{2}}$$

A minimum of this function on Λ is approximately equal to $M(16.58) = 9,509$ (see Figure 5.11).

Thus, choosing the minimum breakdown on the level of $M = 9,509$, we can computer simulate the model $X_{\Lambda}(t)$ of Gaussian process $X(t)$ with spectral function $F(\Lambda) = \ln \frac{2e^{\Lambda}}{1+e^{\Lambda}}$ (see Figure 5.12).

5.1.3. *The accuracy of modeling Gaussian stationary random processes in norms of Orlicz spaces*

Let $X = \{X(t), t \in \mathbf{T}\}$ be Gaussian centered stationary real continuous in the mean square random process, where $\mathbf{T}$ is some parametric set. The definitions and properties of this process and its model $X_{\Lambda}(t)$ are described in section 5.4.

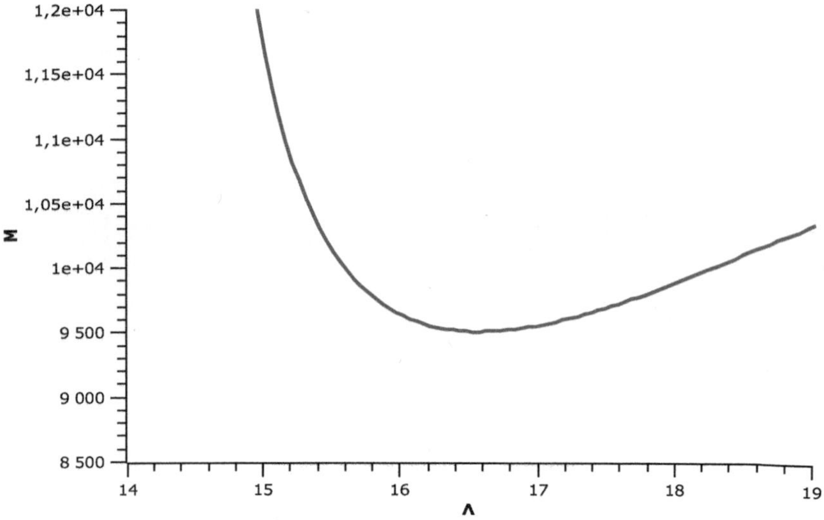

Figure 5.11. *Graph of function* $M \geq \Lambda \left(\frac{8 \ln \frac{2e^{\Lambda}}{1+e^{\Lambda}} \ln 200}{0.0001 + 32 \ln 200 \ln \frac{e^{\Lambda}}{1+e^{\Lambda}}} \right)^{\frac{1}{2}}$

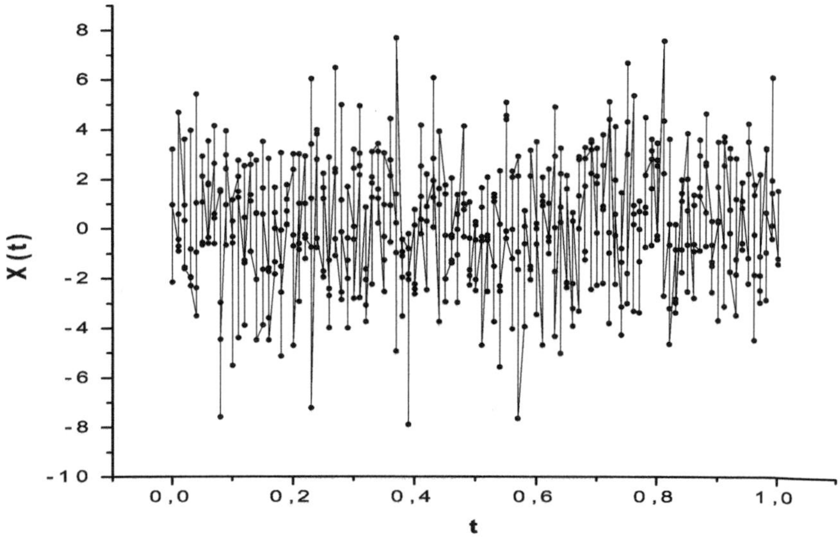

Figure 5.12. *The model of Gaussian random processes in space* $L_p([0,1])$, $p \geq 1$ *with spectral density* $f(\lambda) = \frac{1}{1+e^{\lambda}}$

DEFINITION 5.2.– *A random process $X_\Lambda(t)$ approximates process $X(t)$ with reliability $1-\beta$, $0<\beta<1$ and accuracy $\delta>0$ in the Orlicz space $L_U(\Omega)$, if there exists a partition of Λ, that the next inequality*

$$P\{\|X(t)-X_\Lambda(t)\|_{L_U}>\delta\}\le\beta$$

is fulfilled.

THEOREM 5.5.– [KOZ 88] Let $U = \{U(x), x\in R\}$ be C-function such that the function $G_U(t)=\exp\left\{\left(U^{(-1)}(t-1)\right)^2\right\}$ is convex for $t\ge 1$. Then, with probability 1 $X\in L_U(\mathbf{T})$ and for all ε such that

$$\varepsilon\ge\max(\mu(\mathbf{T}),1)\cdot\tau\left(2+\left(U^{(-1)}(1)\right)^{-2}\right)^{\frac{1}{2}}$$

we have

$$P\{\|X\|_{L_U}>\varepsilon\}\le\sqrt{e}\frac{\varepsilon U^{(-1)}(1)}{\hat{\mu}(\mathbf{T})\cdot\tau}\cdot\exp\left\{-\frac{\varepsilon^2(U^{(-1)}(1))^2}{2(\hat{\mu}(\mathbf{T}))^2\cdot\tau^2}\right\}, \qquad [5.10]$$

where $\hat{\mu}(\mathcal{T})=max(\mu(\mathbf{T}),1)$.

THEOREM 5.6.– Let the partition of Λ in the model $X_\Lambda(t)$ be such that inequality

$$\tau\left(\eta_\Lambda(t)\right)\le\tau(\Lambda,T)$$

holds, where $\tau(\Lambda,T)$ is defined in [4.7].

$$\tau(\Lambda,T)\le\frac{\delta}{\hat{T}\cdot\left(2+\left(U^{(-1)}(1)\right)^{-2}\right)^{\frac{1}{2}}}, \qquad [5.11]$$

$$\tau(\Lambda,T)\le\frac{\delta U^{(-1)}(1)}{\hat{T}x(\beta)}, \qquad [5.12]$$

where $x(\beta)>1$ is a root of the equation $\sqrt{e}x\cdot\exp\left\{-\frac{x^2}{2}\right\}=\beta$ with $\hat{T}=max(T,1)$. Then, the model approximates Gaussian process $X(t)$ with reliability $1-\beta$, $0<\beta<1$ and accuracy $\delta>0$ in Orlicz space $L_U([0,T])$, where C -function U satisfies the conditions of theorem 5.5 ($\mu(\cdot)$ is Lebesgue measure).

PROOF.– The statement of this theorem follows from theorem 5.5. Indeed, let $\tau(\Lambda,T)$ be such that $\delta>\hat{T}\cdot\tau(\Lambda,T)\cdot\left(2+\left(U^{(-1)}(1)\right)^{-2}\right)^{\frac{1}{2}}$ (condition [5.11]). Then

$$U^{(-1)}(1)>\left(2+\left(U^{(-1)}(1)\right)^{-2}\right)^{\frac{1}{2}},$$

$$\frac{\delta U^{(-1)}(1)}{\hat{T} \cdot \tau(\Lambda, T)} \geq \frac{\delta}{\hat{T}\tau(\Lambda, T)\left(2 + \left(U^{(-1)}(1)\right)^{-2}\right)^{\frac{1}{2}}} \geq 1.$$

The function $f(x) = \sqrt{e}x \exp\left\{-\frac{x^2}{2}\right\}$ decreases as $x > 1$, $f(1) = 1$. Thus, from [5.10] it follows that

$$P\left\{\|\eta_\Lambda(t)\|_{L_U} > \delta\right\} \leq \sqrt{e}\frac{\delta U^{(-1)}(1)}{\hat{T} \cdot \tau(\Lambda, T)} \cdot \exp\left\{-\frac{\delta^2 (U^{(-1)}(1))^2}{2\hat{T}^2 \cdot \tau^2(\Lambda, T)}\right\} = \beta$$

holds if the condition $\frac{\delta U^{(-1)}(1)}{\hat{T} \cdot \tau(\Lambda, T)} \geq x(\beta)$ is satisfied, that is [5.12]. □

5.2. The accuracy and reliability of the model stationary random processes in the uniform metric

In section 5.2.1, the estimates of supremum norm for sub-Gaussian random processes with bounded spectrum are obtained. Then, they will be used in the investigation of selection conditions of the partition $[0, \Lambda]$ (on which a spectral function is defined) such that there exists Gaussian process for the model of a random process that is approximated by the model with desired accuracy and reliability. These issues are considered in [TEG 01].

In section 5.2.2, the norms of sub-Gaussian processes are estimated. Using the theory of $L_p(\Omega)$-processes and preliminary estimates, the conditions on partition Λ of the set $[0, \infty]$ are found such that for model there exists a Gaussian random process, which is approached with desired accuracy and reliability in uniform metric.

5.2.1. *The accuracy of simulation of stationary Gaussian processes with bounded spectrum*

Let $X(t)$ be a Gaussian centered stationary real continuous in mean square stochastic process with bounded spectrum, it means that the covariance function has the form:

$$r(\tau) = \mathbf{E}X(t + \tau)X(t) = \int_0^\Lambda \cos \lambda t dF(\lambda),$$

where $F(\lambda)$ is continuous spectral function of the process.

DEFINITION 5.3.– *A random process $X_\Lambda(t)$ approximates Gaussian process $X(t)$ with the reliability of $1-\beta$, $0<\beta<1$ and accuracy $\delta>0$ in the space $C([0,T])$, if there exists such partition of Λ, that inequality*

$$P\left\{\sup_{0\le t\le T}|X(t)-X_\Lambda(t)|>\delta\right\}\le\beta$$

holds.

The process $X(t)$ has such representation

$$X(t)=\int_0^\Lambda \cos\lambda t d\eta_1(\lambda)+\int_0^\Lambda \sin\lambda t d\eta_2(\lambda),$$

where $\eta_1(\lambda)$ and $\eta_2(\lambda)$ are independent centered Gaussian random processes that $E(\eta_i(\lambda_2)-\eta_i(\lambda_1))^2=F(\lambda_2)-F(\lambda_1)$ as $\lambda_1<\lambda_2$, $i=1,2$.

As a model of random process, we can take

$$X_\Lambda(t)=\sum_{k=0}^{M}[\eta_{k1}\cos\zeta_k t+\eta_{k2}\sin\zeta_k t],$$

where the components are described in section 5.4.

For arbitrary $t,s\in[0,T]$, we consider the difference

$$\eta_\Lambda(t)-\eta_\Lambda(s)=\sum_{k=0}^{M}\left[\int_{\lambda_k}^{\lambda_{k+1}}(\cos\lambda t-\cos\zeta_k t-\cos\lambda s+\cos\zeta_k s)d\eta_1(\lambda)\right.$$
$$\left.+\int_{\lambda_k}^{\lambda_{k+1}}(\sin\lambda t-\sin\zeta_k t-\sin\lambda s+\sin\zeta_k s)d\eta_2(\lambda)\right],$$

where the process $\eta_\Lambda(t)$ is defined in [4.6]. The following lemma holds true.

LEMMA 5.3.– For $m=0,1,\ldots$

$$\mathbf{E}\left(\int_{\lambda_k}^{\lambda_{k+1}}(\cos\lambda t-\cos\zeta_k t-\cos\lambda s+\cos\zeta_k s)d\eta_1(\lambda)\right)^{2m+1}=0,$$

$$\mathbf{E}\left(\int_{\lambda_k}^{\lambda_{k+1}}(\sin\lambda t-\sin\zeta_k t-\sin\lambda s+\sin\zeta_k s)d\eta_2(\lambda)\right)^{2m+1}=0,$$

$$\mathbf{E}\left(\int_{\lambda_k}^{\lambda_{k+1}}(\cos\lambda t-\cos\zeta_k t-\cos\lambda s+\cos\zeta_k s)d\eta_1(\lambda)\right)^{2m}\leq V_{km},$$

$$\mathbf{E}\left(\int_{\lambda_k}^{\lambda_{k+1}}(\sin\lambda t-\sin\zeta_k t-\sin\lambda s+\sin\zeta_k s)d\eta_2(\lambda)\right)^{2m}\leq V_{km},$$

where

$$V_{km}=4^{2m}\Delta_{2m}\mathbf{E}\left(\int_{\lambda_k}^{\lambda_{k+1}}\left(\left|\sin\frac{(s-t)(\lambda-\zeta_k)}{4}\right|+\left|\sin\frac{\zeta_k(s-t)}{2}\right|\right.\right.$$
$$\left.\left.\times\left|\sin\frac{(t+s)(\zeta_k-\lambda)}{4}\right|\right)^2dF(\lambda)\right)^m,\quad \Delta_{2m}=\frac{(2m)!}{2^m m!}.$$

PROOF.–

$$|\cos\lambda t-\cos\zeta_k t-\cos\lambda s+\cos\zeta_k s|$$
$$=\left|2\sin\frac{\lambda(s-t)}{2}\sin\frac{\lambda(s+t)}{2}-2\sin\frac{\zeta_k(s-t)}{2}\sin\zeta_k\frac{(s+t)}{2}\right|$$
$$=\left|2\sin\frac{\lambda(s+t)}{2}\left(\sin\frac{\lambda(s-t)}{2}-\sin\frac{\zeta_k(s-t)}{2}\right)\right.$$
$$\left.+2\sin\frac{\zeta_k(s-t)}{2}\left(\sin\frac{\lambda(s+t)}{2}-\sin\frac{\zeta_k(s+t)}{2}\right)\right|$$
$$=\left|4\sin\frac{\lambda(s+t)}{2}\sin\frac{(s-t)(\lambda-\zeta_k)}{4}\cos\frac{(s-t)(\lambda+\zeta_k)}{4}\right.$$
$$\left.+4\sin\frac{\zeta_k(s-t)}{2}\sin\frac{(s+t)(\lambda-\zeta_k)}{4}\cos\frac{(s+t)(\lambda+\zeta_k)}{4}\right|$$
$$\leq 4\left(\left|\sin\frac{(s-t)(\lambda-\zeta_k)}{4}\right|+\left|\sin\frac{\zeta_k(s-t)}{2}\right|\cdot\left|\sin\frac{(t+s)(\zeta_k-\lambda)}{4}\right|\right),$$

$$\begin{aligned}
&|\sin\lambda t-\sin\zeta_k t-\sin\lambda s+\sin\zeta_k s|\\
&=\left|2\sin\frac{\lambda(t-s)}{2}\cdot\cos\frac{\lambda(t+s)}{2}-2\sin\frac{\zeta_k(t-s)}{2}\cdot\cos\frac{\zeta_k(t+s)}{2}\right|\\
&=2\left|\cos\frac{\lambda(t+s)}{2}\left(\sin\frac{\lambda(t-s)}{2}-\sin\frac{\zeta_k(t-s)}{2}\right)\right.\\
&\qquad\left.+\sin\frac{\zeta_k(t-s)}{2}\left(\cos\frac{\lambda(t+s)}{2}-\cos\frac{\zeta_k(t+s)}{2}\right)\right|\\
&=4\left|\cos\frac{\lambda(t+s)}{2}\sin\frac{(t-s)(\lambda-\zeta_k)}{4}\cos\frac{(t-s)(\lambda+\zeta_k)}{4}\right.\\
&\qquad\left.+\sin\frac{\zeta_k(t-s)}{2}\sin\frac{(\zeta_k-\lambda)(t+s)}{4}\sin\frac{(\zeta_k+\lambda)(t+s)}{4}\right|\\
&\le 4\left(\left|\sin\frac{(t-s)(\lambda-\zeta_k)}{4}\right|+\left|\sin\frac{\zeta_k(t-s)}{2}\right|\cdot\left|\sin\frac{(t+s)(\zeta_k-\lambda)}{4}\right|\right).
\end{aligned}$$

Since for centered Gaussian random variable ξ, we have $\mathbf{E}\xi=0$, $\mathbf{E}\xi^2=\sigma^2$, $\mathbf{E}\xi^{2k+1}=0$, $\mathbf{E}\xi^{2k}=\Delta_{2k}\sigma^{2k}$, $k=1,2,\ldots$, $\Delta_{2k}=\frac{(2k)!}{2^k k!}$, then

$$\begin{aligned}
&\mathbf{E}\left(\int_{\lambda_k}^{\lambda_{k+1}}(\cos\lambda t-\cos\lambda s-\cos\zeta_k t+\cos\zeta_k s)d\eta_1(\lambda)\right)^{2m}\\
&\le\Delta_{2m}\mathbf{E}\left(\int_{\lambda_k}^{\lambda_{k+1}}(\cos\lambda t-\cos\lambda s-\cos\zeta_k t+\cos\zeta_k s)^2dF(\lambda)\right)^{m}\\
&\le 4^{2m}\Delta_{2m}\mathbf{E}\left(\int_{\lambda_k}^{\lambda_{k+1}}\left(\left|\sin\frac{(s-t)(\lambda-\zeta_k)}{4}\right|+\left|\sin\frac{\zeta_k(s-t)}{2}\right|\right.\right.\\
&\qquad\times\left.\left.\left|\sin\frac{(t+s)(\zeta_k-\lambda)}{4}\right|\right)^2dF(\lambda)\right)^{m}.
\end{aligned}$$

Similarly, for the sinuses:

$$\mathbf{E}\left(\int_{\lambda_k}^{\lambda_{k+1}}(\sin\lambda t-\sin\lambda s-\sin\zeta_k t+\sin\zeta_k s)d\eta_2(\lambda)\right)^{2m}$$

$$\leq 4^{2m}\Delta_{2m}\mathbf{E}\left(\int\limits_{\lambda_k}^{\lambda_{k+1}}\left(\left|\sin\frac{(t-s)(\lambda-\zeta_k)}{4}\right|+\left|\sin\frac{\zeta_k(t-s)}{2}\right|\right.\right.$$

$$\left.\left.\times\left|\sin\frac{(t+s)(\zeta_k-\lambda)}{4}\right|\right)^2 dF(\lambda)\right)^m. \qquad \square$$

For the process $\eta_\Lambda(t)$ at $[0,T]$ find the estimates of $\sigma_0 = \sup\limits_{0\leq t\leq T}\tau(\eta_\Lambda(t))$ and $\sigma(h) = \sup\limits_{|t-s|\leq h}\tau(\eta_\Lambda(t)-\eta_\Lambda(s))$.

Estimate σ_0. From lemma 1.7, it follows that

$$\tau^2(\eta_\Lambda(t))$$

$$\leq \sum_{k=0}^{M}\tau^2\left(\int\limits_{\lambda_k}^{\lambda_{k+1}}(\cos\lambda t-\cos\zeta_k t)d\eta_1(\lambda)+\int\limits_{\lambda_k}^{\lambda_{k+1}}(\sin\lambda t-\sin\zeta_k t)d\eta_2(\lambda)\right)$$

$$\leq \sum_{k=0}^{M}\left[\tau\left(\int\limits_{\lambda_k}^{\lambda_{k+1}}(\cos\lambda t-\cos\zeta_k t)d\eta_1(\lambda)\right)+\tau\left(\int\limits_{\lambda_k}^{\lambda_{k+1}}(\sin\lambda t-\sin\zeta_k t)d\eta_2(\lambda)\right)\right]^2.$$

Since the random values $\chi_{k1} = \int\limits_{\lambda_k}^{\lambda_{k+1}}(\cos\lambda t - \cos\zeta_k t)d\eta_1(\lambda)$, and $\chi_{k2} = \int\limits_{\lambda_k}^{\lambda_{k+1}}(\sin\lambda t-\sin\zeta_k t)d\eta_2(\lambda)$ are such that their odd moments equal zero, then by of the corresponding theorem about the necessary and sufficient conditions for the existence of a sub-Gaussian random variable [BUL 00], we will have

$$\tau(\chi_{ki})\leq\Theta_1(\chi_{ki})=\sup_{m\geq 1}\left[\frac{1}{\Delta_{2m}}E\chi_{ki}^{2m}\right]^{\frac{1}{2m}}, i=1,2.$$

From lemma 4.1, we have

$$\mathbf{E}\chi_{ki}^{2m}\leq 4^m\Delta_{2m}\int\limits_{\lambda_k}^{\lambda_{k+1}}\left(\int\limits_{\lambda_k}^{\lambda_{k+1}}\left|\sin\frac{t(u-\lambda)}{2}\right|^2 dF(\lambda)\right)^m dF_k(u)$$

$$\leq 4^m\Delta_{2m}b_k^{2m}\int\limits_{\lambda_k}^{\lambda_{k+1}}\left(\int\limits_{\lambda_k}^{\lambda_{k+1}}\frac{t^2|u-\lambda|^2}{4}dF_k(\lambda)\right)^m dF_k(u)$$

$$\leq 4^m \Delta_{2m} t^{2m} \frac{1}{4^m} |\lambda_{k+1} - \lambda_k|^{2m} \left(F(\lambda_{k+1}) - F(\lambda_k)\right)^m$$
$$= t^{2m} \Delta_{2m} |\lambda_{k+1} - \lambda_k|^{2m} \left(F(\lambda_{k+1}) - F(\lambda_k)\right)^m.$$

Then,

$$\tau(\chi_{ki}) \leq \sup_{m\geq 1} \left[t^{2m}|\lambda_{k+1} - \lambda_k|^{2m}(F(\lambda_{k+1}) - F(\lambda_k))^m\right]^{\frac{1}{2m}}$$
$$= t|\lambda_{k+1} - \lambda_k|(F(\lambda_{k+1}) - F(\lambda_k))^{\frac{1}{2}}.$$

So,

$$\tau^2(\eta_\Lambda(t)) \leq 4\sum_{k=0}^{M} \tau^2(\chi_{ki}) \leq 4t^2 \sum_{k=0}^{M} |\lambda_{k+1} - \lambda_k|^2 (F(\lambda_{k+1}) - F(\lambda_k)).$$

Namely,

$$\tau(\eta_\Lambda(t)) \leq 2t\left(\sum_{k=0}^{M} |\lambda_{k+1} - \lambda_k|^2 (F(\lambda_{k+1}) - F(\lambda_k))\right)^{\frac{1}{2}},$$

So,

$$\sigma_0 \leq 2T\left(\sum_{k=0}^{M} |\lambda_{k+1} - \lambda_k|^2 (F(\lambda_{k+1}) - F(\lambda_k))\right)^{\frac{1}{2}} = b_0.$$

If we take $\lambda_{k+1} - \lambda_k = \frac{\Lambda}{M}$, then

$$b_0 = 2T\frac{\Lambda}{M}\left(\sum_{k=0}^{M} (F(\lambda_{k+1}) - F(\lambda_k))\right)^{\frac{1}{2}} = 2T\frac{\Lambda}{M}\left(F(\Lambda)\right)^{\frac{1}{2}}.$$

Estimate $\sigma(h)$. Consider the value

$$\omega_{k1} = \int_{\lambda_k}^{\lambda_{k+1}} (\cos\lambda t - \cos\zeta_k t - \cos\lambda s + \cos\zeta_k s)d\eta_1(\lambda),$$

$$\omega_{k2} = \int_{\lambda_k}^{\lambda_{k+1}} (\sin\lambda t - \sin\zeta_k t - \sin\lambda s + \sin\zeta_k s)d\eta_2(\lambda).$$

As in estimation of $\tau(\eta_\Lambda(t))$, the inequality

$$\tau^2(\eta_\Lambda(t)-\eta_\Lambda(s)) \le 2\sum_{k=0}^{M}\left(\tau^2(\omega_{k1})+\tau^2(\omega_{k2})\right) \le 2\sum_{k=0}^{M}\left(\Theta_1^2(\omega_{k1})+\Theta_1^2(\omega_{k2})\right)$$

is obtained, where $\Theta_1(\omega_{ki}) = \sup\limits_{m\ge 1}\left(\frac{2^m m!}{(2m)!}E\omega_{ki}^{2m}\right)^{\frac{1}{2m}}$. Thus, by lemma 5.3

$$\tau^2(\eta_\Lambda(t)-\eta_\Lambda(s))$$

$$\le 4^3\sum_{k=0}^{M}\sup_{m\ge 1}\left[\int_{\lambda_k}^{\lambda_{k+1}}\left(\int_{\lambda_k}^{\lambda_{k+1}}\left(\left|\sin\frac{(s-t)(\lambda-u)}{4}\right|\right.\right.\right.$$

$$\left.\left.\left.+\left|\sin\frac{u(s-t)}{2}\right|\left|\sin\frac{(\lambda-u)(t+s)}{4}\right|\right)^2 dF(\lambda)\right)^m dF_k(u)\right]^{\frac{1}{m}}$$

$$\le 4^3\sum_{k=0}^{M}\sup_{m\ge 1}\left[b_k^{2m}\int_{\lambda_k}^{\lambda_{k+1}}\left(\int_{\lambda_k}^{\lambda_{k+1}}\left(\frac{|s-t||\lambda-u|}{4}\right.\right.\right.$$

$$\left.\left.\left.+\frac{|u||s-t|}{2}\cdot\frac{|\lambda-u|(t+s)}{4}\right)^2 dF_k(\lambda)\right)^m dF_k(u)\right]^{\frac{1}{m}}$$

$$\le 4^3|s-t|^2\sum_{k=0}^{M}\sup_{m\ge 1}\left[b_k^{2m}\int_{\lambda_k}^{\lambda_{k+1}}\left(\int_{\lambda_k}^{\lambda_{k+1}}\frac{|\lambda-u|}{4^2}\right.\right.$$

$$\left.\left.\times\left(1+\frac{u(t+s)}{2}\right)^2 dF_k(\lambda)\right)^m dF_k(u)\right]^{\frac{1}{m}}$$

$$\le 4|s-t|^2\sum_{k=0}^{M}b_k^2|\lambda_{k+1}-\lambda_k|^2\left(1+\frac{\lambda_{k+1}(t+s)}{2}\right).$$

If we put $\lambda_{k+1}-\lambda_k=\frac{\Lambda}{M}$, then we get

$$\tau(\eta_\Lambda(t)-\eta_\Lambda(s)) \le 2|s-t|\left(\sum_{k=0}^{M}b_k^2\frac{\Lambda^2}{M^2}\left(1+\frac{\Lambda(t+s)}{2}\right)\right)^{\frac{1}{2}}$$

$$\le 2|t-s|(1+\Lambda T)\frac{\Lambda}{M}(F(\lambda))^{\frac{1}{2}}.$$

$$\sigma(h) \leq 2h(1+\Lambda T)\frac{\Lambda}{M}(F(\Lambda))^{\frac{1}{2}}. \quad [5.13]$$

THEOREM 5.7.– Let the model $X_\Lambda(t)$ with a partition Λ be such that under $\delta > 8\tilde{I}(\varepsilon_0)$ the relationship

$$2\exp\left\{-\frac{1}{2\varepsilon_0^2}\left(\delta-\sqrt{8\delta\tilde{I}(\varepsilon_0)}\right)^2\right\} \leq \beta$$

holds, where $\varepsilon_0 = \sup\limits_{0\leq t\leq T} \tau(\eta_\Lambda(t)) = \sigma_0$, $\eta_\Lambda(t) = X(t) - X_\Lambda(t)$,

$$\tilde{I}(\varepsilon_0) \leq \frac{1}{\sqrt{2}}\int\limits_0^{\varepsilon_0}\sqrt{\ln\left(\frac{T\Lambda(1+\Lambda T)}{\varepsilon M}\sqrt{F(\lambda)}+1\right)}d\varepsilon < \infty.$$

Then, the model approximates Gaussian random process $X(t)$ with the reliability $1-\beta$, $0<\beta<1$ and accuracy $\delta > 0$ in space $C([0,T])$.

PROOF.– This theorem follows from entropy characteristics [BUL 00]. Indeed, at $\delta > 8I(\varepsilon_0)$ for sub-Gaussian process $\eta_\Lambda(t)$ the inequality [TEG 01].

$$P\left\{\sup_{0\leq t\leq T}|\eta_\Lambda(t)| > \delta\right\} \leq 2\exp\left\{-\frac{1}{2\varepsilon_0^2}\left(\delta-\sqrt{8\delta\tilde{I}(\varepsilon_0)}\right)^2\right\}$$

holds, where

$$\tilde{I}(\varepsilon_0) = \frac{1}{\sqrt{2}}\int\limits_0^{\varepsilon_0}\sqrt{H(\varepsilon)}d\varepsilon = \frac{1}{\sqrt{2}}\int\limits_0^{\varepsilon_0}\sqrt{\ln\left(\frac{T}{2\sigma^{(-1)}(\varepsilon)}+1\right)}d\varepsilon < \infty,$$

$H(\varepsilon)$ is metric entropy of a compact set $[0,T]$,

$$\sigma(h) = \sup_{|t-s|<h} \tau(\eta_\Lambda(t)-\eta_\Lambda(s)).$$

From the previous estimates for $\sigma(h)$, we have

$$\sigma^{(-1)}(h) = \frac{Mh}{2\Lambda\sqrt{F(\Lambda)}(1+\Lambda T)},$$

then

$$\tilde{I}(\varepsilon_0) = \frac{1}{\sqrt{2}}\int\limits_0^{\varepsilon_0}\sqrt{\ln\left(\frac{T\Lambda(1+T\Lambda)}{\varepsilon M}\sqrt{F(\Lambda)}+1\right)}d\varepsilon,$$

which can be made as small as it needs by choosing Λ and M. It would be such a partition Λ that the condition (by definition 5.3)

$$2\exp\left\{-\frac{1}{2\varepsilon_0^2}(\delta-\sqrt{8\delta\tilde{I}(\varepsilon_0)})^2\right\}\leq\beta$$

is fulfilled. □

EXAMPLE 5.9.– At first, we estimate the integral $\tilde{I}(\varepsilon_0)$ in theorem 5.7

$$\tilde{I}(\varepsilon_0)\leq\frac{1}{\sqrt{2}}\int_0^{\varepsilon_0}\sqrt{\ln\left(\frac{T\Lambda(1+\Lambda T)}{\varepsilon M}\sqrt{F(\Lambda)}+1\right)}d\varepsilon$$

$$\leq\frac{1}{\sqrt{2}}\int_0^{\varepsilon_0}\sqrt{\frac{T\Lambda(1+\Lambda T)}{\varepsilon M}}\sqrt{F(\Lambda)}d\varepsilon=\sqrt{\frac{2T\Lambda(1+\Lambda T)\sqrt{F(\Lambda)}\varepsilon_0}{M}}.$$

Now you can find a partition of $[0,\Lambda]$, by which we construct a model $X_\Lambda(t)$ Gaussian process $X(t)$. From theorem 5.7 at $\delta>8\tilde{I}(\varepsilon_0)$,

$$M>\frac{16T\Lambda}{\delta}\sqrt{F(\Lambda)(1+T\Lambda)} \qquad [5.14]$$

be correct ratio

$$2\exp\left\{-\frac{1}{2\varepsilon_0^2}\left(\delta-\sqrt{8\delta\sqrt{\frac{2T\Lambda(1+\Lambda T)\sqrt{F(\Lambda)}\varepsilon_0}{M}}}\right)^2\right\}\leq\beta$$

$$2\exp\left\{-\frac{M^2}{8T^2\Lambda^2F(\Lambda)}\left(\delta-\sqrt{\frac{16\delta T\Lambda}{M}}(F(\Lambda)(1+\Lambda T))^{\frac{1}{4}}\right)^2\right\}\leq\beta$$

$$\frac{\delta M}{T\Lambda\sqrt{8F(\Lambda)}}-\frac{\sqrt{2\delta M}}{\sqrt{T\Lambda}}\left(\frac{1+\Lambda T}{F(\Lambda)}\right)^{\frac{1}{4}}\geq\sqrt{\ln\frac{2}{\beta}}$$

We obtain

$$M\geq\frac{T\Lambda\sqrt{F(\Lambda)}}{\delta}\left(2(1+\Lambda T)^{\frac{1}{4}}+\sqrt{4\sqrt{1+\Lambda T}+2\sqrt{2\ln\frac{2}{\beta}}}\right)^2 \qquad [5.15]$$

Under given conditions [5.14], we can conclude that for all

$$M > \frac{T\Lambda\sqrt{F(\Lambda)}}{\delta}\left(2(1+\Lambda T)^{\frac{1}{4}} + \sqrt{4\sqrt{1+\Lambda T} + 2\sqrt{2\ln\frac{2}{\beta}}}\right)^2$$

the model $X\Lambda(t)$ approximates Gaussian process $X(t)$ of reliability $1 - \beta$ and accuracy $\delta > 0$ in space $C([0,T])$.

1) Let $T = 1$, $\delta = 0.01$, $\beta = 0.01$, $F(\Lambda) = 1 - e^{-\Lambda}$. Then, for $\Lambda = 2.01$, it is performed $M = 7,429$. Model $X_\Lambda(t)$ of Gaussian process will be as follows (see Figure 5.13).

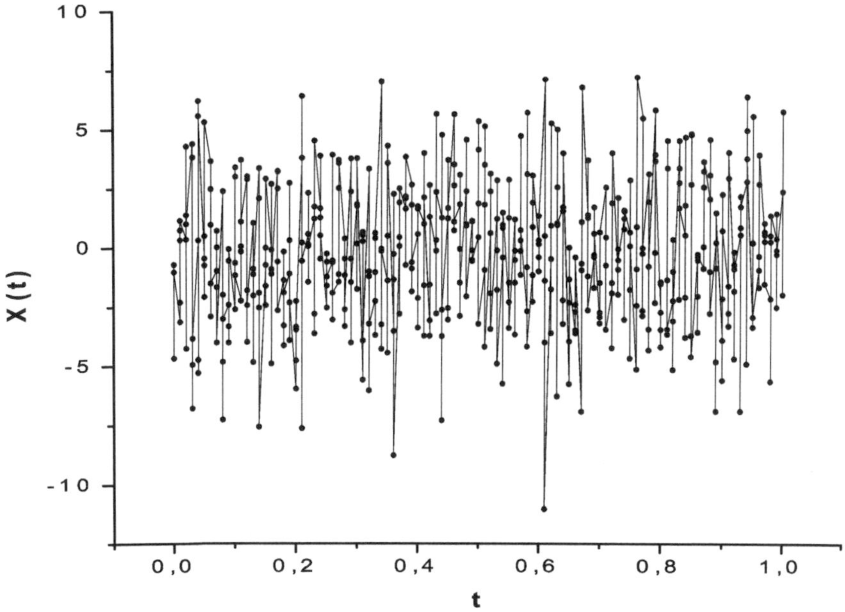

Figure 5.13. *Model of Gaussian random processes in space $C([0,1])$ with spectral density $f(\lambda) = e^{-\lambda}$*

2) Let $T = 1$, $\delta = 0.01$, $\beta = 0.01$, $F(\Lambda) = \frac{1}{9}(1-(1+\Lambda)^{-9})$. Then, for $\Lambda = 2.01$ we have that $M = 2,661$. The model $X_\Lambda(t)$ of Gaussian process will be as follows (see Figure 5.14).

3) Let $T = 1$, $\delta = 0.01$, $F(\Lambda) = \ln\frac{2e^\Lambda}{1+e^\Lambda}$. Then, at $\Lambda = 2.01$ have that $M = 6,013$. Model $X_\Lambda(t)$ of Gaussian process will be as follows (see Figure 5.15).

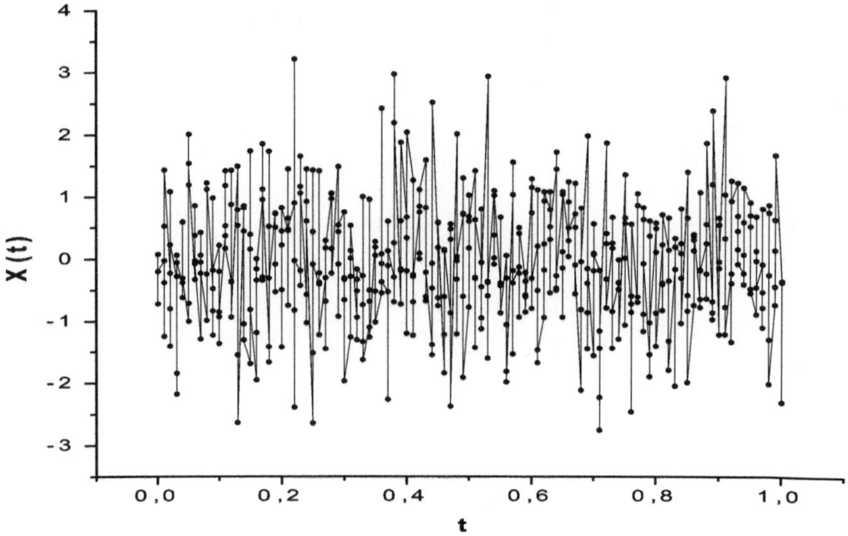

Figure 5.14. *Model of Gaussian random process in space $C([0,1])$ with spectral density $f(\lambda) = \frac{1}{(1+\lambda)^{10}}$*

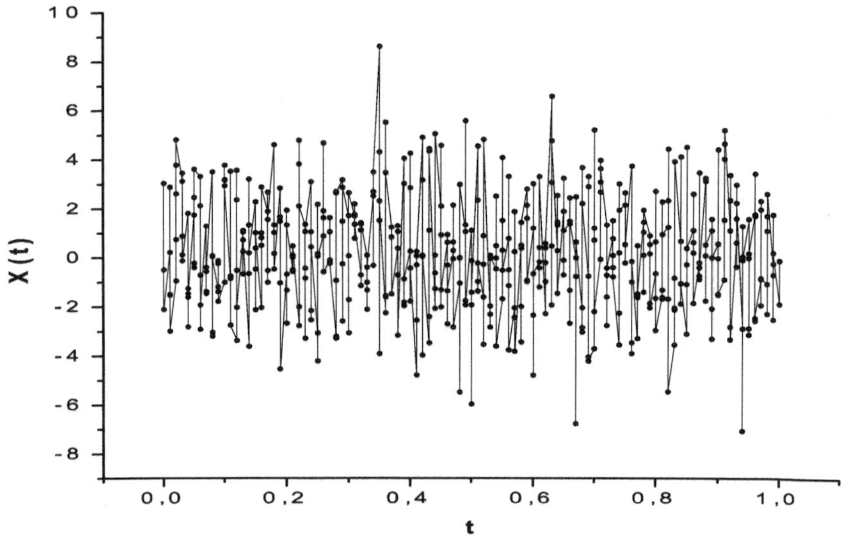

Figure 5.15. *Model of Gaussian random process in space $C([0,1])$ with spectral density $f(\lambda) = \frac{1}{1+e^{\lambda}}$*

5.2.2. *Application of $L_p(\Omega)$-processes theory in simulation of Gaussian stationary random processes*

Consider $L_p(\Omega)$-processes in Orlicz space of random variables generated by the function $U(x) = |x|^p$, $x \in R$, $p \geq 2$. Random process in this space is called $L_p(\Omega)$-process. The norm is defined as

$$\|X(t)\|_U = \|X(t)\|_{L_p} = (\mathbf{E}|X(t)|^p)^{\frac{1}{p}}.$$

Let $X = \{X(t), t \in \mathbf{T}\}$ be Gaussian stationary centered continuous in mean square random process with covariance function

$$\mathbf{E}X(t+\tau)X(t) = r(\tau) = \int_0^\infty \cos \lambda\tau dF(\lambda).$$

The representation of Gaussian random process $X(t)$ and its model $X_\Lambda(t)$ are described in section 5.4.

Consider sub-Gaussian process $\eta_\Lambda(t) = X(t) - X_\Lambda(t)$. It is defined by the expression [4.6]. Further, we will need the following assertion.

LEMMA 5.4.– [MAC 88] Let $\|\xi\|_{L_p} = (\mathbf{E}|\xi|^p)^{\frac{1}{p}}$, $1 \leq p < \infty$, $\xi_i \in L_p$ is a sequence of independent random variables with $\mathbf{E}\xi_i = 0$, $i = \overline{1, \infty}$. Then

$$\|\sum_{i=1}^n \xi_i\|_{L_p}^2 \leq C_p \left(\sum_{i=1}^n \|\xi_i\|_{L_p}^2 \right),$$

where

$$C_p = 8 \left(\frac{G(p+1)}{2\sqrt{\pi}} \right)^{\frac{2}{p}}.$$

LEMMA 5.5.– If $\int\limits_0^\infty \lambda^p dF(\lambda) < \infty$, $p \geq 2$, then for sub-Gaussian random process $\eta\Lambda(t)$ the inequality

$$\|\eta_\Lambda(t)\|_{L_p} \leq 2C_p^{\frac{1}{2}} \tilde{\Delta}_p^{\frac{1}{p}} T \left[\left(\frac{\lambda_M}{M} \right)^2 F(\lambda_M) + 4b_M^{2-\frac{4}{p}} \left(\int_{\lambda_M}^\infty u^p dF(u) \right)^{\frac{2}{p}} \right]^{\frac{1}{2}} \qquad [5.16]$$

holds.

PROOF.– From lemma 5.4 follows that

$$\|\eta_\Lambda(t)\|^2_{L_p}$$

$$\leq C_p \sum_{k=0}^{M} \left\| \int_{\lambda_k}^{\lambda_{k+1}} (\cos \lambda t - \cos \zeta_k t) d\eta_1(\lambda) + \int_{\lambda_k}^{\lambda_{k+1}} (\sin \lambda t - \sin \zeta_k t) d\eta_2(\lambda) \right\|^2_{L_p}$$

$$\leq 2C_p \sum_{k=0}^{M} \left\| \int_{\lambda_k}^{\lambda_{k+1}} (\cos \lambda t - \cos \zeta_k t) d\eta_1(\lambda) \right\|^2_{L_p} + \left\| \int_{\lambda_k}^{\lambda_{k+1}} (\sin \lambda t - \sin \zeta_k t) d\eta_2(\lambda) \right\|^2_{L_p},$$

From Fubini's theorem and lemma 4.1, we have

$$\left\| \int_{\lambda_k}^{\lambda_{k+1}} (\cos \lambda t - \cos \zeta_k t) d\eta_1(\lambda) \right\|^2_{L_p} = \left(\mathbf{E} \left| \int_{\lambda_k}^{\lambda_{k+1}} (\cos \lambda t - \cos \zeta_k t) d\eta_1(\lambda) \right|^p \right)^{\frac{2}{p}}$$

$$= \left(\mathbf{E}\mathbf{E}_{\zeta_k} \left| \int_{\lambda_k}^{\lambda_{k+1}} (\cos \lambda t - \cos \zeta_k t) d\eta_1(\lambda) \right|^p \right)^{\frac{2}{p}}$$

$$= \left(\widetilde{\Delta}_p \mathbf{E} \left| \int_{\lambda_k}^{\lambda_{k+1}} (\cos \lambda t - \cos \zeta_k t)^2 dF(\lambda) \right|^{\frac{p}{2}} \right)^{\frac{2}{p}}$$

$$\leq \widetilde{\Delta}_p^{\frac{2}{p}} \left(\mathbf{E} \left(\int_{\lambda_k}^{\lambda_{k+1}} \left| 2 \sin \frac{t(\zeta_k - \lambda)}{2} \right|^2 dF(\lambda) \right)^{\frac{p}{2}} \right)^{\frac{2}{p}} = Y_{kp},$$

where

$$\widetilde{\Delta}_p = \frac{1}{\sqrt{2\pi}} \int_{-\infty}^{\infty} |t|^p e^{-\frac{t^2}{2}} dt.$$

Similarly

$$\left\| \int_{\lambda_k}^{\lambda_{k+1}} (\sin \lambda t - \sin \zeta_k t) d\eta_2(\lambda) \right\|_{L_p}^2 = \left(\mathbf{E} \left| \int_{\lambda_k}^{\lambda_{k+1}} (\sin \lambda t - \sin \zeta_k t) d\eta_2(\lambda) \right|^p \right)^{\frac{2}{p}}$$

$$\leq \left(\widetilde{\Delta}_p \mathbf{E} \left| \int_{\lambda_k}^{\lambda_{k+1}} (\sin \lambda t - \sin \zeta_k t)^2 dF(\lambda) \right|^{\frac{p}{2}} \right)^{\frac{2}{p}}$$

$$\leq \widetilde{\Delta}_p^{\frac{2}{p}} \left(\mathbf{E} \left(\int_{\lambda_k}^{\lambda_{k+1}} \left| 2 \sin \frac{t(\zeta_k - \lambda)}{2} \right|^2 dF(\lambda) \right)^{\frac{p}{2}} \right)^{\frac{2}{p}} = Y_{kp},$$

$$Y_{kp} \leq 4 \widetilde{\Delta}_p^{\frac{2}{p}} b_k^2 \left(\mathbf{E} \left(\int_{\lambda_k}^{\lambda_{k+1}} \left| \sin \frac{t(\zeta_k - \lambda)}{2} \right|^2 dF_k(\lambda) \right)^{\frac{p}{2}} \right)^{\frac{2}{p}}$$

$$= 4 \widetilde{\Delta}_p^{\frac{2}{p}} b_k^2 \left(\int_{\lambda_k}^{\lambda_{k+1}} \left(\int_{\lambda_k}^{\lambda_{k+1}} \left| \sin \frac{t(u - \lambda)}{2} \right|^2 dF_k(\lambda) \right)^{\frac{p}{2}} dF_k(u) \right)^{\frac{2}{p}}$$

$$\leq 4 \widetilde{\Delta}_p^{\frac{2}{p}} b_k^2 \left(\int_{\lambda_k}^{\lambda_{k+1}} \int_{\lambda_k}^{\lambda_{k+1}} \left| \sin \frac{t(u - \lambda)}{2} \right|^p dF_k(\lambda) dF_k(u) \right)^{\frac{2}{p}}$$

$$\leq 4 \widetilde{\Delta}_p^{\frac{2}{p}} b_k^2 \left(\int_{\lambda_k}^{\lambda_{k+1}} \int_{\lambda_k}^{\lambda_{k+1}} \frac{t^p |u - \lambda|^p}{2^p} dF_k(\lambda) dF_k(u) \right)^{\frac{2}{p}} = \widetilde{\Delta}_p^{\frac{2}{p}} b_k^2 t^2 \left(E|\theta_k|^p \right)^{\frac{2}{p}},$$

where θ_k is such random variable that $\theta_k = \theta_{k1} - \theta_{k2}$, θ_{k1}, θ_{k2} are independent identically distributed random variables with cdf

$$F_k(\lambda) = \frac{F(\lambda) - F(\lambda_k)}{F(\lambda_{k+1}) - F(\lambda_k)}.$$

If $k < M$, then

$$(\mathbf{E}|\theta_k|^p)^{\frac{2}{p}} = \left(\int_{\lambda_k}^{\lambda_{k+1}} \int_{\lambda_k}^{\lambda_{k+1}} |u-\lambda|^p dF_k(\lambda) dF_k(u) \right)^{\frac{2}{p}} \le |\lambda_{k+1} - \lambda_k|^2.$$

If $k = M$, then

$$\begin{aligned}
(\mathbf{E}|\theta_M|^p)^{\frac{2}{p}} &\le \left(\int_{\lambda_M}^{\infty} \int_{\lambda_M}^{\infty} |\lambda - u|^p dF_M(\lambda) dF_M(u) \right)^{\frac{2}{p}} \\
&\le \left(\int_{\lambda_M}^{\infty} \int_{\lambda_M}^{\infty} |\lambda + u|^p dF_M(\lambda) dF_M(u) \right)^{\frac{2}{p}} \\
&\le \left(\left(\int_{\lambda_M}^{\infty} \int_{\lambda_M}^{\infty} \lambda^p dF_M(\lambda) dF_M(u) \right)^{\frac{1}{p}} + \left(\int_{\lambda_M}^{\infty} \int_{\lambda_M}^{\infty} u^p dF_M(\lambda) dF_M(u) \right)^{\frac{1}{p}} \right)^2 \\
&= \left(2 \left(\int_{\lambda_M}^{\infty} u^p dF_M(u) \right)^{\frac{1}{p}} \right)^2 = \frac{4}{b_M^{\frac{4}{p}}} \left(\int_{\lambda_M}^{\infty} u^p dF(u) \right)^{\frac{2}{p}}.
\end{aligned}$$

Then

$$\begin{aligned}
\|\eta_\Lambda(t)\|_{L_p}^2 &\le 4C_p \sum_{k=0}^{M} Y_{kp} \\
&= 4C_p \widetilde{\Delta}_p^{\frac{2}{p}} t^2 \left(\sum_{k=0}^{M-1} b_k^2 |\lambda_{k+1} - \lambda_k|^2 + 4 b_M^{2-\frac{4}{p}} \left(\int_{\lambda_M}^{\infty} u^p dF(u) \right)^{\frac{2}{p}} \right).
\end{aligned}$$

If we take $|\lambda_{k+1} - \lambda_k| = \dfrac{\lambda_M}{M}$, then

$$\|\eta_\Lambda(t)\|_{L_p}^2 = 4C_p \widetilde{\Delta}_p^{\frac{2}{p}} T^2 \left[\left(\frac{\lambda_M}{M} \right)^2 F(\lambda_M) + 4 b_M^{2-\frac{4}{p}} \left(\int_{\lambda_M}^{\infty} u^p dF(u) \right)^{\frac{2}{p}} \right].$$

□

LEMMA 5.6.– If $\int_0^\infty \lambda^p dF(\lambda) < \infty, p \geq 2$, then the following inequality carries out

$$\|\eta_\Lambda(t) - \eta_\Lambda(s)\|_{L_p} \leq 2C_p^{\frac{1}{2}} \widetilde{\Delta}_p^{\frac{1}{p}} |s-t| \left[\left(\frac{\lambda_M}{M}\right)^2 (1+\lambda_M T)^2 F(\lambda_M) + 16 b_M^{2-\frac{4}{p}} \left(\int_{\lambda_M}^\infty u^p dF(u) \right)^{\frac{2}{p}} \right]^{\frac{1}{2}}. \quad [5.17]$$

PROOF.– From lemma 5.4, we have

$$\|\eta_\Lambda(t) - \eta_\Lambda(s)\|_{L_p}^2 \leq 2C_p \sum_{k=0}^M \left(\left\| \int_{\lambda_k}^{\lambda_{k+1}} (\cos\lambda t - \cos\zeta_k t - \cos\lambda s + \cos\zeta_k s) d\eta_1(\lambda) \right\|_{L_p}^2 + \left\| \int_{\lambda_k}^{\lambda_{k+1}} (\sin\lambda t - \sin\zeta_k t - \sin\lambda s + \sin\zeta_k s) d\eta_2(\lambda) \right\|_{L_p}^2 \right);$$

From Fubini's theorem and lemma 5.3 follows

$$\left\| \int_{\lambda_k}^{\lambda_{k+1}} (\cos\lambda t - \cos\zeta_k t - \cos\lambda s + \cos\zeta_k s) d\eta_1(\lambda) \right\|_{L_p}^2$$

$$= \left(\mathbf{E} \left| \int_{\lambda_k}^{\lambda_{k+1}} (\cos\lambda t - \cos\zeta_k t - \cos\lambda s + \cos\zeta_k s) d\eta_1(\lambda) \right|^p \right)^{\frac{2}{p}}$$

$$= \left(\widetilde{\Delta}_p \mathbf{E} \left| \int_{\lambda_k}^{\lambda_{k+1}} (\cos\lambda t - \cos\zeta_k t - \cos\lambda s + \cos\zeta_k s)^2 dF(\lambda) \right|^{\frac{p}{2}} \right)^{\frac{2}{p}}$$

$$\leq 16 \left(\widetilde{\Delta}_p \mathbf{E} \left(\int_{\lambda_k}^{\lambda_{k+1}} \left(\left| \sin\frac{(s-t)(\lambda-\zeta_k)}{4} \right| \right.\right.\right.$$

$$+\left|\sin\frac{(s-t)\zeta_k}{2}\right|\cdot\left|\sin\frac{(\zeta_k-\lambda)(t+s)}{4}\right|\Bigg)^2 dF(\lambda)\Bigg)^{\frac{p}{2}}\Bigg)^{\frac{2}{p}}=W_{kp}.$$

A similar estimate for sine:

$$\left\|\int\limits_{\lambda_k}^{\lambda_{k+1}}(\sin\lambda t-\sin\zeta_k t-\sin\lambda s+\sin\zeta_k s)d\eta_2(\lambda)\right\|_{L_p}^2$$

$$\leq 16\left(\widetilde{\Delta}_p\mathbf{E}\left(\int\limits_{\lambda_k}^{\lambda_{k+1}}\left(\left|\sin\frac{(s-t)(\lambda-\zeta_k)}{4}\right|\right.\right.\right.$$

$$+\left|\sin\frac{(s-t)\zeta_k}{2}\right|\cdot\left|\sin\frac{(\zeta_k-\lambda)(t+s)}{4}\right|\Bigg)^2 dF(\lambda)\Bigg)^{\frac{p}{2}}\Bigg)^{\frac{2}{p}}=W_{kp}.$$

Then, $\left\|\eta_\Lambda(t)-\eta_\Lambda(s)\right\|_{L_p}^2\leq 4C_p\sum\limits_{k=0}^{M}W_{kp}.$

If $k<M$, then

$$W_{kp}\leq 16\widetilde{\Delta}_p^{\frac{2}{p}}b_k^2\left(\mathbf{E}\left(\int\limits_{\lambda_k}^{\lambda_{k+1}}\left(\left|\sin\frac{(s-t)(\lambda-\zeta_k)}{4}\right|\right.\right.\right.$$

$$+\left|\sin\frac{(s-t)\zeta_k}{2}\right|\cdot\left|\sin\frac{(\zeta_k-\lambda)(t+s)}{4}\right|\Bigg)^2 dF_k(\lambda)\Bigg)^{\frac{p}{2}}\Bigg)^{\frac{2}{p}}$$

$$\leq 16\widetilde{\Delta}_p^{\frac{2}{p}}b_k^2\left(\int\limits_{\lambda_k}^{\lambda_{k+1}}\left(\int\limits_{\lambda_k}^{\lambda_{k+1}}\left(\frac{|s-t|\cdot|\lambda-u|}{4}\right.\right.\right.$$

$$+\frac{u|s-t|}{2}\frac{|u-\lambda|(s+t)}{4}\Bigg)^2 dF_k(\lambda)\Bigg)^{\frac{p}{2}}dF_k(u)\Bigg)^{\frac{2}{p}}$$

$$=\widetilde{\Delta}_p^{\frac{2}{p}}b_k^2|s-t|^2\left(\int\limits_{\lambda_k}^{\lambda_{k+1}}\left(\int\limits_{\lambda_k}^{\lambda_{k+1}}|\lambda-u|^2\left(1+\frac{(s+t)u}{2}\right)^2 dF_k(\lambda)\right)^{\frac{p}{2}}dF_k(u)\right)^{\frac{2}{p}}$$

$$\leq \widetilde{\Delta}_p^{\frac{2}{p}} b_k^2 |s-t|^2 \cdot |\lambda_{k+1} - \lambda_k|^2 \left(1 + \frac{(s+t)\lambda_{k+1}}{2}\right)^2$$

$$\leq \widetilde{\Delta}_p^{\frac{2}{p}} b_k^2 |s-t|^2 \cdot |\lambda_{k+1} - \lambda_k|^2 (1 + \lambda_M T)^2.$$

If $k = M$, then

$$W_{Mp} \leq$$

$$\leq 16 \widetilde{\Delta}_p^{\frac{2}{p}} b_M^2 \left(\mathbf{E} \left(\int\limits_{\lambda_M}^{\infty} \left(\left| \sin \frac{(s-t)(\lambda - \zeta_k)}{4} \right| + \left| \sin \frac{(s-t)\zeta_k}{2} \right| \right)^2 dF_M(\lambda) \right)^{\frac{p}{2}} \right)^{\frac{2}{p}}$$

$$\leq 16 \widetilde{\Delta}_p^{\frac{2}{p}} b_M^2 \left(\int\limits_{\lambda_M}^{\infty} \left(\int\limits_{\lambda_M}^{\infty} \left(\frac{|s-t| \cdot |\lambda - u|}{4} + \frac{|s-t| u}{2} \right)^2 dF_M(\lambda) \right)^{\frac{p}{2}} dF_M(u) \right)^{\frac{2}{p}}$$

$$\leq \widetilde{\Delta}_p^{\frac{2}{p}} b_M^2 |s-t|^2 \left(\int\limits_{\lambda_M}^{\infty} \left(\int\limits_{\lambda_M}^{\infty} ((u+\lambda) + 2u)^2 \, dF_M(\lambda) \right)^{\frac{p}{2}} dF_M(u) \right)^{\frac{2}{p}}$$

$$\leq \widetilde{\Delta}_p^{\frac{2}{p}} b_M^2 |s-t|^2 \left(\int\limits_{\lambda_M}^{\infty} \int\limits_{\lambda_M}^{\infty} (3u + \lambda)^p \, dF_M(\lambda) dF_M(u) \right)^{\frac{2}{p}}$$

$$\leq \widetilde{\Delta}_p^{\frac{2}{p}} b_M^2 |s-t|^2 \left(\left(\int\limits_{\lambda_M}^{\infty} \int\limits_{\lambda_M}^{\infty} (3u)^p \, dF_M(\lambda) dF_M(u) \right)^{\frac{1}{p}} \right.$$

$$\left. + \left(\int\limits_{\lambda_M}^{\infty} \int\limits_{\lambda_M}^{\infty} \lambda^p dF_M(\lambda) dF_M(u) \right)^{\frac{1}{p}} \right)^2$$

$$= 4^2 \widetilde{\Delta}_p^{\frac{2}{p}} b_M^2 |s-t|^2 \left(\int\limits_{\lambda_M}^{\infty} u^p dF_M(u) \right)^{\frac{2}{p}}.$$

Then, we obtain

$$W_{Mp} = 16\widetilde{\Delta}_p^{\frac{2}{p}} b_M^{2-\frac{4}{p}} |s-t|^2 \left(\int\limits_{\lambda_M}^{\infty} u^p dF(u) \right)^{\frac{2}{p}}$$

Then,

$$\|\eta_\Lambda(t) - \eta_\Lambda(s)\|_{L_p}^2 \leq 4C_p \widetilde{\Delta}_p^{\frac{2}{p}} |s-t|^2 \left(\sum_{k=0}^{M-1} b_k^2 |\lambda_{k+1} - \lambda_k|^2 (1+\lambda_M T)^2 + 16 b_M^{2-\frac{4}{p}} \left(\int\limits_{\lambda_M}^{\infty} u^p dF(u) \right)^{\frac{2}{p}} \right).$$

So,

$$\|\eta_\Lambda(t) - \eta_\Lambda(s)\|_{L_p} \leq L|s-t|,$$

where

$$L = \left(4C_p \widetilde{\Delta}_p^{\frac{2}{p}} \left(\sum_{k=0}^{M-1} b_k^2 |\lambda_{k+1} - \lambda_k|^2 (1+\lambda_M T)^2 + 16 b_M^{2-\frac{4}{p}} \left(\int\limits_{\lambda_M}^{\infty} u^p dF(u) \right)^{\frac{2}{p}} \right) \right)^{\frac{1}{2}}.$$

Note that when $|\lambda_{k+1} - \lambda_k| = \frac{\lambda_M}{M}$, then

$$L \leq 2C_p^{\frac{1}{2}} \widetilde{\Delta}_p^{\frac{1}{p}} \left(\left(\frac{\lambda_M}{M} \right)^2 (1+\lambda_M T)^2 F(\lambda_M) + 16 b_M^{2-\frac{4}{p}} \left(\int\limits_{\lambda_M}^{\infty} u^p dF(u) \right)^{\frac{2}{p}} \right)^{\frac{1}{2}}. \quad [5.18]$$

□

THEOREM 5.8.– If in the model $X_\Lambda(t)$ the partition Λ are such that the following inequalities hold:

$$\int\limits_0^{\infty} \lambda^p dF(\lambda) < \infty, \quad p \geq 2 \quad [5.19]$$

$$\frac{(p+1)^{p+1}}{(p\delta)^p}\left(\frac{p}{p-1}\left(\frac{TL}{2}\right)^{\frac{1}{p}}\varepsilon_0^{1-\frac{1}{p}}+\varepsilon_0\right)^p \leq \beta,$$

where L is defined in [5.18]. Then, the model approximates Gaussian random process $X(t)$ with reliability $1-\beta, 0<\beta<1$ and accuracy $\delta>0$ in uniform metric.

PROOF.– In section 5.4, it is shown that the corresponding process $X(t)$ and its model are separable processes. And since

$$\int_0^\infty (\ln(1+\lambda))^{1+\varepsilon}\, dF(\lambda) \leq \int_0^\infty \lambda^p dF(\lambda) < \infty, \quad p \geq 2,$$

then from theorem 4.1, a separable random process $\eta_\Lambda(t)$ is continuous with probability 1.

If condition [5.19] is satisfied, then from lemma 5.5 it follows that the process $\eta_\Lambda(t)$ is $L_p(\Omega)$-process (because of $\sup_{0\leq t\leq T}\|\eta_\Lambda(t)\|_{L_p}<\infty$).

Then, the entropy characteristics for $L_p(\Omega)$-process yields the inequality:

$$P\left\{\sup_{0\leq t\leq T}|\eta_\Lambda(t)|>\delta\right\} \leq \frac{\widetilde{B}_p^p}{\delta^p},$$

where

$$\widetilde{B}_p = \inf_{0\leq t\leq T}(\mathbf{E}|\eta_\Lambda(t)|^p)^{\frac{1}{p}} + \inf_{0<\theta<1}\frac{1}{\theta(1-\theta)}\int_0^{\theta 2\varepsilon_0} N^{\frac{1}{p}}(\varepsilon)d\varepsilon,$$

$\varepsilon_0 = \sup_{0\leq t\leq T}\|\eta_\Lambda(t)\|_{L_p}$.

Since [TEG 01] $N(\varepsilon)=\frac{T}{2\sigma^{(-1)}(\varepsilon)}+1$, $\sigma(h)=\sup_{|t-s|<h}\|\eta_\Lambda(t)-\eta_\Lambda(s)\|_{L_p}$.

In our case $\sigma(h)=hL$, where L is defined in [5.18],

$$\sigma^{(-1)}(h)=\frac{h}{L}, \quad \inf_{0\leq t\leq T}(\mathbf{E}|\eta_\Lambda(t)|^p)^{\frac{1}{p}}=0,$$

Then

$$B_p = \inf_{0<\theta<1}\frac{1}{\theta(1-\theta)}\int_0^{2\theta\varepsilon_0}\left(\frac{TL}{2\varepsilon}+1\right)^{\frac{1}{p}} d\varepsilon$$

$$\leq \inf_{0<\theta<1}\frac{1}{\theta(1-\theta)}\left[\left(\frac{TL}{2}\right)^{\frac{1}{p}}(2\theta\varepsilon_0)^{1-\frac{1}{p}}\frac{1}{1-\frac{1}{p}}+2\theta\varepsilon_0\right]$$

$$\leq \inf_{0<\theta<1} \frac{\theta^{1-\frac{1}{p}}}{\theta(1-\theta)} \left[\left(\frac{TL}{2}\right)^{\frac{1}{p}} (2\varepsilon_0)^{1-\frac{1}{p}} \frac{p}{p-1} + 2\varepsilon_0 \right]$$

$$= \frac{(p+1)^{1+\frac{1}{p}}}{p} \left(\frac{p}{p-1} \left(\frac{TL}{2}\right)^{\frac{1}{p}} (2\varepsilon_0)^{1-\frac{1}{p}} + 2\varepsilon_0 \right),$$

$$P\{ \sup_{0<t\leq T} |\eta_\Lambda(t)| > \delta \} \leq \frac{(p+1)^{p+1}}{(p\delta)^p} \left(\frac{p}{p-1} \left(\frac{TL}{2}\right)^{\frac{1}{p}} (2\varepsilon_0)^{1-\frac{1}{p}} + 2\varepsilon_0 \right)^p.$$

From [5.16] and [5.18] follows that we can take such λ_M and M, that ε_0 and L will be made as small as it needs. Then, there exists a partition of Λ such that by definition 5.3 inequality

$$\frac{(p+1)^{p+1}}{(p\delta)^p} \left(\frac{p}{p-1} \left(\frac{TL}{2}\right)^{\frac{1}{p}} (2\varepsilon_0)^{1-\frac{1}{p}} + 2\varepsilon_0 \right)^p \leq \beta$$

holds true. □

5.3. Application of $Sub_\varphi(\Omega)$ space theory to find the accuracy of modeling for stationary Gaussian processes

In the previous section, we have proved that the model approximates Gaussian process under condition $\int_0^\infty \lambda^\varepsilon dF(\lambda) < \infty$, as $\varepsilon \geq 2$. In this section, under more restrictive conditions the estimates are found that significantly improve estimates of the previous section. The theory of spaces $Sub_\varphi(\Omega)$ random variables is used. Note also that new inequalities for norms of random variables with spaces $Sub_\varphi(\Omega)$ are obtained. These issues are considered in [KOZ 02].

Let $X = \{X(t),\ t \in R\}$ be Gaussian stationary real centered in mean square continuous random process. The model construction $X_\Lambda(t)$ of approximated process is described in section 4.

The following statement is needed. This section will consider the spaces $Sub_\varphi(\Omega)$, generated by the functions $\varphi_p(x)$, $p \geq 2$

$$\varphi_p(x) = \begin{cases} |x|^p, \text{ as} |x| > 1 \\ |x|^2, \text{ as} |x| < 1 \end{cases}. \qquad [5.20]$$

Recall that for $p = 2$ space $Sub_{\varphi_2}(\Omega)$ is called space of sub-Gaussian random variables.

Now prove the theorem that improves the appropriate theorem from [KOZ 85a].

THEOREM 5.9.– Let ξ be a random variable such that $\mathbf{E}\xi^{2k+1} = 0$ as $k = 0, 1, 2, \dots$, and the condition

$$S_{\varphi_p}(\xi) = \sup_{n\geq 1}(2n)^{\frac{1}{p}} \frac{\left(\mathbf{E}\xi^{2n}\right)^{\frac{1}{2n}}}{\left((2n)!\right)^{\frac{1}{2n}}} < \infty$$

is fulfilled, then $\xi \in Sub_{\varphi_p}(\Omega)$ and inequality

$$\tau_{\varphi_p} \leq 2^{\frac{1}{2}-\frac{1}{2p}} S_{\varphi_p}(\xi)$$

holds true.

PROOF.– For all $\lambda > 0$, we have

$$\mathbf{E}\exp\{\lambda\xi\} = \sum_{k=0}^{\infty} \frac{\lambda^k \mathbf{E}\xi^k}{k!} = 1 + \sum_{n=1}^{\infty} \frac{\lambda^{2n}\mathbf{E}\xi^{2n}}{(2n)!} = S(\lambda),$$

$$\begin{aligned} S(\lambda) &= 1 + \sum_{n=1}^{\infty} \left(\frac{\lambda}{(2n)^{\frac{1}{p}}}\right)^{2n} \left(\frac{\left(\mathbf{E}\xi^{2n}\right)^{\frac{1}{2n}}}{((2n)!)^{\frac{1}{2n}}}(2n)^{\frac{1}{p}}\right)^{2n} \\ &\leq 1 + \sum_{n=1}^{\infty} \left(\frac{\lambda}{(2n)^{\frac{1}{p}}}\right)^{2n} \left(S_{\varphi_p}(\xi)\right)^{2n}. \end{aligned}$$

Note that $S_{\varphi_p} = S$, so

$$S(\lambda) = 1 + \sum_{n=1}^{\infty} \left(\frac{\lambda S}{(2n)^{\frac{1}{p}}}\right)^{2n}.$$

Let γ be any number such that $0 < \gamma < \frac{1}{\sqrt{2}}$, $\lambda_1 = \frac{2^{\frac{1}{p}}\gamma}{S}$. First, consider such λ that $0 \leq |\lambda| \leq \lambda_1$, namely $|\lambda| \leq \frac{2^{\frac{1}{p}}\gamma}{S}$. Then

$$\begin{aligned} S(\lambda) &\leq 1 + \sum_{n=1}^{\infty} \left(\frac{\lambda S}{2^{\frac{1}{p}}}\right)^{2n} = 1 + \left(\frac{\lambda S}{2^{\frac{1}{p}}}\right)^2 \left(1 - \left(\frac{\lambda S}{2^{\frac{1}{p}}}\right)^2\right)^{-1} \\ &\leq 1 + \left(\frac{\lambda S}{2^{\frac{1}{p}}}\right)^2 (1-\gamma^2)^{-1} = 1 + \left(\frac{\lambda S}{2^{\frac{1}{p}}\sqrt{1-\gamma^2}}\right)^2 \\ &\leq \exp\left\{\left(\frac{\lambda S}{2^{\frac{1}{p}}\sqrt{1-\gamma^2}}\right)^2\right\}. \end{aligned} \qquad [5.21]$$

Since

$$\frac{|\lambda|S}{2^{\frac{1}{p}}\sqrt{1-\gamma^2}} \le \frac{\lambda_1 S}{2^{\frac{1}{p}}\sqrt{1-\gamma^2}} = \frac{\gamma}{\sqrt{1-\gamma^2}} \le \left(\frac{\gamma^2}{1-\gamma^2}\right)^{\frac{1}{2}} \le 1,$$

then as $0 \le |\lambda| < \lambda_1$ from [5.20] and [5.21], it follows that the inequality

$$S(\lambda) \le \exp\left\{\varphi_p\left(\frac{\lambda S}{2^{\frac{1}{p}}\sqrt{1-\gamma^2}}\right)\right\} \qquad [5.22]$$

holds. Consider now the case $|\lambda| > \lambda_1$. Denote under n_λ such integer that $1 \le n_\lambda \le \frac{1}{2}\left(\frac{|\lambda|S}{\gamma}\right)^p$ and $n_\lambda + 1 > \frac{1}{2}\left(\frac{|\lambda|S}{\gamma}\right)^p$. There exists n_λ because of

$$\frac{1}{2}\left(\frac{|\lambda|S}{\gamma}\right)^p > \frac{1}{2}\left(\frac{\lambda_1 S}{\gamma}\right)^p = 1.$$

Put

$$A_1(\lambda) = \sum_{n=1}^{n_\lambda}\left(\frac{\lambda S}{(2n)^{\frac{1}{p}}}\right)^{2n}, \quad A_2(\lambda) = \sum_{n=n_\lambda+1}^{\infty}\left(\frac{\lambda S}{(2n)^{\frac{1}{p}}}\right)^{2n}.$$

For $n \le n_\lambda$

$$\left(\frac{|\lambda|S}{\gamma}\right)(2n)^{-\frac{1}{p}} \le \left(\frac{|\lambda|S}{\gamma}\right)^p (2n)^{-1}.$$

So,

$$A_1(\lambda) = \sum_{n=1}^{n_\lambda}\left(\frac{\frac{|\lambda|S}{\gamma}\gamma}{(2n)^{\frac{1}{p}}}\right)^{2n} \le \sum_{n=1}^{n_\lambda}\left(\frac{\left(\frac{|\lambda|S}{\gamma}\right)^p\gamma}{2n}\right)^{2n} = \sum_{n=1}^{n_\lambda}\frac{\left(\left(\frac{|\lambda|S}{\gamma}\right)^p\gamma\right)^{2n}}{(2n)^{2n}};$$

$$A_2(\lambda) \le \sum_{n=n_\lambda+1}^{\infty}\left(\frac{|\lambda|S}{(2(n_\lambda+1))^{\frac{1}{p}}}\right)^{2n} = \left(\frac{|\lambda|S}{(2(n_\lambda+1))^{\frac{1}{p}}}\right)^{2(n_\lambda+1)}$$

$$= \left(1-\left(\frac{|\lambda|S}{(2(n_\lambda+1))^{\frac{1}{p}}}\right)^2\right)^{-1}. \qquad [5.23]$$

Since $n_\lambda + 1 > \frac{1}{2}\left(\frac{|\lambda|S}{\gamma}\right)^p$, then $\frac{2(n_\lambda+1)}{(|\lambda|S)^p} > \frac{1}{\gamma^p}$, it means that $\gamma > \frac{|\lambda|S}{(2(n_\lambda+1))^{\frac{1}{p}}}$.

Hence, [5.23] yields

$$A_2(\lambda) \le \frac{\gamma^{2(n_\lambda+1)}}{1-\gamma^2} \le \frac{\gamma^4}{1-\gamma^2} \le \gamma^2\frac{\gamma^2}{1-\gamma^2} \le \gamma^2 \le 2\gamma \le \left(\frac{\lambda S}{\gamma}\right)^p\gamma$$

(since $\left(\frac{\lambda S}{\gamma}\right)^p > 2$). So, under $|\lambda| > \lambda_1$ the following relationship holds

$$\begin{aligned} S(\lambda) &= 1 + \left(\tfrac{|\lambda|S}{\gamma}\right)^p \gamma + \textstyle\sum_{n=1}^{n_\lambda} \frac{\left(\left(\frac{|\lambda|S}{\gamma}\right)^p \gamma\right)^{2n}}{(2n)^{2n}} \\ &\leq 1 + \sum_{k=1}^{\infty} \frac{\left(\left(\frac{|\lambda|S}{\gamma}\right)^p \gamma\right)^k}{k!} \leq \exp\left\{\left(\tfrac{|\lambda|S}{\gamma}\right)^p \gamma\right\} \\ &= \exp\left\{\left(\frac{|\lambda|S}{\gamma 2^{\frac{1}{2p}}}\right)^p 2^{\frac{1}{2}}\gamma\right\} \leq \exp\left\{\left(\frac{|\lambda|S}{\gamma 2^{\frac{1}{2p}}}\right)^p\right\}. \end{aligned} \qquad [5.24]$$

Since $\frac{|\lambda|S}{\gamma 2^{\frac{1}{2p}}} \geq \frac{\lambda_1 S}{\gamma 2^{\frac{1}{2p}}} \geq 1$, then from [5.24] and [5.20], it follows that

$$S(\lambda) \leq \exp\left\{\varphi_p\left(\frac{|\lambda|S}{\gamma 2^{\frac{1}{2p}}}\right)\right\} \qquad [5.25]$$

as $|\lambda| > \lambda_1$.

[5.22] and [5.25] imply

$$\begin{aligned} S(\lambda) &\leq \exp\left\{\varphi\left(\frac{|\lambda|S}{2^{\frac{1}{2p}}} \inf_{0\leq\gamma\leq\frac{1}{\sqrt{2}}} \max\left(\frac{1}{\gamma}, \frac{1}{2^{\frac{1}{2p}}\sqrt{1-\gamma^2}}\right)\right)\right\} \\ &= \exp\left\{\varphi_p\left(|\lambda|S 2^{\frac{1}{2}-\frac{1}{2p}}\right)\right\}. \end{aligned}$$

□

LEMMA 5.7.– Let ξ be some random variable, $\alpha > 0$, $b > 0$, $S > 0$, then the inequality

$$\mathbf{E}|\xi|^S \leq b^S \left(\frac{S}{\alpha}\right)^{\frac{S}{\alpha}} \exp\left\{-\frac{S}{\alpha}\right\} \mathbf{E}\exp\left\{\frac{|\xi|^\alpha}{b^\alpha}\right\} \qquad [5.26]$$

is fulfilled.

PROOF.– As $x > 0$

$$x^S \leq e^{x^\alpha}\left(\frac{S}{\alpha}\right)^{\frac{S}{\alpha}} e^{-\frac{S}{\alpha}}$$

(It follows from $\max\limits_{x>0} \frac{x^s}{e^{x^\alpha}} = \left(\frac{S}{\alpha}\right)^{\frac{S}{\alpha}} e^{-\frac{S}{\alpha}}$). Put $x = \frac{|\xi|}{b}$, then

$$|\xi|^s \leq b^s \exp\left\{\left(\frac{|\xi|}{b}\right)^\alpha\right\}\left(\frac{s}{\alpha}\right)^{\frac{s}{\alpha}} e^{-\frac{s}{\alpha}}.$$

If we take the mathematical expectation of the right-hand and left-hand sides of above inequality, we get [5.26]. □

LEMMA 5.8.– As $u > 0, v > 0, \alpha \geq 1, 0 \leq \gamma \leq \alpha$

$$\left|\sin \frac{u}{v}\right| \leq \left(\frac{\ln(e^{\alpha-1}+u)}{\ln(e^{\alpha-1}+v)}\right)^{\gamma}.$$

PROOF.– If $u > v$, then inequality is trivial. Let $u < v$, since a function $f(u) = \frac{\ln(e^{\alpha-1}+u)}{u}$ monotonically decreases as $u > 0$, then

$$\left|\sin \frac{u}{v}\right| \leq \frac{|u|}{|v|} \leq \left(\frac{\ln(e^{\alpha-1}+u)}{\ln(e^{\alpha-1}+v)}\right)^{\gamma},$$

Really,

$$f'(u) = \frac{\gamma\left(\ln(e^{\alpha-1}+u)\right)^{\gamma-1}\frac{u}{e^{\alpha-1}+u} - \left(\ln(e^{\alpha-1}+u)\right)^{\gamma}}{u^2}$$
$$= \frac{\gamma u\left(\ln(e^{\alpha-1}+u)\right)^{\gamma-1} - (e^{\alpha-1}+u)\left(\ln(e^{\alpha-1}+u)\right)^{\gamma}}{(e^{\alpha-1}+u)u^2} \leq 0$$

since

$$g(u) = \gamma u \leq (e^{\alpha-1}+u)(\ln(e^{\alpha-1}+u)) = r(u),$$

and $q'(u) = \gamma$; $r'(u) = \ln(e^{\alpha-1}+u)+1$; $r'(u) > r'(0) = \alpha \geq \gamma = q'(u)$. □

COROLLARY 5.1.– An inequality

$$\left|\sin \frac{u}{v}\right| \leq \left(\frac{\ln(1+|u|)}{\ln(1+|v|)}\right)^{\alpha} \quad [5.27]$$

holds as $0 < \alpha \leq 1$.

PROOF.– If $|u| \geq |v|$, then the inequality is trivial. If $|u| < |v|$, it is enough to prove [5.27] as $\alpha = 1$. So, since a function $f(v) = \dfrac{\ln(1+v)}{v}$ monotonically decreases with respect to $v > 0$, then as $|u| < |v|$

$$\left|\sin \frac{u}{v}\right| \leq \frac{|u|}{|v|} \leq \frac{\ln(1+|u|)}{\ln(1+|v|)}.$$

□

Consider the random process $\eta_\Lambda(t)$ that is described in [4.6].

REMARK 5.1.– In theorem 4.2, it was proved that the stochastic process η_Λ is sub-Gaussian that belongs to the space $Sub_{\varphi_2}(\Omega)$, where $\varphi_2(x) = x^2$. From this follows that $\eta_\Lambda(t) \in Sub_{\varphi_p}(\Omega)$, where $p \geq 2$.

THEOREM 5.10.– Suppose that for some $\alpha > 2$ the condition

$$\int_0^\infty \exp\left\{(\ln(1+\lambda))^\alpha\right\} dF(\lambda) < \infty \qquad [5.28]$$

is satisfied, then inequality

$$\tau^2_{\varphi_p}(\eta_\Lambda(t)) \leq \frac{1}{\ln^2\left(1+\frac{1}{t}\right)} C_\alpha \sum_{k=0}^{M} b_k^2 d_k^2 \qquad [5.29]$$

holds, where p is such number that $\frac{1}{p} + \frac{1}{\alpha} = \frac{1}{2}$, $\left(p = \frac{2\alpha}{\alpha-2}\right)$, $C_\alpha = 32 \cdot 2^{1-\frac{1}{p}} (e\alpha)^{-\frac{2}{\alpha}} e^{\frac{26}{24}}$, $b_k^2 = F(\lambda_{k+1}) - F(\lambda_k)$, d_k is the Luxembourg norm of the random variable $\ln(1+\theta_k)$ in Orlicz space $L_U(\Omega)$, where $U(x) = \exp\{|x|^\alpha\} - 1$, $\theta_k = \frac{|\theta_{k1}-\theta_{k2}|}{2}$, θ_{k1}, θ_{k2} are independent random variables identically distributed with function of distribution $F_k(\lambda) = \frac{F(\lambda)-F(\lambda_k)}{F(\lambda_{k+1}-F(\lambda_k))}$, $F(\lambda)$ is a spectral function of the process.

PROOF.– Since in $\eta_\Lambda(t)$ all items with different k are independent, then it follows that for $p \geq 2$

$$\begin{aligned}
&\tau^2_{\varphi_p}(\eta_\Lambda(t)) \\
&\leq \sum_{k=0}^{M} \tau^2_{\varphi_p}\left[\int_{\lambda_k}^{\lambda_{k+1}} (\cos\lambda t - \cos\zeta_k t) d\eta_1(\lambda) + \int_{\lambda_k}^{\lambda_{k+1}} (\sin\lambda t - \sin\zeta_k t) d\eta_2(\lambda)\right] \\
&\leq 2\sum_{k=0}^{M}\left(\tau^2_{\varphi_p}\left(\int_{\lambda_k}^{\lambda_{k+1}} (\cos\lambda t - \cos\zeta_k t) d\eta_1(\lambda)\right)\right. \qquad [5.30] \\
&\quad \left. + \tau^2_{\varphi_p}\left(\int_{\lambda_k}^{\lambda_{k+1}} (\sin\lambda t - \sin\zeta_k t) d\eta_2(\lambda)\right)\right).
\end{aligned}$$

By theorem 5.9 and [5.30], we have

$$\tau^2_{\varphi_p}(\eta_\Lambda(t)) \le 2\cdot 2^{1-\frac{1}{p}} \sum_{k=0}^{M} S^2_{\varphi_p}\left(\int\limits_{\lambda_k}^{\lambda_{k+1}} (\cos\lambda t - \cos\zeta_k t)d\eta_1(\lambda)\right)$$
$$+ S^2_{\varphi_p}\left(\int\limits_{\lambda_k}^{\lambda_{k+1}} (\sin\lambda t - \sin\zeta_k t)d\eta_2(\lambda)\right). \quad [5.31]$$

From lemma 4.1 follows that ($b_k^2 = F(\lambda_{k+1}) - F(\lambda_k)$)

$$I_k = \mathbf{E}\left(\int\limits_{\lambda_k}^{\lambda_{k+1}} (\cos\lambda t - \cos\zeta_k t)d\eta_1(\lambda)\right)^{2m}$$
$$\le 4^m\Delta_{2m}b_k^{2m}\mathbf{E}\left(\int\limits_{\lambda_k}^{\lambda_{k+1}}\left(\sin\frac{t(\zeta_k-\lambda)}{2}\right)^2 dF_k(\lambda)\right)^m$$
$$= 4^m\Delta_{2m}b_k^{2m}\left(\int\limits_{\lambda_k}^{\lambda_{k+1}}\left(\int\limits_{\lambda_k}^{\lambda_{k+1}}\left(\sin\frac{t(u-\lambda)}{2}\right)^2 dF_k(u)\right)^m dF_k(\lambda)\right)$$
$$= 4^m\Delta_{2m}b_k^{2m}\left(\int\limits_{\lambda_k}^{\lambda_{k+1}}\int\limits_{\lambda_k}^{\lambda_{k+1}}\left(\sin\frac{t(u-\lambda)}{2}\right)^{2m} dF_k(u)dF_k(\lambda)\right).$$

From corollary 5.1 follows that

$$I_k \le 4^m\Delta_{2m}b_k^{2m}\left(\int\limits_{\lambda_k}^{\lambda_{k+1}}\int\limits_{\lambda_k}^{\lambda_{k+1}}\left(\frac{\ln\left(1+\frac{|u-\lambda|}{2}\right)}{\ln\left(1+\frac{1}{t}\right)}\right)^{2m} dF_k(u)dF_k(\lambda)\right)$$
$$= \frac{4^m\Delta_{2m}b_k^{2m}}{\left(\ln(1+\frac{1}{t})\right)^{2m}}\mathbf{E}\left(\ln(1+\theta_k)\right)^{2m}, \quad [5.32]$$

where θ_k is random variable that $\theta_k = \frac{|\theta_{k1}-\theta_{k2}|}{2}$, where θ_{k1}, θ_{k2} are independent identically distributed random variables with function of distribution $F_k(\lambda) = \frac{F(\lambda)-F(\lambda_k)}{F(\lambda_{k+1})-F(\lambda_k)}$.

From lemma 5.7 for any $d_k > 0, \alpha > 2$ follows inequality ($S = 2m$)

$$\mathbf{E}\left(\ln(1+\theta_k)\right)^{2m} \le d_k^{2m}\left(\frac{2m}{\alpha}\right)^{\frac{2m}{\alpha}}\exp\left\{-\frac{2m}{\alpha}\right\}\mathbf{E}\exp\left\{\frac{(\ln(1+\theta_k))^\alpha}{d_k^\alpha}\right\}. \quad [5.33]$$

Since for $\alpha > 2$ the function $U(x) = \exp\{|x|^\alpha\} - 1$ is N-Orlicz function then from the conditions of theorem follows that random variable $\ln(1+\theta_k)$ belongs to Orlicz space $L_U(\Omega)$. Put $d_k = \|\ln(1+\theta_k)\|_\alpha$, then $\mathbf{E}\exp\left\{\frac{(\ln(1+\theta_k))^\alpha}{d_k^\alpha}\right\} \le 2$. Thus, for each $m = 1, 2, \ldots$ of [5.33], the following inequality holds true

$$\mathbf{E}\left(\ln(1+\theta_k)\right)^{2m} \le 2d_k^{2m}\left(\frac{2m}{\alpha}\right)^{\frac{2m}{\alpha}} \exp\left\{-\frac{2m}{\alpha}\right\}. \qquad [5.34]$$

So, from [5.34] and [5.32] follows

$$I_k \le \frac{4^m \Delta_{2m} b_k^{2m}}{\left(\ln(1+\frac{1}{t})\right)^{2m}} \cdot 2d_k^{2m}\left(\frac{2m}{\alpha}\right)^{\frac{2m}{\alpha}} \exp\left\{-\frac{2m}{\alpha}\right\}. \qquad [5.35]$$

From [5.35] and definitions $S_{\varphi_p}(\bullet)$, we have

$$S_{\varphi_p}\left(\int_{\lambda_k}^{\lambda_{k+1}} (\cos\lambda t - \cos\zeta_k t)\,d\eta_1(\lambda)\right) \le \sup_{m\ge1}(2m)^{\frac{1}{p}}\frac{(I_k)^{\frac{1}{2m}}}{((2m)!)^{\frac{1}{2m}}}$$

$$\le \sup_{m\ge1}\frac{(2m)^{\frac{1}{p}}}{((2m)!)^{\frac{1}{2m}}}\frac{2((2m)!)^{\frac{1}{2m}}\sqrt{b_k^2}\,2^{\frac{1}{2m}}d_k}{\sqrt{2}(m!)^{\frac{1}{2m}}\ln\left(1+\frac{1}{t}\right)}\left(\frac{2m}{\alpha}\right)^{\frac{1}{\alpha}}\exp\left\{-\frac{1}{\alpha}\right\}$$

$$= \sup_{m\ge1}\frac{\sqrt{2}b_k d_k 2^{\frac{1}{p}+\frac{1}{2m}+\frac{1}{\alpha}}m^{\frac{1}{p}+\frac{1}{\alpha}}}{(m!)^{\frac{1}{2m}}\ln\left(1+\frac{1}{t}\right)\alpha^{\frac{1}{\alpha}}\exp\left\{\frac{1}{\alpha}\right\}}. \qquad [5.36]$$

From Stirling's formula follows inequality $\left(\frac{1}{m!}\right)^{\frac{1}{2m}} \le \frac{1}{m^{\frac{1}{2}}}e^{\frac{1}{2}+\frac{1}{24}}$. Moreover, [5.36] implies $\left(\frac{1}{p}+\frac{1}{\alpha}-\frac{1}{2}=0\right)$ that

$$S_{\varphi_p}\left(\int_{\lambda_k}^{\lambda_{k+1}} (\cos\lambda t - \cos\zeta_k t)\,d\eta_1(\lambda)\right)$$

$$\le \frac{\sqrt{2}}{\ln\left(1+\frac{1}{t}\right)}b_k d_k 2^{\frac{1}{p}+\frac{1}{\alpha}+\frac{1}{2}}\left(\frac{1}{e\alpha}\right)^{\frac{1}{\alpha}}e^{\frac{13}{24}} = \frac{2\sqrt{2}}{\ln\left(1+\frac{1}{t}\right)}b_k d_k (e\alpha)^{-\frac{1}{\alpha}}e^{\frac{13}{24}}.$$

Similarly, we prove that

$$\mathbf{E}\left(\int\limits_{\lambda_k}^{\lambda_{k+1}}(\sin\lambda t-\sin\zeta_k t)d\eta_2(\lambda)\right)^{2m}$$

$$\leq 4^m\Delta_{2m}b_k^{2m}\left(\int\limits_{\lambda_k}^{\lambda_{k+1}}\int\limits_{\lambda_k}^{\lambda_{k+1}}\left(\sin\frac{t(u-\lambda)}{2}\right)^{2m}dF_k(\lambda)dF_k(u)\right)=I_k,$$

$$S_{\varphi_p}\left(\int\limits_{\lambda_k}^{\lambda_{k+1}}(\sin\lambda t-\sin\zeta_k t)d\eta_2(\lambda)\right)$$

$$\leq\sup_{m\geq 1}(2m)^{\frac{1}{p}}\frac{(I_k)^{\frac{1}{2m}}}{((2m)!)^{\frac{1}{2m}}}\leq\sup_{m\geq 1}\frac{2\sqrt{2}}{\ln\left(1+\frac{1}{t}\right)}b_k d_k(e\alpha)^{-\frac{1}{\alpha}}e^{\frac{13}{24}}.$$

From [5.31] follows

$$\tau^2_{\varphi_p}(\eta_\Lambda(t))\leq 32\cdot 2^{1-\frac{1}{p}}(e\alpha)^{-\frac{2}{\alpha}}e^{\frac{26}{24}}\sum_{k=0}^{M}b_k^2 d_k^2\frac{1}{\ln^2\left(1+\frac{1}{t}\right)}.$$

□

THEOREM 5.11.– Suppose that for some $\alpha>2$, condition [5.28] is satisfied, then the inequality

$$\tau^2_{\varphi_p}(\eta_\Lambda(t)-\eta_\Lambda(s))\leq\frac{1}{\ln^2\left(1+\frac{1}{|t-s|}\right)}\widetilde{C}_\alpha\sum_{k=0}^{M}b_k^2\hat{d}_k^2$$

holds, where $p=\frac{2\alpha}{\alpha-2}$, $\widetilde{C}_\alpha=64\cdot 2^{1-\frac{1}{p}}(e\alpha)^{-\frac{2}{\alpha}}e^{\frac{26}{24}}$, $b_k^2=F(\lambda_{k+1})-F(\lambda_k)$, $\hat{d}_k$ is the Luxembourg norm of random variable

$$\ln\left(1+\frac{|\theta_{k1}-\theta_{k2}|}{4}\right)+\ln\left(1+\frac{\theta_{k1}}{2}\right)\frac{\ln\left(1+\frac{|\theta_{k1}-\theta_{k2}|}{4}\right)}{\ln\left(1+\frac{1}{2T}\right)}$$

in Orlicz space $L_U(\Omega)$, where $U(x)=\exp\{|x|^\alpha\}-1$, θ_{k1} and θ_{k2} are independent identically distributed random variables with cdf $F_k(\lambda)=\frac{F(\lambda)-F(\lambda_k)}{F(\lambda_{k+1})-F(\lambda_k)}$.

PROOF.– Denote

$$\omega_{k1}=\int\limits_{\lambda_k}^{\lambda_{k+1}}(\cos\lambda t-\cos\zeta_k t-\cos\lambda s+\cos\zeta_k s)d\eta_1(\lambda),$$

$$\omega_{k2} = \int\limits_{\lambda_k}^{\lambda_{k+1}} (\sin \lambda t - \sin \zeta_k t - \sin \lambda s + \sin \zeta_k s) d\eta_2(\lambda).$$

From lemma 1.7 follows

$$\tau_{\varphi_p}^2(\eta_\Lambda(t) - \eta_\Lambda(t)) \le 2 \sum_{k=0}^{M} \left[\tau_{\varphi_p}^2(\omega_{k1}) + \tau_{\varphi_p}^2(\omega_{k2})\right].$$

By theorem 5.9, we have

$$\tau_{\varphi_p}^2(\eta_\Lambda(t) - \eta_\Lambda(s)) \le 2 \cdot 2^{1-\frac{1}{p}} \sum_{k=0}^{M} \left(S_{\varphi_p}^2(\omega_{k1}) + S_{\varphi_p}^2(\omega_{k2})\right). \qquad [5.37]$$

From lemma 5.3 follows that

$$\begin{aligned} I_k &= \mathbf{E}\left(\omega_{k1}\right)^{2m} \\ &\le 4^{2m}\Delta_{2m}\mathbf{E}\left(\int\limits_{\lambda_k}^{\lambda_{k+1}}\left(\left|\sin\frac{(s-t)(\lambda-\zeta_k)}{4}\right|\right.\right. \\ &\quad \left.\left. + \left|\sin\frac{\zeta_k(s-t)}{2}\right|\cdot\left|\sin\frac{(t+s)(\zeta_k-\lambda_k)}{4}\right|\right)^2 dF(\lambda)\right)^m \\ &= 4^{2m}\Delta_{2m}b_k^{2m}\left(\int\limits_{\lambda_k}^{\lambda_{k+1}}\int\limits_{\lambda_k}^{\lambda_{k+1}}\left(\left|\sin\frac{(s-t)(\lambda-u)}{4}\right|\right.\right. \\ &\quad \left.\left. + \left|\sin\frac{u(s-t)}{2}\right|\cdot\left|\sin\frac{(t+s)(u-\lambda)}{4}\right|\right)^{2m} dF_k(u)dF_k(\lambda)\right). \qquad [5.38] \end{aligned}$$

By corollary 5.1

$$\begin{aligned} I_k &\le 4^{2m}\Delta_{2m}b_k^{2m}\left(\int\limits_{\lambda_k}^{\lambda_{k+1}}\int\limits_{\lambda_k}^{\lambda_{k+1}}\left(\frac{\ln\left(1+\frac{|u-\lambda|}{4}\right)}{\ln\left(1+\frac{1}{|s-t|}\right)}\right.\right. \\ &\quad \left.\left. + \frac{\ln\left(1+\frac{u}{2}\right)}{\ln\left(1+\frac{1}{|s-t|}\right)}\frac{\ln\left(1+\frac{|u-\lambda|}{4}\right)}{\ln\left(1+\frac{1}{(t+s)}\right)}\right)^{2m} dF_k(u)dF_k(\lambda)\right) \qquad [5.39] \end{aligned}$$

$$= \frac{4^{2m}\Delta_{2m} b_k^{2m}}{\left(\ln\left(1+\frac{1}{|s-t|}\right)\right)^{2m}} \mathbf{E}\left(\ln\left(1+\frac{|\theta_{k1}-\theta_{k2}|}{4}\right)\right.$$

$$\left.+\frac{\ln\left(1+\frac{\theta_{k1}}{2}\right)\cdot\ln\left(1+\frac{|\theta_{k1}-\theta_{k2}|}{4}\right)}{\ln\left(1+\frac{1}{2T}\right)}\right)^{2m},$$

where θ_{k1}, θ_{k2} are independent identically distributed random variables with cdf $F_k(x) = \frac{F(x)-F(\lambda_k)}{F(\lambda_{k+1})-F(\lambda_k)}$; $t, s \in [0,T]$.

It follows from lemma 5.7 that for arbitrary $\hat{d}_k > 0$, $\alpha > 2$

$$\mathbf{E}\left(\ln\left(1+\frac{|\theta_{k1}-\theta_{k2}|}{4}\right)+\frac{\ln\left(1+\frac{\theta_{k1}}{2}\right)\ln\left(1+\frac{|\theta_{k1}-\theta_{k2}|}{4}\right)}{\ln\left(1+\frac{1}{2T}\right)}\right)^{2m}$$

$$\leq \hat{d}_k^{2m}\left(\left(\frac{2m}{\alpha}\right)^{\frac{2m}{\alpha}}\exp\left\{-\frac{2m}{\alpha}\right\}\mathbf{E}\exp\left\{\frac{L^\alpha}{\hat{d}_k^\alpha}\right\}\right), \quad [5.40]$$

where

$$L = \ln\left(1+\frac{|\theta_{k1}-\theta_{k2}|}{4}\right)+\frac{\ln\left(1+\frac{\theta_{k1}}{2}\right)\ln\left(1+\frac{|\theta_{k1}-\theta_{k2}|}{4}\right)}{\ln\left(1+\frac{1}{2T}\right)}.$$

Since as $\alpha > 2$ the function $U(x) = \exp\{|x|^\alpha\} - 1$ is N-Orlicz function, then the condition of the theorem implies that random variable L belongs to the Orlicz space $L_u(\Omega)$. Put $\hat{d}_k = \|L\|_\alpha$, then

$$\mathbf{E}\exp\left\{\frac{L^\alpha}{\hat{d}^\alpha}\right\} \leq 2.$$

Hence, for each $m = 1, 2, \ldots$ from [5.40] follows that

$$\mathbf{E}\left(\ln\left(1+\frac{|\theta_{k1}-\theta_{k2}|}{4}\right)+\frac{\ln\left(1+\frac{\theta_{k1}}{2}\right)\ln\left(1+\frac{|\theta_{k1}-\theta_{k2}|}{4}\right)}{\ln\left(1+\frac{1}{2T}\right)}\right)$$

$$\leq 2\hat{d}_k^{2m}\left(\frac{2m}{\alpha}\right)^{\frac{2m}{\alpha}}\exp\left\{-\frac{2m}{\alpha}\right\}. \quad [5.41]$$

So, [5.39] and [5.41] imply

$$I_k \leq \frac{4^{2m}\Delta_{2m} b_k^{2m}}{\left(\ln\left(1+\frac{1}{|s-t|}\right)\right)^{2m}} 2\hat{d}_k^{2m}\left(\frac{2m}{\alpha}\right)^{\frac{2m}{\alpha}}\exp\left\{-\frac{2m}{\alpha}\right\}. \quad [5.42]$$

Therefore, from [5.42] and definitions of $S_{\varphi_p}(\bullet)$ follows that

$$S_{\varphi_p}\left(\int\limits_{\lambda_k}^{\lambda_{k+1}} (\cos\lambda t - \cos\zeta_k t - \cos\lambda s + \cos\zeta_k s)d\eta_1(\lambda)\right)$$

$$\leq \sup_{m\geq 1}(2m)^{\frac{1}{p}}\frac{(I_k)^{\frac{1}{2m}}}{((2m)!)^{\frac{1}{2m}}}$$

$$\leq \sup_{m\geq 1}\frac{(2m)^{\frac{1}{p}}4\left(\frac{(2m)!}{2^m m!}\right)^{\frac{1}{2m}}b_k 2^{\frac{1}{2m}}}{((2m)!)^{\frac{1}{2m}}\ln\left(1+\frac{1}{|s-t|}\right)}\hat{d}_k\left(\frac{2m}{\alpha}\right)^{\frac{1}{\alpha}}\exp\left\{-\frac{1}{\alpha}\right\}$$

$$= \sup_{m\geq 1}\frac{2\sqrt{2}b_k\hat{d}_k 2^{\frac{1}{p}+\frac{1}{2m}+\frac{1}{\alpha}}m^{\frac{1}{p}+\frac{1}{\alpha}}}{(m!)^{\frac{1}{2m}}\ln\left(1+\frac{1}{|s-t|}\right)\alpha^{\frac{1}{\alpha}}\exp\left\{\frac{1}{\alpha}\right\}}.$$

By Stirling's formula, we have

$$\left(\frac{1}{m!}\right)^{\frac{1}{2m}} \leq \frac{1}{\sqrt{m}}e^{\frac{1}{2}+\frac{1}{24}}.$$

Then, using $\frac{1}{p}+\frac{1}{\alpha}-\frac{1}{2}=0$, we obtain

$$S_{\varphi_p}\left(\int\limits_{\lambda_k}^{\lambda_{k+1}} (\cos\lambda t - \cos\zeta_k t - \cos\lambda s + \cos\zeta_k s)d\eta_1(\lambda)\right)$$

$$\leq \frac{2\sqrt{2}}{\ln\left(1+\frac{1}{|s-t|}\right)}b_k\hat{d}_k 2^{\frac{1}{p}+\frac{1}{\alpha}+\frac{1}{2}}(e\alpha)^{-\frac{1}{\alpha}}e^{\frac{13}{24}} = \frac{4\sqrt{2}}{\ln\left(1+\frac{1}{|s-t|}\right)}b_k\hat{d}_k(e\alpha)^{-\frac{1}{\alpha}}e^{\frac{13}{24}}.$$

Similarly, we obtain

$$S_{\varphi_p}\left(\int\limits_{\lambda_k}^{\lambda_{k+1}} (\sin\lambda t - \sin\zeta_k t - \sin\lambda s + \sin\zeta_k s)d\eta_2(\lambda)\right)$$

$$\leq \frac{4\sqrt{2}}{\ln\left(1+\frac{1}{|s-t|}\right)}b_k\hat{d}_k(e\alpha)^{-\frac{1}{\alpha}}e^{\frac{13}{24}}.$$

From [5.37] follows that

$$\tau_{\varphi_p}^2(\eta_\Lambda(t)-\eta_\Lambda(s)) \leq 64\cdot 2^{1-\frac{1}{p}}(e\alpha)^{-\frac{2}{\alpha}}e^{\frac{26}{24}}\sum_{k=0}^{M} b_k^2\hat{d}_k^2\frac{1}{\ln^2\left(1+\frac{1}{|s-t|}\right)}.$$

□

REMARK 5.2.– The value d_k from theorem 5.10 can be evaluated in this way. Consider in [5.33] $\mathbf{E}\exp\left\{\frac{(\ln(1+\theta_k))^\alpha}{d_k^\alpha}\right\}$. Let $k < M$. Since

$$L_k = \int\limits_{\lambda_k}^{\lambda_{k+1}}\int\limits_{\lambda_k}^{\lambda_{k+1}} \exp\left\{\left(\frac{\ln\left(1+\frac{|u-v|}{2}\right)}{S}\right)^\alpha\right\} dF_k(u)dF_k(v)$$
$$\leq \exp\left\{\left(\frac{\ln\left(1+\frac{\lambda_{k+1}-\lambda_k}{2}\right)}{S}\right)^\alpha\right\},$$

then $L_k \leq 2$, when $S \geq \frac{\ln\left(1+\frac{\lambda_{k+1}-\lambda_k}{2}\right)}{(\ln 2)^{\frac{1}{\alpha}}}$. Thus, by the Luxembourg norm

$$d_k \leq \frac{\ln\left(1+\frac{\lambda_{k+1}-\lambda_k}{2}\right)}{(\ln 2)^{\frac{1}{\alpha}}}.$$

As $k = M$ from the theory of Orlicz spaces

$$d_M \leq \left(\int\limits_{\lambda_M}^{\infty}\int\limits_{\lambda_M}^{\infty} \exp\left\{\left(\ln\left(1+\frac{|u-v|}{2}\right)\right)^\alpha\right\} dF_M(u)dF_M(v)\right)^{\frac{1}{\alpha}}$$
$$= \left(\int\limits_{\lambda_M}^{\infty}\left(\int\limits_{\lambda_M}^{v} \exp\left\{\left(\ln\left(1+\frac{|u-v|}{2}\right)\right)^\alpha\right\} dF_M(u)\right.\right.$$
$$\left.\left. + \int\limits_{v}^{\infty} \exp\left\{\left(\ln\left(1+\frac{|u-v|}{2}\right)\right)^\alpha\right\} dF_M(u)\right) dF_M(v)\right)^{\frac{1}{\alpha}}$$
$$\leq \left(\int\limits_{\lambda_M}^{\infty}\left(\int\limits_{\lambda_M}^{v} \exp\left\{\left(\ln\left(1+\frac{v-\lambda_M}{2}\right)\right)^\alpha\right\} dF_M(u)\right.\right.$$
$$\left.\left. + \int\limits_{v}^{\infty} \exp\left\{\left(\ln\left(1+\frac{u-\lambda_M}{2}\right)\right)^\alpha\right\} dF_M(u)\right) dF_M(v)\right)^{\frac{1}{\alpha}}$$
$$\leq \left(\int\limits_{\lambda_M}^{\infty} dF_M(u) \int\limits_{\lambda_M}^{\infty} \exp\left\{\left(\ln\left(1+\frac{v-\lambda_M}{2}\right)\right)^\alpha\right\}\right.$$

$$+\int_{\lambda_M}^{\infty} dF_M(v)\int_{\lambda_M}^{\infty}\exp\left\{\left(\ln\left(1+\frac{u-\lambda_M}{2}\right)\right)^{\alpha}\right\}dF_M(v)\right)^{\frac{1}{\alpha}}$$

$$\leq\left(2\int_{\lambda_M}^{\infty}\exp\left\{(\ln(1+u))^{\alpha}\right\}dF_M(u)\right)^{\frac{1}{\alpha}}=\left(\frac{2}{b_M^2}\int_{\lambda_M}^{\infty}\exp\left\{(\ln(1+u))^{\alpha}\right\}dF(u)\right)^{\frac{1}{\alpha}}.$$

So,

$$\sum_{k=0}^{M} b_k^2 d_k^2 \leq \sum_{k=0}^{M-1} b_k^2\left(\frac{\ln\left(1+\frac{\lambda_{k+1}-\lambda_k}{2}\right)}{(\ln 2)^{\frac{1}{\alpha}}}\right)^2$$
$$+\, b_M^{2-\frac{4}{\alpha}}\left(2\int_{\lambda_M}^{\infty}\exp\left\{(\ln(1+u))^{\alpha}\right\}dF(u)\right)^{\frac{2}{\alpha}}.$$

From the inequality above, condition [5.28] and $\alpha>2$ imply that the sum $\sum_{k=0}^{M} b_k^2 d_k^2$ can be made as small as it needs by selecting sufficiently large λ_M and sufficiently small $\max_{0\leq k\leq M-1}(\lambda_{k+1}-\lambda_k)$. For example, when $\lambda_M=\Lambda$, $\lambda_{k+1}-\lambda_k=\frac{\Lambda}{M}$ at $k=0,1,\ldots,M-1$, $\sum_{k=0}^{M} b_k^2\leq F(+\infty)$, then

$$\sum_{k=0}^{M} b_k^2 d_k^2 \leq \left(\frac{\ln\left(1+\frac{\Lambda}{2M}\right)}{(\ln 2)^{\frac{1}{\alpha}}}\right)^2 F(+\infty)$$
$$+\left(2\int_{\Lambda}^{\infty}\exp\left\{(\ln(1+u))^{\alpha}\right\}dF(u)\right)^{\frac{2}{\alpha}}(F(+\infty)-F(\Lambda))^{2-\frac{4}{\alpha}}.$$

Similarly, we estimate the value $\hat{d}_k$ from theorem 5.11. Consider in [5.40]

$$\mathbf{E}\exp\left\{\frac{\left(\ln\left(1+\frac{|\theta_{k1}-\theta_{k2}|}{4}\right)+\frac{\ln\left(1+\frac{\theta_{k1}}{2}\right)\ln\left(1+\frac{|\theta_{k1}-\theta_{k2}|}{4}\right)}{\ln\left(1+\frac{1}{2T}\right)}\right)^{\alpha}}{\hat{d}_k^{\alpha}}\right\}.$$

Let $k < M$.

$$\tilde{L}_k =$$

$$= \int\limits_{\lambda_k}^{\lambda_{k+1}} \int\limits_{\lambda_k}^{\lambda_{k+1}} \exp\left\{\left(\frac{\ln\left(1+\frac{|u-v|}{4}\right) + \frac{\ln\left(1+\frac{u}{2}\right)}{\ln\left(1+\frac{1}{2T}\right)} \ln\left(1+\frac{|u-v|}{4}\right)}{S}\right)^{\alpha}\right\}$$

$$dF_k(u)dF_k(v)$$

$$\leq \exp\left\{\left(\frac{\ln\left(1+\frac{\lambda_{k+1}-\lambda_k}{4}\right) + \frac{\ln\left(1+\frac{\lambda_{k+1}}{2}\right)}{\ln\left(1+\frac{1}{2T}\right)} \ln\left(1+\frac{\lambda_{k+1}-\lambda_k}{4}\right)}{S}\right)^{\alpha}\right\}.$$

$\tilde{L}_k \leq 2$ when

$$S \geq \frac{\ln\left(1+\frac{\lambda_{k+1}-\lambda_k}{4}\right) + \frac{\ln\left(1+\frac{\lambda_{k+1}}{2}\right)}{\ln\left(1+\frac{1}{2T}\right)} \ln\left(1+\frac{\lambda_{k+1}-\lambda_k}{4}\right)}{(\ln 2)^{\frac{1}{\alpha}}}.$$

Thus, by the norm of Luxembourg

$$\hat{d}_k \leq \frac{\ln\left(1+\frac{\lambda_{k+1}-\lambda_k}{4}\right) + \frac{\ln\left(1+\frac{\lambda_{k+1}}{2}\right)}{\ln\left(1+\frac{1}{2T}\right)} \ln\left(1+\frac{\lambda_{k+1}-\lambda_k}{4}\right)}{(\ln 2)^{\frac{1}{\alpha}}}.$$

When $k = M$, then we obtain

$$\hat{d}_M \leq \left(\int\limits_{\lambda_M}^{\infty} \int\limits_{\lambda_M}^{\infty} \exp\left\{\left[\ln\left(1+\frac{|u-v|}{4}\right)\right.\right.\right.$$

$$\left.\left.\left.+\frac{\ln\left(1+\frac{u}{2}\right)}{\ln\left(1+\frac{1}{2T}\right)} \ln\left(1+\frac{|u-v|}{4}\right)\right]^{\alpha}\right\} dF_M(u)dF_M(v)\right)^{\frac{1}{\alpha}}$$

$$\leq \left(\int\limits_{\lambda_M}^{\infty} \int\limits_{\lambda_M}^{\infty} \exp\left\{\left[\ln\left(1+\frac{u}{2}\right)\right.\right.\right.$$

$$\left.\left.\left.+\frac{\ln^2\left(1+\frac{u}{2}\right)}{\ln\left(1+\frac{1}{2T}\right)}\right]^{\alpha}\right\} dF_M(u)dF_M(v)\right)^{\frac{1}{\alpha}}$$

$$\leq \left(\frac{2}{b_M^2} \int_{\lambda_M}^{\infty} \exp \left\{ \left(\ln\left(1+\frac{u}{2}\right) + \frac{\ln^2\left(1+\frac{u}{2}\right)}{\ln\left(1+\frac{1}{2T}\right)} \right)^{\alpha} \right\} dF(u) \right)^{\frac{1}{\alpha}}.$$

So,

$$\sum_{k=0}^{M} b_k^2 \hat{d}_k^2$$

$$\leq \sum_{k=0}^{M-1} b_k^2 \left(\frac{\ln\left(1+\frac{(\lambda_{k+1}-\lambda_k)}{4}\right) + \frac{\ln\left(1+\frac{\lambda_{k+1}}{2}\right)}{\ln\left(1+\frac{1}{2T}\right)} \ln\left(1+\frac{(\lambda_{k+1}-\lambda_k)}{4}\right)}{(\ln 2)^{\frac{1}{\alpha}}} \right)^2$$

$$+ b_M^{2-\frac{4}{\alpha}} \left(2 \int_{\lambda_M}^{\infty} \exp \left\{ \left(\ln\left(1+\frac{u}{2}\right) + \frac{\ln^2\left(1+\frac{u}{2}\right)}{\ln\left(1+\frac{1}{2T}\right)} \right)^{\alpha} \right\} dF(u) \right)^{\frac{2}{\alpha}}.$$

REMARK 5.3.– The norms can always be estimated more accurately by approximating methods. The next corollary follows from last two theorems.

COROLLARY 5.2.– Suppose that for some $\alpha > 2$, condition [5.28] is fulfilled, then for separable process $\eta_\Lambda(t)$, which belongs to the space $Sub_{\varphi_p}(\Omega)$, where

$$\varphi_p(u) = \begin{cases} u^2, \ |u| < 1, \\ |u|^p, \ |u| > 1, \end{cases} \quad p = \frac{2\alpha}{\alpha - 2},$$

inequalities

$$\tau_{\varphi_p}(\eta_\Lambda(t)) \leq \frac{1}{\ln\left(1+\frac{1}{t}\right)} C_\Lambda \qquad [5.43]$$

holds true, where $C_\Lambda = \left(C_\alpha \sum_{k=0}^{M} b_k^2 d_k^2 \right)^{\frac{1}{2}}$, C_α, b_k^2, d_k^2 is defined by formula [5.29], and

$$\tau_{\varphi_p}(\eta_\Lambda(t) - \eta_\Lambda(s)) \leq \frac{1}{\ln\left(1+\frac{1}{|t-s|}\right)} L_\Lambda, \qquad [5.44]$$

where $L_\Lambda = \left(\widetilde{C}_\alpha \sum_{k=0}^{M} b_k^2 \hat{d}_k^2 \right)^{\frac{1}{2}}$, $\widetilde{C}_\alpha$, b_k^2, $\hat{d}_k^2$ is defined in theorem 5.11.

THEOREM 5.12.– Suppose that for some $\alpha > 2$ condition [5.28] is satisfied. Then, for any $T \geq 1$, $\lambda > 0$ and $\delta > 0$ such that $\delta < \frac{1}{\gamma_T}\min(L_\Lambda, \kappa_T)$, where $\kappa_T = \frac{L_\Lambda}{\ln(1+\frac{2}{T})}$, $\gamma_T = \frac{C_\Lambda}{\ln(1+\frac{1}{T})}$, C_Λ is defined in [5.43], L_Λ from [5.44], an inequality

$$\mathbf{E}\exp\left\{\lambda \sup_{0\leq t\leq T} |\eta_\Lambda(t)|\right\} \leq 2\tilde{G}(\lambda,\delta), \qquad [5.45]$$

where

$$\tilde{G}(\lambda,\delta) = \exp\left\{\varphi_p\left(\frac{\lambda\gamma_T}{1-\delta}\right) + 2\lambda B_\delta\right\},$$

$$B_\delta = \frac{1}{(1-\delta)\delta}\left((\ln T)^{\frac{\alpha+2}{2\alpha}}\delta\gamma_T + L_\Lambda^{\frac{\alpha-2}{2\alpha}}(\delta\gamma_T)^{\frac{\alpha-2}{2\alpha}}\frac{2\alpha}{\alpha-2}\right),$$

where $\varphi_p(x)$ is defined in [5.20], $p = \frac{2\alpha}{\alpha-2}$.

PROOF.– The theorem follows from entropy characteristics and corollary 5.2. Indeed, we put $\mathcal{T} = [0,T]$, $\rho(t,s) = |t-s|$, $X(t) = \eta_\Lambda(t)$, $\varphi(x) = \varphi_p(x)$. From corollary 5.2, we obtain that

$$\sigma(h) = \frac{L_\Lambda}{\ln\left(1+\frac{1}{h}\right)}, \quad \gamma_0 = \frac{C_\Lambda}{\ln\left(1+\frac{1}{T}\right)} = \gamma_T,$$

$$\kappa = \kappa_T,\ \sigma^{(-1)}(u) = \left(\exp\left\{\frac{L_\Lambda}{u}\right\} - 1\right)^{-1}.$$

It is clear that $N(\varepsilon) \leq \frac{T}{2\varepsilon} + 1$, then

$$H(\sigma^{(-1)}(u)) = \ln\left(N(\sigma^{(-1)}(u))\right) \leq \ln\left(\frac{T}{2}\left(\exp\left\{\frac{L_\Lambda}{u}\right\} - 1\right) + 1\right)$$
$$\leq \ln\left(T\exp\left\{\frac{L_\Lambda}{u}\right\} - (T-1)\right) \leq \ln\left(T\exp\left\{\frac{L_\Lambda}{u}\right\}\right) = \frac{L_\Lambda}{u} + \ln T. \qquad [5.46]$$

Since $\delta\gamma_T \leq \kappa_T$, then we can put $\beta = \gamma_T$, then from [5.46], we obtain

$$G(\lambda,\delta) \leq \exp\left\{\varphi_p\left(\frac{\lambda\gamma_T}{1-\delta}\right) + 2\lambda\left[\frac{1}{(1-\delta)\delta}\int_0^{\delta\gamma_T}\zeta_\varphi(u)du\right]\right\}. \qquad [5.47]$$

Since $\dfrac{\delta\gamma_T}{L_\Lambda} < 1$, then

$$\int_0^{\delta\gamma_T} \zeta_\varphi(u)du = \int_0^{\delta\gamma_T} \left(\frac{L_\Lambda}{u} + \ln T\right)^{1-\frac{1}{p}} du \leq L_\Lambda^{\frac{p-1}{p}} (\delta\gamma_T)^{\frac{1}{p}} p + (\ln T)^{1-\frac{1}{p}} \delta\gamma_T.$$

Last equality, inequality [5.47] implies that

$$\begin{aligned}
&\mathbf{E} \exp\left\{\lambda \sup_{0\leq t\leq T} |\eta_\Lambda(t)|\right\} \\
&\leq 2\exp\left\{\varphi_p\left(\frac{\lambda\gamma_T}{1-\delta}\right) + 2\lambda\left[\frac{1}{(1-\delta)\delta}\left(L_\Lambda^{\frac{p-1}{p}} (\delta\gamma_T)^{\frac{1}{p}} p + (\ln T)^{1-\frac{1}{p}} \delta\gamma_T\right)\right]\right\} \\
&= 2\exp\left\{\varphi_p\left(\frac{\lambda\gamma_T}{1-\delta}\right) + 2\lambda\frac{1}{(1-\delta)\delta}\left((\ln T)^{\frac{\alpha+2}{2\alpha}} \delta\gamma_T + \right.\right. \\
&\left.\left. + \ L_\Lambda^{\frac{\alpha-2}{2\alpha}} (\delta\gamma_T)^{\frac{\alpha-2}{2\alpha}} \frac{2\alpha}{\alpha-2}\right)\right\}.
\end{aligned}$$

□

COROLLARY 5.3.– Assume that the conditions of theorem 5.12 are satisfied, then for any $T > 0$, $\varepsilon > 2B_\delta$, $\delta > 0$, $\delta < \frac{L_\Lambda}{\gamma_T}$; $\delta < \frac{\kappa_T}{\gamma_T}$ an inequality

$$P\left\{\sup_{0\leq t\leq T} |\eta_\Lambda(t)| > \varepsilon\right\} \leq 2\exp\left\{-\varphi_p^*\left(\frac{\varepsilon - 2B_\delta}{\gamma_T}(1-\delta)\right)\right\} \qquad [5.48]$$

holds, where $B_\delta = \dfrac{1}{(1-\delta)\delta}\left((\ln T)^{\frac{\alpha+2}{2\alpha}} \delta\gamma_T + L_\Lambda^{\frac{\alpha-2}{2\alpha}} (\delta\gamma_T)^{\frac{\alpha-2}{2\alpha}} \dfrac{2\alpha}{\alpha-2}\right)$, $p = \frac{2\alpha}{\alpha-2}$, $\varphi_p^*(u)$, $u > 0$ is a Young–Fenchel transform of the function $\varphi_p(u)$, $\varphi_p^*(u) = \sup\limits_{v>0}(uv - \varphi_p(v))$.

PROOF.– From the Chebyshev inequality and [5.45] it follows that

$$\begin{aligned}
P\left\{\sup_{0\leq t\leq T} |\eta_\Lambda(t)| > \varepsilon\right\} &\leq 2\exp\left\{\varphi_p\left(\frac{\lambda\gamma_T}{1-\delta}\right) + 2\lambda B_\delta\right\} \cdot \exp\{-\lambda\varepsilon\} \\
&= 2\exp\left\{-\left(\frac{\lambda\gamma_T}{1-\delta}\frac{\varepsilon - 2B_\delta}{\gamma_T}(1-\delta) - \varphi_p\left(\frac{\lambda\gamma_T}{1-\delta}\right)\right)\right\}.
\end{aligned}$$

If in the right-hand side of this inequality infinimum $\frac{\lambda\gamma_T}{1-\delta}$ is taken with respect to $\varepsilon \geq 2B_\delta$, we obtain [5.48].

Find the exact form of the function $\varphi^*(u)$:

$$\varphi_p^*(u) = \sup_{v>0}\{uv - \varphi_p(v)\} = \sup_{|v|<1}\{uv - v^2\} = u\frac{u}{2} - \frac{u^2}{4} = \frac{u^2}{4}, \quad as\ 0 \le u < 2,$$

$$\varphi_p^*(u) = \sup_{v>1}\{uv - v^p\} = u\left(\frac{u}{p}\right)^{\frac{1}{p-1}} - \left(\frac{u}{p}\right)^{\frac{p}{p-1}} = \frac{p-1}{p^{\frac{p}{p-1}}}u^{\frac{p}{p-1}}, \quad as\ u > p,$$

$\varphi_p^*(u) = u - 1$, as $v = 1$ and $2 \le u \le p$.

So

$$\varphi_p^*(u) = \begin{cases} \frac{u^2}{4}, & 0 \le u < 2 \\ u - 1, & 2 \le u \le p \\ \frac{p-1}{p^{\frac{p}{p-1}}}u^{\frac{p}{p-1}}, & u > p \end{cases}.$$

□

COROLLARY 5.4.– The model $X_\Lambda(t)$ approximates the process $X(t)$ with given accuracy $\varepsilon > B_\delta$ and reliability $1-\kappa$, $\kappa > 0$ in uniform metric if for the partition Λ the inequality

$$2\exp\left\{-\varphi_p^*\left(\frac{\varepsilon - 2B_\delta}{\gamma_T}(1-\delta)\right)\right\} \le \kappa$$

holds, where $B_\delta = \frac{1}{(1-\delta)\delta}\left((\ln T)^{\frac{\alpha+2}{2\alpha}}\delta\gamma_T + L_\Lambda^{\frac{\alpha-2}{2\alpha}}(\delta\gamma_T)^{\frac{\alpha-2}{2\alpha}}\frac{2\alpha}{\alpha-2}\right)$, $p = \frac{2\alpha}{\alpha-2}$. From corollary 5.3, we obtain the following statement.

COROLLARY 5.5.– If in [5.48] we put $\delta = \left(\frac{2A(\alpha,T)}{\varepsilon}\right)^{\frac{2\alpha}{3\alpha+2}}$, where

$$A(\alpha,T) = L_\Lambda^{\frac{\alpha+2}{2\alpha}}(\gamma_T)^{\frac{\alpha-2}{2\alpha}}\frac{2\alpha}{\alpha-2}$$

we find, that for any

$$\varepsilon > 2A(\alpha,T)\cdot\max\left(\left(\frac{\gamma_T}{L_\Lambda}\right)^{\frac{3\alpha+2}{2\alpha}}, \left(\frac{\gamma_T}{\kappa_T}\right)^{\frac{3\alpha+2}{2\alpha}}, 2\right)$$

the inequality

$$P\left\{\sup_{0\le t\le T}|\eta_{N,\Lambda}(t)| \ge \varepsilon\right\}$$

$$\le 2\exp\left\{\varphi_p^*\left(\frac{\varepsilon - \frac{2\delta^{\frac{\alpha-2}{2\alpha}}}{(1-\delta)\delta}\left((\ln T)^{\frac{\alpha+2}{2\alpha}}\delta^{\frac{\alpha+2}{2\alpha}}\gamma_T + A(\alpha,T)\right)}{\gamma_T}(1-\delta)\right)\right\}$$

$$= 2\exp\left\{\varphi_p^*\left(\frac{\varepsilon(1-\delta)}{\gamma_T} - 2(\ln T)^{\frac{\alpha+2}{2\alpha}} - \frac{2A(\alpha,T)}{\gamma_T \delta^{\frac{\alpha+2}{2\alpha}}}\right)\right\}$$

$$= 2\exp\left\{\varphi_p^*\left(\frac{\varepsilon - \varepsilon\left(\frac{2A(\alpha,T)}{\varepsilon}\right)^{\frac{2\alpha}{3\alpha+2}} - (2A(\alpha,T))^{\frac{2\alpha}{3\alpha+2}}\varepsilon^{\frac{\alpha+2}{3\alpha+2}}}{\gamma_T} - 2(\ln T)^{\frac{\alpha+2}{2\alpha}}\right)\right\}$$

$$= 2\; exp\left\{\varphi_p^*\left(\frac{\varepsilon - 2\varepsilon^{\frac{\alpha+2}{3\alpha+2}}(2A(\alpha,T))^{\frac{2\alpha}{3\alpha+2}}}{\gamma_T} - 2(\ln T)^{\frac{\alpha+2}{2\alpha}}\right)\right\}$$

holds true.

5.4. Generalized model of Gaussian stationary processes

In this section, the model of Gaussian stationary random process is constructed such that the correlation function of the process does not coincide with the correlation function of the model.

With high modeling accuracy, the estimates are worse than in previous cases. But this method does not require the additional restrictions on spectral function of the process. The estimates hold true only under constraints that ensure sample continuity of the process with probability 1. The results of this section are described in [TEG 02].

Let $X = \{X(t), t \in \mathbf{T}\}$ be Gaussian stationary real centered continuous in mean square random process with covariance function

$$\mathbf{E}X(t+\tau)X(t) = r(\tau) = \int_0^\infty \cos\lambda\tau dF(\lambda),$$

where $F(\lambda)$ is continuous spectral function of the process.

Random process $X(t)$ can be represented as

$$X(t) = \int_0^\infty \cos\lambda t d\eta_1(\lambda) + \int_0^\infty \sin\lambda t d\eta_2(\lambda),$$

where $\eta_1(\lambda)$, $\eta_2(\lambda)$ are defined in [4.2]

$$X(t) = X_\Lambda(t) + X^\Lambda(t),$$

where $X_\Lambda(t) = \int_0^\Lambda \cos\lambda t d\eta_1(\lambda) + \int_0^\Lambda \sin\lambda t d\eta_2(\lambda)$ with covariance function

$$r_\Lambda(\tau) = \mathbf{E}X_\Lambda(t+\tau)X_\Lambda(t) = \int_0^\Lambda \cos\lambda\tau dF(\lambda),$$

$$X^\Lambda(t) = \int_\Lambda^\infty \cos\lambda t d\eta_1(\lambda) + \int_\Lambda^\infty \sin\lambda t d\eta_2(\lambda)$$

with covariance function

$$r^\Lambda(\tau) = \mathbf{E}X^\Lambda(t+\tau)X^\Lambda(t) = \int_\Lambda^\infty \cos\lambda\tau dF(\lambda).$$

According to the model of $X(t)$, we take the model type

$$X_\Lambda^M(t) = \sum_{k=0}^{M}(\eta_{k1}\cos\zeta_k t + \eta_{k2}\sin\zeta_k t),$$

where $\Lambda = \{\lambda_0, \dots, \lambda_M\}$ a partition of the set $[0, \Lambda]$ that $\lambda_0 = 0$, $\lambda_k < \lambda_{k+1}$, $\lambda_M = \Lambda$, $\eta_{k1}, \eta_{k2}, \zeta_k$ are independent random variables such that $\mathbf{E}\eta_{k1} = \mathbf{E}\eta_{k2} = 0$,

$$\mathbf{E}\eta_{k1}^2 = \mathbf{E}\eta_{k2}^2 = F(\lambda_{k+1}) - F(\lambda_k) = b_k^2, \quad k = 1, \dots, M,$$

ζ_k are random variables taking values on the segments $[\lambda_k, \lambda_{k+1}]$ and have the following distribution function

$$P\{\zeta_k < \lambda\} = F_k(\lambda) = \frac{F(\lambda) - F(\lambda_k)}{F(\lambda_{k+1}) - F(\lambda_k)}.$$

Show now under which conditions the partition Λ should be chosen such that for model X_Λ^M there exists a centered Gaussian process $X(t)$, that is approximated by the model in the space $C([0,T])$ with given accuracy and reliability.

Note that

$$\eta_\Lambda(t) = X_\Lambda^M(t) - X_\Lambda(t)$$
$$= \sum_{k=0}^{M}\left[\int_{\lambda_k}^{\lambda_{k+1}} (\cos\lambda t - \cos\zeta_k t)d\eta_1(\lambda) + \int_{\lambda_k}^{\lambda_{k+1}} (\sin\lambda t - \sin\zeta_k t)d\eta_2(\lambda)\right]. \qquad [5.49]$$

The main statements for sub-Gaussian process $\eta_\Lambda(t)$ are given in section 5.4.

The following theorems are carried out.

THEOREM 5.13.– For sub-Gaussian random process $\eta\Lambda(t)$, the inequality

$$\tau(\eta_\Lambda(t)) \leq 4\frac{\ln^\gamma\left(1+\frac{\Lambda}{M}\right)}{\ln^\gamma\left(1+\frac{2}{t}\right)}(F(\Lambda))^{\frac{1}{2}}$$

holds, where $\Lambda = \{\lambda_0, \lambda_1, ...\lambda_M\}$, $\lambda_0 = 0, \lambda_k \leq \lambda_{k+1}, \lambda_M = \Lambda$, $F(\Lambda)$ are spectral function.

PROOF.– From properties $Sub(\Omega)$- space and lemma 4.1 follows

$$\tau^2\left(\int\limits_{\lambda_k}^{\lambda_{k+1}}(cos\lambda t - cos\zeta_k t)d\eta_1(\lambda)\right) \leq \Theta_1^2\left(\int\limits_{\lambda_k}^{\lambda_{k+1}}(cos\lambda t - cos\zeta_k t)d\eta_1(\lambda)\right)$$

$$\leq \sup_{m\geq 1}\left[\frac{1}{\Delta_{2m}}\mathbf{E}\left(\int\limits_{\lambda_k}^{\lambda_{k+1}}(cos\lambda t - cos\zeta_k t)d\eta_1(\lambda)\right)^{2m}\right]^{\frac{1}{m}}$$

$$\leq \sup_{m\geq 1} b_k^2\left(\int\limits_{\lambda_k}^{\lambda_{k+1}}\left(\int\limits_{\lambda_k}^{\lambda_{k+1}}\left(2sin\frac{t(u-\lambda)}{2}\right)^2 dF_k(\lambda)\right)^m dF_k(u)\right)^{\frac{1}{m}}$$

$$\leq \sup_{m\geq 1} 4b_k^2\left(\int\limits_{\lambda_k}^{\lambda_{k+1}}\int\limits_{\lambda_k}^{\lambda_{k+1}}\left(\frac{\ln^\gamma(1+u-\lambda)}{\ln^\gamma\left(1+\frac{2}{t}\right)}\right)^{2m} dF_k(\lambda)dF_k(u)\right)^{\frac{1}{m}}$$

$$\leq 4\frac{\ln^{2\gamma}(1+\lambda_{k+1}-\lambda_k)}{\ln^{2\gamma}\left(1+\frac{2}{t}\right)}(F(\lambda_{k+1}) - F(\lambda_k)) = I_k.$$

Similarly, we obtain:

$$\tau^2\left(\int\limits_{\lambda_k}^{\lambda_{k+1}}(\sin\lambda t - \sin\zeta_k t)d\eta_2(\lambda)\right) \leq \Theta_1^2\left(\int\limits_{\lambda_k}^{\lambda_{k+1}}(\sin\lambda t - \sin\zeta_k t)d\eta_2(\lambda)\right)$$

$$\leq \sup_{m\geq 1}\left[\frac{1}{\Delta_{2m}}\mathbf{E}\left(\int\limits_{\lambda_k}^{\lambda_{k+1}}(\sin\lambda t - \sin\zeta_k t)d\eta_2(\lambda)\right)^{2m}\right]^{\frac{1}{m}}$$

$$\leq \sup_{m\geq 1} b_k^2 \left(\int\limits_{\lambda_k}^{\lambda_{k+1}} \left(\int\limits_{\lambda_k}^{\lambda_{k+1}} \left(2\sin\frac{t(u-\lambda)}{2} \right)^2 dF_k(\lambda) \right)^m dF_k(u) \right)^{\frac{1}{m}}$$

$$\leq \sup_{m\geq 1} 4b_k^2 \left(\int\limits_{\lambda_k}^{\lambda_{k+1}} \int\limits_{\lambda_k}^{\lambda_{k+1}} \left(\frac{\ln^\gamma(1+u-\lambda)}{\ln^\gamma\left(1+\frac{2}{t}\right)} \right)^{2m} dF_k(\lambda)dF_k(u) \right)^{\frac{1}{m}}$$

$$\leq 4\frac{\ln^{2\gamma}(1+\lambda_{k+1}-\lambda_k)}{\ln^{2\gamma}\left(1+\frac{2}{t}\right)} (F(\lambda_{k+1})-F(\lambda_k)) = I_k.$$

$$\tau^2 \left(\int\limits_{\lambda_k}^{\lambda_{k+1}} (\cos\lambda t - \cos\zeta_k t) d\eta_1(\lambda) + \int\limits_{\lambda_k}^{\lambda_{k+1}} (\sin\lambda t - \sin\zeta_k t) d\eta_2(\lambda) \right)$$

$$\leq \left(\tau\left(\int\limits_{\lambda_k}^{\lambda_{k+1}} (\cos\lambda t - \cos\zeta_k t) d\eta_1(\lambda) \right) + \tau\left(\int\limits_{\lambda_k}^{\lambda_{k+1}} (\sin\lambda t - \sin\zeta_k t) d\eta_2(\lambda) \right) \right)^2$$

$$\leq 4I_k. \qquad [5.50]$$

Since the summands of sums [5.49] for different k are independent, then [5.50] and lemma 1.7 imply

$$\tau^2(\eta_\Lambda(t)) \leq 4\sum_{k=0}^{M} I_k,$$

$$\tau(\eta_\Lambda(t)) \leq 2\left(\sum_{k=0}^{M} 4\frac{\ln^{2\gamma}(1+\lambda_{k+1}-\lambda_k)}{\ln^{2\gamma}(1+\frac{2}{t})} \cdot (F(\lambda_{k+1})-F(\lambda_k)) \right)^{\frac{1}{2}}.$$

If we put $\lambda_{k+1}-\lambda_k = \frac{\Lambda}{M}$, then

$$\tau(\eta_\Lambda(t)) \leq 4\ln^{-\gamma}\left(1+\frac{2}{t}\right)\ln^\gamma\left(1+\frac{\Lambda}{M}\right)(F(\Lambda))^{\frac{1}{2}}, \ \gamma > 0.$$

□

For any $t, s \in \mathcal{T}$

$$\eta_\Lambda(t) - \eta_\Lambda(s)$$

$$= \sum_{k=0}^{M} \left[\int\limits_{\lambda_k}^{\lambda_{k+1}} (\cos\lambda t - \cos\zeta_k t - \cos\lambda s + \cos\zeta_k s) d\eta_1(\lambda) \right.$$

$$+ \int_{\lambda_k}^{\lambda_{k+1}} (\sin \lambda t - \sin \zeta_k t - \sin \lambda s + \sin \zeta_k s) d\eta_2(\lambda) \Bigg] .$$

THEOREM 5.14.– The following inequality is fulfilled.

$$\tau\left(\eta_\Lambda(t) - \eta_\Lambda(s)\right)$$
$$\leq 8\left(F(\Lambda)\right)^{\frac{1}{2}} \frac{1}{\ln^\gamma\left(1+\frac{2}{|t-s|}\right)} \cdot \ln^\gamma\left(1+\frac{\Lambda}{M}\right)\left(1+\frac{\ln^\gamma(1+\Lambda)}{\ln^\gamma\left(1+\frac{2}{T}\right)}\right),$$

$t, s \in [0, T]$.

PROOF.– Denote

$$\omega_{k1} = \int_{\lambda_k}^{\lambda_{k+1}} (\cos \lambda t - \cos \zeta_k t - \cos \lambda s + \cos \zeta_k s) d\eta_1(\lambda),$$

$$\omega_{k2} = \int_{\lambda_k}^{\lambda_{k+1}} (\sin \lambda t - \sin \zeta_k t - \sin \lambda s + \sin \zeta_k s) d\eta_2(\lambda).$$

By lemma 1.7, we have

$$\tau^2(\eta_\Lambda(t) - \eta_\Lambda(s)) \leq 2\sum_{k=0}^{M}(\tau^2(\omega_{k1}) + \tau^2(\omega_{k2})) \leq 2\sum_{k=0}^{M}(\Theta_1^2(\omega_{k1}) + \Theta_1^2(\omega_{k2})),$$

where

$$\Theta_1(\omega_{ki}) = \sup_{m\geq 1}\left(\frac{1}{\Delta_{2m}} E\omega_{ki}^{2m}\right)^{\frac{1}{2m}}, \quad i = 1, 2.$$

From lemma 5.3 follows that

$$\tau^2\left(\int_{\lambda_k}^{\lambda_{k+1}} (\cos \lambda t - \cos \zeta_k t - \cos \lambda s + \cos \zeta_k s)\, d\eta_1(\lambda)\right)$$

$$\leq 16 \sup_{m\geq 1}\left(\mathbf{E}\left(\int_{\lambda_k}^{\lambda_{k+1}}\left(\left|\sin\frac{(s-t)(\lambda-\zeta_k)}{4}\right| + \right.\right.\right.$$

$$+\left|\sin\frac{\zeta_k(s-t)}{2}\right|\cdot\left|\sin\frac{(\lambda-\zeta_k)(t+s)}{4}\right|\Bigg)^2 dF(\lambda)\Bigg)^m\Bigg)^{\frac{1}{m}}$$

$$\leq 16b_k^2\sup_{m\geq 1}\left(\int_{\lambda_k}^{\lambda_{k+1}}\int_{\lambda_k}^{\lambda_{k+1}}\left(\frac{\ln^\gamma(1+|\lambda-u|)}{\ln^\gamma(1+\frac{4}{|s-t|})}\right.\right.$$

$$\left.\left.+\frac{\ln^\gamma(1+u)}{\ln^\gamma(1+\frac{2}{|s-t|})}\frac{\ln^\gamma(1+|\lambda-u|)}{\ln\left(1+\frac{4}{s+t}\right)}\right)^{2m} dF_k(\lambda)dF_k(u)\right)^{\frac{1}{m}}$$

$$\leq 16\left(\frac{\ln^\gamma(1+\lambda_{k+1}-\lambda_k)}{\ln^\gamma\left(1+\frac{2}{|s-t|}\right)}+\frac{\ln^\gamma(1+\lambda_{k+1})}{\ln^\gamma\left(1+\frac{2}{|s-t|}\right)}\frac{\ln^\gamma(1+\lambda_{k+1}-\lambda_k)}{\ln^\gamma\left(1+\frac{4}{s+t}\right)}\right)^2$$

$$\times\left(F(\lambda_{k+1})-F(\lambda_k)\right)$$

$$\leq 16\left(F(\lambda_{k+1})-F(\lambda_k)\right)\frac{\ln^{2\gamma}(1+\lambda_{k+1}-\lambda_k)}{\ln^{2\gamma}\left(1+\frac{2}{|s-t|}\right)}\left(1+\frac{\ln^\gamma(1+\lambda_{k+1})}{\ln^\gamma\left(1+\frac{2}{T}\right)}\right)^2=J_k.$$

Similarly, we obtain

$$\tau^2\left(\int_{\lambda_k}^{\lambda_{k+1}}(\sin\lambda t-\sin\zeta_k t-\sin\lambda s+\sin\zeta_k s)\,d\eta_2(\lambda)\right)$$

$$\leq 16\sup_{m\geq 1}\left(\mathbf{E}\left(\int_{\lambda_k}^{\lambda_{k+1}}\left(\left|\sin\frac{(s-t)(\lambda-\zeta_k)}{4}\right|\right.\right.\right.$$

$$+\left|\sin\frac{\zeta_k(s-t)}{2}\right|\cdot\left|\sin\frac{(\lambda-\zeta_k)(t+s)}{4}\right|\Bigg)^2 dF(\lambda)\Bigg)^m\Bigg)^{\frac{1}{m}}$$

$$\leq 16b_k^2\sup_{m\geq 1}\left(\int_{\lambda_k}^{\lambda_{k+1}}\int_{\lambda_k}^{\lambda_{k+1}}\left(\frac{\ln^\gamma(1+|\lambda-u|)}{\ln^\gamma(1+\frac{4}{|s-t|})}\right.\right.$$

$$\left.\left.+\frac{\ln^\gamma(1+u)}{\ln^\gamma(1+\frac{2}{|s-t|})}\frac{\ln^\gamma(1+|\lambda-u|)}{\ln\left(1+\frac{4}{s+t}\right)}\right)^{2m} dF_k(\lambda)dF_k(u)\right)^{\frac{1}{m}}$$

$$\leq 16\left(F(\lambda_{k+1}) - F(\lambda_k)\right) \frac{\ln^{2\gamma}(1+\lambda_{k+1}-\lambda_k)}{\ln^{2\gamma}\left(1+\frac{2}{|s-t|}\right)} \left(1+\frac{\ln^{\gamma}(1+\lambda_{k+1})}{\ln^{\gamma}\left(1+\frac{2}{T}\right)}\right)^2 = J_k.$$

Then, $\tau^2\left(\eta_\Lambda(t) - \eta_\Lambda(s)\right) \leq 4\sum\limits_{k=0}^{M} J_k.$

If we put $\lambda_{k+1} - \lambda_k = \frac{\Lambda}{M}$, then

$$\tau\left(\eta_\Lambda(t) - \eta_\Lambda(s)\right) \leq 8(F(\Lambda))^{\frac{1}{2}} \frac{\ln^{\gamma}\left(1+\frac{\Lambda}{M}\right)}{\ln^{\gamma}\left(1+\frac{2}{|s-t|}\right)} \left(1+\frac{\ln^{\gamma}(1+\Lambda)}{\ln^{\gamma}\left(1+\frac{2}{T}\right)}\right).$$

□

THEOREM 5.15.– If condition

$$\int\limits_0^\infty \ln^{2\varepsilon}(1+u)dF(u) < \infty, \quad 0 < \varepsilon \leq 1$$

holds, then

$$\tau\left(X^\Lambda(t) - X^\Lambda(s)\right) \leq \frac{2\tilde{Q}^{\frac{1}{2}}}{\ln^{\gamma}(1+\frac{1}{|t-s|})},$$

where $\tilde{Q} = \int\limits_\Lambda^\infty \ln^{2\gamma}(1+\lambda)dF(\lambda)$, $0 < \gamma \leq 1$.

PROOF.–

$$r^\Lambda(\tau) = \mathbf{E}X^\Lambda(t+\tau)X^\Lambda(t) = \int\limits_\Lambda^\infty \cos\lambda\tau dF(\lambda),$$

$$\begin{aligned}
\mathbf{E}\left|X^\Lambda(t) - X^\Lambda(s)\right|^2 &= 2\left(r^\Lambda(0) - r^\Lambda(t-s)\right)\\
&= 2\left[(F(+\infty) - F(\Lambda)) - \int\limits_\Lambda^\infty \cos\lambda(t-s)dF(\lambda)\right]\\
&= 2\int\limits_\Lambda^\infty (1-\cos\lambda(t-s))\, dF(\lambda)\\
&= 4\int\limits_\Lambda^\infty \sin^2\frac{\lambda(t-s)}{2} dF(\lambda) \leq 4\int\limits_\Lambda^\infty \frac{\ln^{2\gamma}(1+\lambda)}{\ln^{2\gamma}(1+\frac{2}{|t-s|})} dF(\lambda).
\end{aligned}$$

Then,

$$\tau\left(X^{\Lambda}(t)-X^{\Lambda}(s)\right)\leq\left(\mathbf{E}\left|X^{\Lambda}(t)-X^{\Lambda}(s)\right|^{2}\right)^{\frac{1}{2}}$$

$$=2\left(\int\limits_{\Lambda}^{\infty}\frac{\ln^{2\gamma}(1+\lambda)}{\ln^{2\gamma}(1+\frac{2}{|t-s|})}dF(\lambda)\right)^{\frac{1}{2}}=\frac{2}{\ln^{\gamma}(1+\frac{2}{|t-s|})}\left(\int\limits_{\Lambda}^{\infty}\ln^{2\gamma}(1+\lambda)dF(\lambda)\right)^{\frac{1}{2}}.$$

□

THEOREM 5.16.– If the random process $X_{\Lambda}^{M}(t)$ has a partition Λ such that the inequalities

$$\int\limits_{0}^{\infty}\ln^{2\gamma}(1+\lambda)dF(\lambda)<\infty\,,0<\gamma\leq1\,,$$

$$2\exp\left\{-\frac{1}{2\varepsilon_{01}^{2}}\left(\alpha\delta-\sqrt{8\alpha\delta I_{1}(\varepsilon_{01})}\right)^{2}\right\}$$
$$+2\exp\left\{-\frac{1}{2\varepsilon_{02}^{2}}\left((1-\alpha)\delta-\sqrt{8(1-\alpha)\delta I_{2}(\varepsilon_{02})}\right)^{2}\right\}\leq\beta \quad [5.51]$$

as $\delta>8\max\left(I_{1}(\varepsilon_{01}),I_{2}(\varepsilon_{02})\right)$ holds, where $0<\alpha<1$, $\varepsilon_{01}=\sup\limits_{0\leq t\leq T}\tau(X_{\Lambda}(t)-X_{\Lambda}^{M}(t))$, $\varepsilon_{02}=\sup\limits_{0\leq t\leq T}\tau(X^{\Lambda}(t))$,

$$I_{1}(\varepsilon_{01})=\frac{1}{\sqrt{2}}\int\limits_{0}^{\varepsilon_{01}}\left(\ln\left(\frac{T}{4}\exp\left\{\left(\frac{8\sqrt{F(\Lambda)}\ln^{\gamma}\left(1+\frac{\Lambda}{M}\right)\left(1+\frac{\ln^{\gamma}(1+\Lambda)}{\ln^{\gamma}\left(1+\frac{2}{T}\right)}\right)}{\varepsilon}\right)^{\frac{1}{\gamma}}\right\}\right.\right.$$
$$\left.\left.-\frac{T}{4}+1\right)\right)^{\frac{1}{2}}d\varepsilon,$$

$$I_{2}(\varepsilon_{02})$$
$$=\frac{1}{\sqrt{2}}\int\limits_{0}^{\varepsilon_{02}}\left(\ln\left(\frac{T}{4}\exp\left\{\left(\frac{2\int\limits_{\Lambda}^{\infty}\ln^{2\gamma}(1+\lambda)dF(\lambda)}{\varepsilon}\right)^{\frac{1}{\gamma}}\right\}-\frac{T}{4}+1\right)\right)^{\frac{1}{2}}d\varepsilon,$$

then the model approximates Gaussian random process $X(t)$ in the space $C([0,T])$ with reliability $1-\beta$, $0<\beta<1$ and accuracy $\delta>0$.

PROOF.– From entropy characteristics [BUL 00], we have

$$P\left\{\sup_{0\le t\le T}|X(t)-X_\Lambda^M(t)|\ge\delta\right\}$$

$$\le P\left\{\sup_{0\le t\le T}|X_\Lambda(t)-X_\Lambda^M(t)|>\alpha\delta\right\}+P\left\{\sup_{0\le t\le T}|X^\Lambda(t)|>(1-\alpha)\delta\right\}$$

$$\le 2\exp\left\{-\frac{1}{2\varepsilon_{01}^2}\left(\alpha\delta-\sqrt{8\alpha\delta I_1(\varepsilon_{01})}\right)^2\right\}+$$

$$+2\exp\left\{-\frac{1}{2\varepsilon_{02}^2}\left((1-\alpha)\sqrt{8(1-\alpha)\delta I_2(\varepsilon_{02})}\right)^2\right\}.$$

From theorem 5.14 follows that $\sigma_1(h)=\sup\limits_{|t-s|<h}\tau(\eta_\Lambda(t)-\eta_\Lambda(s))\le\frac{\tilde G}{\ln^\gamma\left(1+\frac{2}{h}\right)}$, where

$$\tilde G=8\sqrt{F(\Lambda)}\ln^\gamma\left(1+\frac{\Lambda}{M}\right)\left(1+\frac{\ln^\gamma(1+\Lambda)}{\ln^\gamma\left(1+\frac{2}{T}\right)}\right).$$

$$\sigma_1^{(-1)}(h)=\frac{2}{\exp\left\{\left(\frac{\tilde G}{h}\right)^{\frac{1}{\gamma}}\right\}-1}.$$

Then

$$I_1(\varepsilon_{01})=\frac{1}{\sqrt2}\int_0^{\varepsilon_{01}}\left(\ln\left(\frac{T}{2\sigma_1^{(-1)}(\varepsilon)}+1\right)\right)^{\frac12}d\varepsilon$$

$$=\frac{1}{\sqrt2}\int_0^{\varepsilon_{01}}\left(\ln\left(\frac{T}{4}\exp\left\{\left(\frac{\tilde G}{\varepsilon}\right)^{\frac{1}{\gamma}}\right\}-\frac{T}{4}+1\right)\right)^{\frac12}d\varepsilon,$$

$\varepsilon_{01}=\sup\limits_{0\le t\le T}\tau(\eta_\Lambda(t))$ is defined in theorem 5.13.

From theorem 5.15 follows that $\sigma_2(h)=\sup\limits_{|t-s|<h}(X^\Lambda(t)-X^\Lambda(s))\le\frac{2\tilde Q}{\ln^\gamma\left(1+\frac{2}{h}\right)}$, where $\tilde Q=\int\limits_\Lambda^\infty\ln^{2\gamma}(1+\lambda)dF(\lambda)$, then

$$\sigma_2^{(-1)}(h)=2\left(\exp\left\{\left(\frac{2\tilde Q}{\varepsilon}\right)^{\frac{1}{\gamma}}\right\}-1\right)^{-1},$$

$$I_2(\varepsilon_{02}) = \frac{1}{\sqrt{2}} \int_0^{\varepsilon_{02}} \left(\ln \left(\frac{T}{4} \exp \left\{ \left(\frac{2\tilde{Q}}{\varepsilon} \right)^{\frac{1}{\gamma}} \right\} - \frac{T}{4} + 1 \right) \right)^{\frac{1}{2}} d\varepsilon.$$

where

$$\varepsilon_{02} = \sup_{0 \leq t \leq T} \tau(X^\Lambda(t)) \leq \sup_{0 \leq t \leq T} (\mathbf{E}|X^\lambda(t)|^2)^{\frac{1}{2}} = \sqrt{F(+\infty) - F(\Lambda)}.$$

By definition 5.3, the model $X_\Lambda^M(t)$ approximates the process $X_\Lambda^M(t)$ with reliability of $1 - \beta$, $0 < \beta < 1$ and accuracy $\delta > 0$ in the space $C([0, T])$, if condition [5.51] is fulfilled. □

6

Simulation of Cox Random Processes

In this chapter, we introduce random Cox processes and describe two algorithms of their simulation with some given accuracy and reliability. The cases are considered when an intensity of random Cox processes are generated by log Gaussian or square Gaussian homogeneous and inhomogeneous processes, or fields are considered. The results of this chapter are based on the works of [KOZ 06a, KOZ 07c, KOZ 11, KOZ 07b, POG 07, POG 09, POG 11].

6.1. Random Cox processes

In this section, random Cox processes driven by random intensity are considered. All necessary definitions and properties which will be used during their simulation are described.

Let $\{\mathbf{T}, \mathfrak{B}, \mu\}$ be a measurable space, $\mu(\mathbf{T}) < \infty$.

DEFINITION 6.1.– *[MOL 98] Let* $\{Z(t), t \in \mathbf{T}\}, \mathbf{T} \subset \mathbf{R}$ *not be a negative random process. If* $\{\nu(B), B \in \mathfrak{B}\}$ *under fixed simple function* $Z(t)$ *is Poisson process with intensity function* $\mu(B) = \int_B Z(\cdot, t)dt$*, then* $\nu(B)$ *is said to be a random Cox process driven by process* $Z(t)$.

Let $\{Y(t), t \in \mathbf{T}\}$, $\mathbf{T} \subset \mathbf{R}$ be a homogeneous, Gaussian, mean square continuous random process, $\mathbf{E}Y(t) = 0$, $\mathbf{E}Y(t) \times Y(s) = B(t-s)$. If $Z(t) = \exp\{Y(t)\}$, then $\nu(B)$ is said to be the log Gaussian Cox process or Cox process driven by a log Gaussian process $\exp\{Y(t)\}$. If $Z(t) = Y^2(t)$, then $\nu(B)$ is said to be a square Gaussian Cox process or Cox process driven by a square Gaussian process $Y^2(t)$.

If $\mathbf{T} \subset \mathbf{R}^{\mathbf{n}}$, then $\nu(B)$ is said to be a random Cox process driven by the field.

DEFINITION 6.2.– *[FEL 70] Poisson point process N on $\{\mathbf{T}, \mathfrak{B}\}$ is said to be a point process such that for all $B_k \in \mathfrak{B}$, $k = \overline{1, m}$, $m \in \mathbf{N}$, $B_i \cap B_j = \emptyset$ if $i \neq j$, random variables $N(B_k)$, $k = \overline{1, m}$ are independent and have Poisson distribution with mean $\mu(B_k)$,*

$$\mathbf{P}\{N(B_k) = l\} = \frac{(\mu(B_k))^l}{l!} \exp\{-\mu(B_k)\}.$$

Let ξ_i, $i = 1, 2, \ldots$ be independent random elements with the same distribution, that for any set $B \in \mathfrak{B}$, $\mathbf{P}\{\xi_i \in B\} = \frac{\mu(B)}{\mu(\mathbf{T})}$.

Let Θ be a Poisson random variable, which does not depend on ξ_i. Consider a family of random elements $\xi_1, \xi_2, \ldots, \xi_\Theta$.

We denote $\Pi(B)$ by quantity of elements from $\xi_1, \xi_2, \ldots, \xi_\Theta$, which are contained in $B \in \mathfrak{B}$.

THEOREM 6.1.– $\Pi(B)$, $B \in \mathfrak{B}$ is a Poisson ensemble with density function $\mu(B)$.

PROOF.– Let $B_1, B_2, \ldots, B_m \in \mathfrak{B}$, $B_i \cap B_j = \emptyset$ when $i \neq j$. Since joint distribution of random variables $\Pi(B_1), \Pi(B_2), \ldots, \Pi(B_m)$ given that $\Theta = n$ is polynomial, it follows from the formula of total probability:

$$\begin{aligned}
&\mathbf{P}\{\Pi(B_1) = k_1, \Pi(B_2) = k_2, \ldots, \Pi(B_m) = k_m\} \\
&= \sum_{n=\sum_{i=1}^m k_i}^{\infty} \mathbf{P}\{\Pi(B_1) = k_1, \Pi(B_2) = k_2, \ldots, \Pi(B_m) = k_m / \Theta = n\} \mathbf{P}\{\Theta = n\} \\
&= \sum_{n=\sum_{i=1}^m k_i}^{\infty} \frac{n!}{k_1! \ldots k_m! \left(n - \sum_{i=1}^m k_i\right)!} \prod_{i=1}^m \left(\frac{\mu(B_i)}{\mu(\mathbf{T})}\right)^{k_i} \\
&\times \left(1 - \frac{\sum_{i=1}^m \mu(B_i)}{\mu(\mathbf{T})}\right)^{n-\sum_{i=1}^m k_i} e^{-\mu(\mathbf{T})} \frac{(\mu(\mathbf{T}))^n}{n!} \\
&= \left(\prod_{i=1}^m \frac{(\mu(B_i))^{k_i}}{k_i! \, (\mu(\mathbf{T}))^{k_i}}\right) e^{-\mu(\mathbf{T})} \\
&\times \sum_{n=\sum_{i=1}^m k_i}^{\infty} \frac{1}{\left(n - \sum_{i=1}^m k_i\right)!} \frac{\left(\mu(\mathbf{T}) - \sum_{i=1}^m \mu(B_i)\right)^{n-\sum_{i=1}^m k_i}}{(\mu(\mathbf{T}))^{n-\sum_{i=1}^m k_i}} (\mu(\mathbf{T}))^n \\
&= \left(\prod_{i=1}^m \frac{(\mu(B_i))^{k_i}}{k_i!}\right) e^{-\mu(\mathbf{T})} \sum_{n=\sum_{i=1}^m k_i}^{\infty} \frac{1}{\left(n - \sum_{i=1}^m k_i\right)!} \left(\mu(\mathbf{T}) - \sum_{i=1}^m \mu(B_i)\right)^{n-\sum\limits_{i=1}^m k_i}
\end{aligned}$$

$$= \left(\prod_{i=1}^{m} \frac{(\mu(B_i))^{k_i}}{k_i!}\right) e^{-\mu(\mathbf{T})} \sum_{s=0}^{\infty} \frac{1}{s!} \left(\mu(\mathbf{T}) - \sum_{i=1}^{m} \mu(B_i)\right)^s$$

$$= \prod_{i=1}^{m} \exp\{-\mu(B_i)\} \frac{(\mu(B_i))^{k_i}}{k_i!}.$$

□

Thereby, under fixed realization of the random process $Y(t)$ log Gaussian Cox processes $\nu(B)$ is a Poisson ensemble. This result will be used during simulation of Cox processes.

6.2. Simulation of log Gaussian Cox processes as a demand arrival process in actuarial mathematics

In this section, a simulation of the Cox processes methods density, generated by Log Gaussian process ($\mu(B) = \int\limits_B \exp\{Y(t)\}\, dt$, $\{Y(t), t \in \mathbf{T}\}$ is centered, homogeneous, Gaussian), is described. This method can be used as a method of simulation of demands arrival process in actuarial mathematics, since arrival of demands [TEU 04] often may be considered as a log Gaussian Cox process. The model is constructed on finite domain $\mathbf{T} = [0, T], T \in \mathbf{R}$.

We construct a model of log Gaussian process in three steps. In the first step, we simulate centered, homogeneous, Gaussian random process $Y(t)$. In the second step, we simulate a Poisson random variable with density $\widetilde{\mu}(\mathbf{T}) = \int\limits_0^T \exp\left\{\widetilde{Y}(t)\right\}$, where $\widetilde{Y}(t)$ is model of $Y(t)$. Consequently, we obtain a value of a $\widetilde{\nu}(\mathbf{T})$. Under realization of a process that generates density, it follows from theorem 6.1 that the log Gaussian Cox process $\widetilde{\nu}(\mathbf{T})$ is a Poisson ensemble. In the third step, we have to simulate $\widetilde{\nu}(\mathbf{T})$ independent random variables with function density $\widetilde{G}(x) = \frac{\int_0^x \exp\{\widetilde{Y}(t)\}dt}{\int_{\mathbf{T}} \exp\{\widetilde{Y}(u)\}du}$.

Since the model of continuous random variable with function density $G(x)$ is $G^{-1}(\zeta)$, where $G^{-1}(\cdot)$ is inverse to $G(\cdot)$ function, ζ is uniform on $[0, 1]$ random variable and the model of log Gaussian Cox process is constructed in such a way that the difference between process and their model was as small as possible, we should demand that $\left|G^{(-1)}(\zeta) - \widetilde{G}^{(-1)}(\zeta)\right|$ for all ζ is as small as possible. In other words, the placement of each point of the log Gaussian Cox process must differ from their small simulated analog.

DEFINITION 6.3.– *We say that the model of Cox process* $\{\nu(B), B \in \mathfrak{B}\}$, *driven by log Gaussian process* $\exp\{Y(t)\}$, *approximates that process with accuracy* α, $0 < \alpha < 1$ *and reliability* $1-\gamma$, $0<\gamma<1$, *if the following inequality holds true:*

$$\mathbf{P}\left\{\sup_{0\le\zeta\le 1}\left|G^{(-1)}(\zeta) - \widetilde{G}^{(-1)}(\zeta)\right| > \alpha\right\} < \gamma.$$

LEMMA 6.1.– Let $Y(t)$ be a homogeneous, centered, continuous in mean square, Gaussian process with spectral function $F(\lambda)$, a partition D_Λ of domain $[0,\Lambda]$, $\Lambda \in \mathbf{R}$ is such that $\lambda_{k+1} - \lambda_k = \frac{\Lambda}{N}$, $N \in \mathbf{N}$. Then, for all $p > 1$ we have

$$\left(\mathbf{E}\left|\int_0^T e^{Y(u)-Y(t)}du - \int_0^T e^{\widetilde{Y}(u)-\widetilde{Y}(t)}du\right|^p\right)^{\frac{1}{p}} \le 2^{\frac{1}{p}}\sqrt{v_1}A_{N,t}p^{\frac{1}{2}}\exp\left\{2pv_2B(0) - \frac{1}{2}\right\},$$

where the model is equal to

$$\widetilde{Y}(t) = \sum_{k=0}^{N-1}\left(\cos\lambda_k t\int_{\lambda_k}^{\lambda_{k+1}} d\xi(\lambda) + \sin\lambda_k t\int_{\lambda_k}^{\lambda_{k+1}} d\eta(\lambda)\right)$$

and

$$A_{N,t} = \left(\frac{T^2}{2} + Tt\right)\sqrt{2F(\Lambda)}\frac{\Lambda}{N} + 2T\sqrt{2(F(\infty) - F(\Lambda))},$$

where v_1 and v_2 are such positive numbers, that $\frac{1}{v_1} + \frac{1}{v_2} = 1$.

PROOF.– By virtue of the general Minkowski inequality:

$$\left(\mathbf{E}\left|\int_0^T e^{Y(u)-Y(t)}du - \int_0^T e^{\widetilde{Y}(u)-\widetilde{Y}(t)}du\right|^p\right)^{\frac{1}{p}} \le \int_0^T\left(\mathbf{E}\left|e^{Y(u)-Y(t)} - e^{\widetilde{Y}(u)-\widetilde{Y}(t)}\right|^p\right)^{\frac{1}{p}}du. \quad [6.1]$$

It follows from inequality $|e^x - e^y| \le |x-y|\,e^{\max(x,y)}$ and the *Hölder* inequality that

$$\mathbf{E}\left|e^{Y(u)-Y(t)} - e^{\widetilde{Y}(u)-\widetilde{Y}(t)}\right|^p$$

$$\leq \mathbf{E}\left(\left|Y(u) - Y(t) - \widetilde{Y}(u) + \widetilde{Y}(t)\right|^p e^{p\max\left(Y(u)-Y(t),\widetilde{Y}(u)-\widetilde{Y}(t)\right)}\right)$$

$$\leq \left(\mathbf{E}\left|Y(u) - Y(t) - \widetilde{Y}(u) + \widetilde{Y}(t)\right|^{pv_1}\right)^{\frac{1}{v_1}}$$

$$\left(\mathbf{E}e^{pv_2\max\left(Y(u)-Y(t),\widetilde{Y}(u)-\widetilde{Y}(t)\right)}\right)^{\frac{1}{v_2}}, \quad [6.2]$$

$\frac{1}{v_1} + \frac{1}{v_2} = 1$, $v_1 > 1$. Let us estimate each of the multipliers of the right-hand side of the last inequality. For the estimation of the first multiplier, we proof auxiliary relation. Let ξ be a Gaussian random variable with parameters 0 and σ^2:

$$\mathbf{E}|\xi|^p = \frac{1}{\sqrt{2\pi}\sigma}\int_{-\infty}^{\infty}|x|^p\exp\left\{-\frac{x^2}{2\sigma^2}\right\}dx = \left|\frac{x}{\sigma} = t,\, dx = \sigma dt\right|$$

$$= \sigma^p\frac{1}{\sqrt{2\pi}}\int_{-\infty}^{\infty}|t|^p\exp\left\{-\frac{t^2}{2}\right\}dt = c_p\left(\sigma^2\right)^{\frac{p}{2}},$$

$$c_p = \sqrt{\frac{2}{\pi}}\int_0^{\infty}u^p\exp\left\{-\frac{u^2}{2}\right\}du = \left|\frac{u^2}{2} = v,\, du = \frac{dv}{\sqrt{2v}}\right|$$

$$= \sqrt{\frac{2}{\pi}}\int_0^{\infty}(2v)^{\frac{p}{2}}\exp\{-v\}\frac{1}{\sqrt{2v}}dv$$

$$= \frac{2^{\frac{p}{2}}}{\sqrt{\pi}}\int_0^{\infty}v^{\frac{p+1}{2}-1}\exp\{-v\}dv = \frac{2^{\frac{p}{2}}}{\sqrt{\pi}}\Gamma\left(\frac{p+1}{2}\right).$$

It is follows from the Stirling formula that, $0 < \theta < 1$

$$\Gamma\left(\frac{p+1}{2}\right) = \sqrt{2\pi}\left(\frac{p+1}{2}\right)^{\frac{p+1}{2}-\frac{1}{2}}\exp\left\{-\frac{p+1}{2}\right\}\exp\left\{\frac{\theta}{12\frac{p+1}{2}}\right\}$$

$$\leq \sqrt{2\pi}\left(\frac{p}{2}\right)^{\frac{p}{2}}\left(1+\frac{1}{p}\right)^{\frac{p}{2}}\exp\left\{-\frac{1}{2}+\frac{1}{6(p+1)}\right\}\exp\left\{-\frac{p}{2}\right\}.$$

Finding a maximum of $\ln f(p)$, where $f(p) = \left(1+\frac{1}{p}\right)^{\frac{p}{2}}\exp\left\{-\frac{1}{2}+\frac{1}{6(p+1)}\right\}$ we obtain that under $p > 1$ $f(p) < 1$. Thereby for a Gaussian random variable, the following relations hold true:

$$\mathbf{E}|\xi|^p = c_p\left(\sigma^2\right)^{\frac{p}{2}}, \quad [6.3]$$

$$c_p \leq \sqrt{2}p^{\frac{p}{2}}\exp\left\{-\frac{p}{2}\right\}. \quad [6.4]$$

It follows from the first proved relation that:

$$\mathbf{E}\left|Y(u) - Y(t) - \widetilde{Y}(u) + \widetilde{Y}(t)\right|^{pv_1}$$

$$= c_{pv_1}\left(\mathbf{E}\left|Y(u) - Y(t) - \widetilde{Y}(u) + \widetilde{Y}(t)\right|^2\right)^{\frac{pv_1}{2}}.$$

Using the presentations of process

$$Y(t) = \sum_{k=0}^{N} \int_{\lambda_k}^{\lambda_{k+1}} \cos \lambda t d\xi(\lambda) + \sum_{k=0}^{N} \sin \lambda t d\eta(\lambda), \quad \lambda_{N+1} = +\infty,$$

and its model $\widetilde{Y}(t)$ we obtain that:

$$\begin{aligned}
&\mathbf{E}\left|Y(u) - Y(t) - \widetilde{Y}(u) + \widetilde{Y}(t)\right|^2 \\
&= \mathbf{E}\left|\sum_{k=0}^{N-1} \int_{\lambda_k}^{\lambda_{k+1}} (\cos \lambda u - \cos \lambda_k u - \cos \lambda t + \cos \lambda_k t)\, d\xi(\lambda)\right. \\
&\quad + \sum_{k=0}^{N-1} \int_{\lambda_k}^{\lambda_{k+1}} (\sin \lambda u - \sin \lambda_k u - \sin \lambda t + \sin \lambda_k t)\, d\eta(\lambda) \\
&\quad \left. + \int_{\Lambda}^{\infty} (\cos \lambda u - \cos \lambda t)\, d\xi(\lambda) + \int_{\Lambda}^{\infty} (\sin \lambda u - \sin \lambda t)\, d\eta(\lambda)\right|^2 \\
&= \sum_{k=0}^{N-1} \int_{\lambda_k}^{\lambda_{k+1}} (\cos \lambda u - \cos \lambda_k u - \cos \lambda t + \cos \lambda_k t)^2\, dF(\lambda) \\
&\quad + \sum_{k=0}^{N-1} \int_{\lambda_k}^{\lambda_{k+1}} (\sin \lambda u - \sin \lambda_k u - \sin \lambda t + \sin \lambda_k t)^2\, dF(\lambda) + \\
&\quad + \int_{\Lambda}^{\infty} (\cos \lambda u - \cos \lambda t)^2\, dF(\lambda) + \int_{\Lambda}^{\infty} (\sin \lambda u - \sin \lambda t)^2\, dF(\lambda)
\end{aligned}$$

$$\begin{aligned}
&\leq 4 \sum_{k=0}^{N-1} \int_{\lambda_k}^{\lambda_{k+1}} \left(\left|\sin \tfrac{u(\lambda+\lambda_k)}{2} \sin \tfrac{u(\lambda-\lambda_k)}{2}\right|\right. \\
&\quad \left. + \left|\sin \tfrac{t(\lambda_k+\lambda)}{2} \sin \tfrac{t(\lambda-\lambda_k)}{2}\right|\right)^2 dF(\lambda) \\
&\quad +4 \sum_{k=0}^{N-1} \int_{\lambda_k}^{\lambda_{k+1}} \left(\left|\sin \tfrac{u(\lambda-\lambda_k)}{2} \cos \tfrac{u(\lambda+\lambda_k)}{2}\right|\right. \\
&\quad \left. + \left|\sin \tfrac{t(\lambda-\lambda_k)}{2} \sin \tfrac{t(\lambda_k+\lambda)}{2}\right|\right)^2 dF(\lambda) + \\
&\quad +4 \int_{\Lambda}^{\infty} \left(\sin \tfrac{\lambda(u+t)}{2} \sin \tfrac{\lambda(u-t)}{2}\right)^2 dF(\lambda) \\
&\quad +4 \int_{\Lambda}^{\infty} \left(\sin \tfrac{\lambda(u-t)}{2} \cos \tfrac{\lambda(u+t)}{2}\right)^2 dF(\lambda) \\
&\leq 8 \sum_{k=0}^{N-1} \int_{\lambda_k}^{\lambda_{k+1}} \left(\tfrac{u(\lambda-\lambda_k)}{2} + \tfrac{t(\lambda-\lambda_k)}{2}\right)^2 dF(\lambda) + 8(F(\infty) - F(\Lambda)) \\
&\leq 2\tfrac{\Lambda^2}{N^2}(u+t)^2 F(\Lambda) + 8(F(\infty) - F(\Lambda)),
\end{aligned} \qquad [6.5]$$

since on each of the domains $[\lambda_k, \lambda_{k+1}]$, $\lambda - \lambda_k \leq \lambda_{k+1} - \lambda_k = \frac{\Lambda}{N}$. Thereby,

$$\left(\mathbf{E}\left|Y(u) - Y(t) - \widetilde{Y}(u) + \widetilde{Y}(t)\right|^{pv_1}\right)^{\frac{1}{v_1}} \leq c_{pv_1}^{\frac{1}{v_1}} A_{N,u,t}^{\frac{p}{2}}, \qquad [6.6]$$

where

$$A_{N,u,t} = 2\frac{\Lambda^2}{N^2}(u+t)^2 F(\Lambda) + 8\left(F(\infty) - F(\Lambda)\right). \qquad [6.7]$$

Let us estimate $\mathbf{E}e^{pv_2 \max\left(Y(u)-Y(t), \widetilde{Y}(u)-\widetilde{Y}(t)\right)}$. Note that for the Gaussian random variable $\xi = N(0, \sigma^2)$, the next inequality holds true $\mathbf{E}\exp\{\lambda\xi\} = \exp\left\{\frac{\lambda^2\sigma^2}{2}\right\}$. Using this, we can write the next estimation:

$$\begin{aligned}
&\mathbf{E}\exp\left\{pv_2 \max\left(Y(u) - Y(t), \widetilde{Y}(u) - \widetilde{Y}(t)\right)\right\} \\
&\leq \mathbf{E}\exp\left\{pv_2\left(Y(u) - Y(t)\right)\right\} + \mathbf{E}\exp\left\{pv_2\left(\widetilde{Y}(u) - \widetilde{Y}(t)\right)\right\} \\
&= \exp\left\{\frac{(pv_2)^2}{2}\mathbf{E}\left(Y(u) - Y(t)\right)^2\right\} + \left\{\frac{(pv_2)^2}{2}\mathbf{E}\left(\widetilde{Y}(u) - \widetilde{Y}(t)\right)^2\right\}.
\end{aligned}$$

$$\begin{aligned}
&\mathbf{E}\left(Y(u) - Y(t)\right)^2 = B(0) - 2\mathbf{E}Y(u)Y(t) + B(0) \\
&= 2B(0) - 2B(t-u) = 2\int_0^\infty \left(1 - \cos\lambda(t-u)\right) dF(\lambda) \\
&\qquad\qquad = 4\int_0^\infty \sin^2\frac{\lambda(t-u)}{2} dF(\lambda) \leq 4B(0).
\end{aligned}$$

It is obvious that $\mathbf{E}\left|\widetilde{Y}(u) - \widetilde{Y}(t)\right|^2 \leq 4B(0)$. Then,

$$\mathbf{E}e^{pv_2 \max\left(Y(u)-Y(t), \widetilde{Y}(u)-\widetilde{Y}(t)\right)} \leq 2e^{2(pv_2)^2 B(0)}.$$

Thereby, using estimation [6.6] as the last inequality, it follows from [6.2] that:

$$\mathbf{E}\left|e^{Y(u)-Y(t)} - e^{\widetilde{Y}(u)-\widetilde{Y}(t)}\right|^p \leq c_{pv_1}^{\frac{1}{v_1}} A_{N,u,t}^{\frac{p}{2}} 2^{\frac{1}{v_2}} e^{2p^2 v_2 B(0)},$$

where $A_{N,u,t}$ is defined in [6.7]. It follows from the last estimation and estimation [6.4] for c_p that:

$$\int_0^T \left(\mathbf{E}\left|e^{Y(u)-Y(t)} - e^{\widetilde{Y}(u)-\widetilde{Y}(t)}\right|^p\right)^{\frac{1}{p}} du$$

$$\leq \int_0^T \left(\sqrt{2}\,(pv_1)^{\frac{pv_1}{2}} \exp\left\{-\frac{pv_1}{2}\right\}\right)^{\frac{1}{pv_1}} A_{N,\,u,\,t}^{\frac{1}{2}} 2^{\frac{1}{pv_2}} e^{2pv_2 B(0)} du.$$

Taking the integral in the right-hand side of the above estimation and using inequality $\sqrt{x+y} \leq \sqrt{x} + \sqrt{y}$ after one of them is integrated, it is evident that after elementary transformation of the statement of lemma follows from [6.1]. □

LEMMA 6.2.– Let $Y(t)$ be a homogeneous, centered, continuous mean square Gaussian random process with spectral function $F(\lambda)$. There exists spectral moment $\int_0^\infty \lambda^{2\beta} dF(\lambda)$, $0 < \beta \leq 1$, and partition D_Λ of domain $[0, \Lambda]$, $\Lambda \in \mathbf{R}$ such that $\lambda_{k-1} - \lambda_k = \frac{\Lambda}{N}$, $N \in \mathbf{N}$, that for all $p > 1$ the following estimation holds true:

$$\left(\mathbf{E}\left|\int_0^T e^{Y(u)-Y(t+h)} du - \int_0^T e^{\widetilde{Y}(u)-\widetilde{Y}(t+h)} du\right.\right.$$

$$\left.\left. - \int_0^T e^{Y(u)-Y(t)} du + \int_0^T e^{\widetilde{Y}(u)-\widetilde{Y}(t)} du\right|^p\right)^{\frac{1}{p}} \leq h^\beta G_{N,t,p},$$

where

$$G_{N,t,p} = 2^{\frac{1}{p}}\left(T\sqrt{2P_{N,t}} + \left(2^{7-2\beta}\Lambda^{2\beta}B(0)\right)^{\frac{1}{2}} A_{N,\,t}\right) p\, e^{16pB(0)-\frac{1}{2}},$$

$$P_{N,t} = 2^{5-4\beta}\left(\left(\frac{\Lambda}{N}\right)^\beta + 2^{\beta-1} t \frac{\Lambda^{\beta+1}}{N}\right)^2 F(\Lambda) + 2^{3-2\beta}\int_\Lambda^\infty \lambda^{2\beta} dF(\lambda),$$

$$A_{N,\,t} = \left(\frac{T^2}{2} + Tt\right)\sqrt{2F(\Lambda)\frac{\Lambda}{N}} + 2T\sqrt{2\left(F(\infty) - F(\Lambda)\right)}.$$

PROOF.–

$$\left(\mathbf{E}\left|\int_0^T e^{Y(u)-Y(t+h)} du - \int_0^T e^{\widetilde{Y}(u)-\widetilde{Y}(t+h)} du\right.\right.$$
$$\left.\left. - \int_0^T e^{Y(u)-Y(t)} du + \int_0^T e^{\widetilde{Y}(u)-\widetilde{Y}(t)} du\right|^p\right)^{\frac{1}{p}}$$
$$= \left(\mathbf{E}\left|\int_0^T e^{Y(u)-Y(t)} du e^{Y(t)-Y(t+h)} - \int_0^T e^{Y(u)-Y(t)} du e^{\widetilde{Y}(t)-\widetilde{Y}(t+h)}\right.\right.$$
$$+ \int_0^T e^{Y(u)-Y(t)} du e^{\widetilde{Y}(t)-\widetilde{Y}(t+h)} - \int_0^T e^{Y(u)-Y(t)} du$$
$$\left.\left. - \int_0^T e^{\widetilde{Y}(u)-\widetilde{Y}(t)} du e^{\widetilde{Y}(t)-\widetilde{Y}(t+h)} + \int_0^T e^{\widetilde{Y}(u)-\widetilde{Y}(t)} du\right|^p\right)^{\frac{1}{p}}$$

$$= \left(\mathbf{E} \left| \int_0^T e^{Y(u)-Y(t)} du \left(e^{Y(t)-Y(t+h)} - e^{\widetilde{Y}(t)-\widetilde{Y}(t+h)} \right) \right.\right.$$

$$\left.\left. + \int_0^T e^{Y(u)-Y(t)} du \left(e^{\widetilde{Y}(t)-\widetilde{Y}(t+h)} - 1 \right) - \int_0^T e^{\widetilde{Y}(u)-\widetilde{Y}(t)} du \left(e^{\widetilde{Y}(t)-\widetilde{Y}(t+h)} - 1 \right) \right|^p \right)^{\frac{1}{p}}$$

$$\leq \left(\mathbf{E} \left| \int_0^T e^{Y(u)-Y(t)} du \left(e^{Y(t)-Y(t+h)} - e^{\widetilde{Y}(t)-\widetilde{Y}(t+h)} \right) \right|^p \right)^{\frac{1}{p}}$$

$$+ \left(\mathbf{E} \left| \left(e^{\widetilde{Y}(t)-\widetilde{Y}(t+h)} - 1 \right) \int_0^T \left(e^{Y(u)-Y(t)} - e^{\widetilde{Y}(u)-\widetilde{Y}(t)} \right) du \right|^p \right)^{\frac{1}{p}}. \quad [6.8]$$

Let us estimate separately each of the summands of the right-hand side of [6.8]. Consequently, using the general Minkowsky inequality, inequality $|e^x - e^y| \leq |x - y| \, e^{\max(x,y)}$ and *Hölder* inequality, we obtain:

$$\begin{aligned} & \left(\mathbf{E} \left| \int_0^T e^{Y(u)-Y(t)} du \left(e^{Y(t)-Y(t+h)} - e^{\widetilde{Y}(t)-\widetilde{Y}(t+h)} \right) \right|^p \right)^{\frac{1}{p}} \\ & \leq \int_0^T \left(\mathbf{E} \left| e^{Y(u)-Y(t)} du \left(e^{Y(t)-Y(t+h)} - e^{\widetilde{Y}(t)-\widetilde{Y}(t+h)} \right) \right|^p \right)^{\frac{1}{p}} du \\ & \leq \int_0^T \Big(\mathbf{E} \left| e^{Y(u)-Y(t)} \left| Y(t) - Y(t+h) - \widetilde{Y}(t) + \widetilde{Y}(t+h) \right| \right. \\ & \left. \times e^{\max\left(Y(t)-Y(t+h), \widetilde{Y}(t)-\widetilde{Y}(t+h)\right)} \right|^p \Big)^{\frac{1}{p}} du \\ & \leq \int_0^T \left(\left(\mathbf{E} \left| \Delta_1(Y) \right|^{pr_1} \right)^{\frac{1}{r_1}} \left(\mathbf{E} \left| \Delta_2(Y) \right|^{pr_2} \right)^{\frac{1}{r_2}} \right)^{\frac{1}{p}} du, \end{aligned} \quad [6.9]$$

where

$$\Delta_1(Y) = Y(t) - Y(t+h) - \widetilde{Y}(t) + \widetilde{Y}(t+h),$$

$$\Delta_2(Y) = e^{Y(u)-Y(t)} e^{\max\left(Y(t)-Y(t+h), \widetilde{Y}(t)-\widetilde{Y}(t+h)\right)},$$

$\frac{1}{r_1} + \frac{1}{r_2} = 1$, $r_1 > 1$. By virtue of [6.3]

$$\left(\mathbf{E} \left| \Delta_1(Y) \right|^{pr_1} \right) = \left(\mathbf{E} \left| \Delta_1(Y) \right|^2 \right)^{\frac{pr_1}{2}} c_{pr_1}.$$

Using representations [4.3] and [4.8] of the process $Y(t)$ and their model $\widetilde{Y}(t)$ and inequality $|\sin x| \le |x|^{\beta}$, $0 \le \beta \le 1$, we obtain:

$$
\begin{aligned}
&\mathbf{E}\left|\Delta_1(Y)\right|^2 = \mathbf{E}\left|Y(t+h) - \widetilde{Y}(t+h) - Y(t) + \widetilde{Y}(t)\right|^2 \\
&= \sum_{k=0}^{N-1} \int_{\lambda_k}^{\lambda_{k+1}} \left(\cos\lambda(t+h) - \cos\lambda_k(t+h) - \cos\lambda t + \cos\lambda_k t\right)^2 dF(\lambda) \\
&\quad + \sum_{k=0}^{N-1} \int_{\lambda_k}^{\lambda_{k+1}} \left(\sin\lambda(t+h) - \sin\lambda_k(t+h) - \sin\lambda t + \sin\lambda_k t\right)^2 dF(\lambda) \\
&\quad + \int_{\Lambda}^{\infty} \left(\cos\lambda(t+h) - \cos\lambda t\right)^2 dF(\lambda) + \int_{\Lambda}^{\infty} \left(\sin\lambda(t+h) - \sin\lambda t\right)^2 dF(\lambda) \\
&\le \sum_{k=0}^{N-1} \int_{\lambda_k}^{\lambda_{k+1}} 4\left(\left|\sin\frac{(t+h)(\lambda+\lambda_k)}{2}\right| \left|\sin\frac{(t+h)(\lambda-\lambda_k)}{2} - \sin\frac{(\lambda-\lambda_k)t}{2}\right|\right. \\
&\quad \left. + \left|\sin\frac{(\lambda-\lambda_k)t}{2}\right| \left|\sin\frac{(t+h)(\lambda+\lambda_k)}{2} - \sin\frac{(\lambda+\lambda_k)t}{2}\right|\right)^2 dF(\lambda) \\
&\quad + \sum_{k=0}^{N-1} \int_{\lambda_k}^{\lambda_{k+1}} 4\left(\left|\sin\frac{(t+h)(\lambda-\lambda_k)}{2}\right| \left|\cos\frac{(t+h)(\lambda+\lambda_k)}{2} - \cos\frac{(\lambda+\lambda_k)t}{2}\right|\right. \\
&\quad \left. + \left|\cos\frac{(\lambda+\lambda_k)t}{2}\right| \left|\sin\frac{(t+h)(\lambda-\lambda_k)}{2} - \sin\frac{(\lambda-\lambda_k)t}{2}\right|\right)^2 dF(\lambda) \\
&\quad + \int_{\Lambda}^{\infty} 4\left|\sin\frac{\lambda(2t+h)}{2}\right|^2 \left|\sin\frac{\lambda h}{2}\right|^2 dF(\lambda) + \int_{\Lambda}^{\infty} 4\left|\cos\frac{\lambda(2t+h)}{2}\right|^2 \left|\sin\frac{\lambda h}{2}\right|^2 dF(\lambda) \\
&\le \sum_{k=0}^{N-1} \int_{\lambda_k}^{\lambda_{k+1}} 4\left(\left|2\cos\frac{(2t+h)(\lambda-\lambda_k)}{4} \sin\frac{(\lambda-\lambda_k)h}{4}\right|\right. \\
&\quad \left. + \left|\sin\frac{(\lambda-\lambda_k)t}{2}\right| \left|2\cos\frac{(2t+h)(\lambda+\lambda_k)}{4} \sin\frac{(\lambda+\lambda_k)h}{4}\right|\right)^2 dF(\lambda) \\
&\quad + \sum_{k=0}^{N-1} \int_{\lambda_k}^{\lambda_{k+1}} 4\left(\left|\sin\frac{(t+h)(\lambda-\lambda_k)}{2}\right| \left|2\sin\frac{(2t+h)(\lambda+\lambda_k)}{4} \sin\frac{(\lambda+\lambda_k)h}{4}\right|\right. \\
&\quad \left. + \left|2\cos\frac{(2t+h)(\lambda-\lambda_k)}{4} \sin\frac{(\lambda-\lambda_k)h}{4}\right|\right)^2 dF(\lambda) \\
&\quad + 2^{3-2\beta} h^{2\beta} \int_{\Lambda}^{\infty} \lambda^{2\beta} dF(\lambda) \\
&\le \sum_{k=0}^{N-1} \int_{\lambda_k}^{\lambda_{k+1}} 16\left(\frac{(\lambda-\lambda_k)^{\beta} h^{\beta}}{4^{\beta}} + \frac{(\lambda-\lambda_k)t}{2} \frac{(\lambda+\lambda_k)^{\beta} h^{\beta}}{4^{\beta}}\right)^2 dF(\lambda) \\
&\quad + \sum_{k=0}^{N-1} \int_{\lambda_k}^{\lambda_{k+1}} 16\left(\frac{(t+h)(\lambda-\lambda_k)}{2} \frac{(\lambda+\lambda_k)^{\beta} h^{\beta}}{4^{\beta}} + \frac{(\lambda-\lambda_k)^{\beta} h^{\beta}}{4^{\beta}}\right)^2 dF(\lambda) \\
&\quad + 2^{3-2\beta} h^{2\beta} \int_{\Lambda}^{\infty} \lambda^{2\beta} dF(\lambda) \\
&\le 32 \sum_{k=0}^{N-1} \int_{\lambda_k}^{\lambda_{k+1}} \left(\left(\frac{\Lambda}{N}\frac{h}{4}\right)^{\beta} + \frac{\Lambda}{N}\frac{t}{2}\left(2\Lambda\frac{h}{4}\right)^{\beta}\right)^2 dF(\lambda) + 2^{3-2\beta} h^{2\beta} \int_{\Lambda}^{\infty} \lambda^{2\beta} dF(\lambda),
\end{aligned}
$$

$0 < \beta \leq 1$. Therefore,

$$\mathbf{E}\left|\Delta_1(Y)\right|^2 \leq h^{2\beta} P_{N,t}, \qquad [6.10]$$

where $P_{N,t} = 2^{5-4\beta}\left(\left(\frac{\Lambda}{N}\right)^{\beta} + 2^{\beta-1} t \frac{\Lambda^{\beta+1}}{N}\right)^2 F(\Lambda) + 2^{3-2\beta} \int\limits_{\Lambda}^{\infty} \lambda^{2\beta} dF(\lambda)$, $0 < \beta \leq 1$. Thereby:

$$\left(\mathbf{E}\left|\Delta_1(Y)\right|^{pr_1}\right)^{\frac{1}{r_1}} \leq c_{pr_1}^{\frac{1}{r_1}} h^{p\beta} P_{N,t}^{\frac{p}{2}}. \qquad [6.11]$$

Let us estimate $\mathbf{E}\left|\Delta_2(Y)\right|^{pr_2}$. It follows from the *Hölder* inequality that

$$\mathbf{E}\left|\Delta_2(Y)\right|^{pr_2} \leq \left(\mathbf{E}\exp\left\{pr_2 f_1\left(Y(u) - Y(t)\right)\right\}\right)^{\frac{1}{f_1}}$$
$$\times \left(\mathbf{E}\exp\left\{pr_2 f_2 \max\left(Y(t) - Y(t+h), \widetilde{Y}(t) - \widetilde{Y}(t+h)\right)\right\}\right)^{\frac{1}{f_2}}, \qquad [6.12]$$

$\frac{1}{f_1} + \frac{1}{f_2} = 1$, $f_1 > 1$. Since for Gaussian random variable $\xi = N(0, \sigma^2)$, the next relation holds true $\mathbf{E}\exp\{\lambda\xi\} = \exp\left\{\frac{\lambda^2\sigma^2}{2}\right\}$, and it was shown in lemma 6.1 that, $\mathbf{E}\left|Y(u) - Y(t)\right|^2 \leq 4B(0)$, $\mathbf{E}\left|\widetilde{Y}(u) - \widetilde{Y}(t)\right|^2 \leq 4B(0)$, then:

$$\mathbf{E}\exp\left\{pr_2 f_1\left(Y(u) - Y(t)\right)\right\}$$
$$= \exp\left\{\frac{\left(pr_2 f_1\right)^2}{2}\mathbf{E}\left(Y(u) - Y(u)\right)^2\right\} \leq \exp\left\{2\left(pr_2 f_1\right)^2 B(0)\right\}.$$

$$\mathbf{E}\exp\left\{pr_2 f_2 \max\left(Y(t) - Y(t+h), \widetilde{Y}(t) - \widetilde{Y}(t+h)\right)\right\}$$
$$\leq \exp\left\{\frac{(pr_2 f_2)^2}{2}\mathbf{E}\left(Y(t) - Y(t+h)\right)^2\right\}$$
$$+ \exp\left\{\frac{(pr_2 f_2)^2}{2}\mathbf{E}\left(\widetilde{Y}(t) - \widetilde{Y}(t+h)\right)^2\right\}$$
$$\leq 2\exp\left\{2\left(pr_2 f_2\right)^2 B(0)\right\}.$$

Let us put $f_1 = f_2 = 2$. It is evident by using the two last inequalities and [6.12] that

$$\left(\mathbf{E}\left|\Delta_2(Y)\right|^{pr_2}\right)^{\frac{1}{r_2}} \leq 2^{\frac{1}{2r_2}} \exp\left\{8p^2 r_2 B(0)\right\}.$$

In consideration of [6.11] and the last inequality, it follow from [6.9] that:

$$\left(\mathbf{E}\left|\int_0^T e^{Y(u)-Y(t)}du\left(e^{Y(t)-Y(t+h)}-e^{\widetilde{Y}(t)-\widetilde{Y}(t+h)}\right)\right|^p\right)^{\frac{1}{p}}$$

$$\leq \int_0^T \left(\left(\sqrt{2}\,(pr_1)^{\frac{pr_1}{2}}\exp\left\{-\frac{pr_1}{2}\right\}\right)^{\frac{1}{r_1}} h^{p\beta} P_{N,t}^{\frac{p}{2}} 2^{\frac{1}{2r_2}} e^{8p^2 r_2 B(0)}\right)^{\frac{1}{p}} du$$

$$\leq h^{\beta} 2^{\frac{1}{2p}} T\sqrt{r_1 P_{N,t}} p^{\frac{1}{2}} \exp\left\{8pr_2 B(0)-\frac{1}{2}\right\}. \quad [6.13]$$

Let us estimate the second summand of the right-hand side of [6.8].

$$\left(\mathbf{E}\left|\left(e^{\widetilde{Y}(t)-\widetilde{Y}(t+h)}-1\right)\int_0^T\left(e^{Y(u)-Y(t)}-e^{\widetilde{Y}(u)-\widetilde{Y}(t)}\right)du\right|^p\right)^{\frac{1}{p}}$$

$$\leq \int_0^T\left(\mathbf{E}\left|\left(e^{\widetilde{Y}(t)-\widetilde{Y}(t+h)}-1\right)\left(e^{Y(u)-Y(t)}-e^{\widetilde{Y}(u)-\widetilde{Y}(t)}\right)\right|^p\right)^{\frac{1}{p}} du$$

$$\leq \int_0^T\left(\left(\mathbf{E}\left|e^{\widetilde{Y}(t)-\widetilde{Y}(t+h)}-1\right|^{ps_1}\right)^{\frac{1}{s_1}}\right.$$

$$\left.\left(\mathbf{E}\left|\left(e^{Y(u)-Y(t)}-e^{\widetilde{Y}(u)-\widetilde{Y}(t)}\right)\right|^{ps_2}\right)^{\frac{1}{s_2}}\right)^{\frac{1}{p}} du, \quad [6.14]$$

$\frac{1}{s_1}+\frac{1}{s_2}=1$. Using inequality $|\exp(x)-1|\leq|x|\exp\{|x|\}$, and after the *Hölder* inequality,

$$\mathbf{E}\left|e^{\widetilde{Y}(t)-\widetilde{Y}(t+h)}-1\right|^{ps_1}\leq\mathbf{E}\left(\left|\widetilde{Y}(t)-\widetilde{Y}(t+h)\right|e^{\left|\widetilde{Y}(t)-\widetilde{Y}(t+h)\right|}\right)^{ps_1}$$

$$\leq\mathbf{E}\left(\left|\widetilde{Y}(t)-\widetilde{Y}(t+h)\right|^{ps_1 l_1}\right)^{\frac{1}{l_1}}\left(\mathbf{E}e^{ps_1 l_2\left|\widetilde{Y}(t)-\widetilde{Y}(t+h)\right|}\right)^{\frac{1}{l_2}},$$

$\frac{1}{l_1}+\frac{1}{l_2}=1$. Since

$$\mathbf{E}\left|\Delta_2(Y)\right|^2=\mathbf{E}\left|\widetilde{Y}(t+h)-\widetilde{Y}(t)\right|^2$$

$$=\mathbf{E}\left(\sum_{k=0}^{N-1}\int_{\lambda_k}^{\lambda_{k+1}}\cos\lambda_k(t+h)\,d\xi(\lambda)+\sum_{k=0}^{N-1}\int_{\lambda_k}^{\lambda_{k+1}}\sin\lambda_k(t+h)\,d\eta(\lambda)\right.$$

$$\left.-\sum_{k=0}^{N-1}\int_{\lambda_k}^{\lambda_{k+1}}\cos\lambda_k t\,d\xi(\lambda)-\sum_{k=0}^{N-1}\int_{\lambda_k}^{\lambda_{k+1}}\sin\lambda_k t d\eta(\lambda)\right)^2$$

$$= \mathbf{E}\left(\sum_{k=0}^{N-1}\int\limits_{\lambda_k}^{\lambda_{k+1}} -2\sin\frac{\lambda_k(2t+h)}{2}\sin\frac{\lambda_k h}{2}d\xi(\lambda)+\right.$$

$$\left.+\sum_{k=0}^{N-1}\int\limits_{\lambda_k}^{\lambda_{k+1}} 2\sin\frac{\lambda_k h}{2}\cos\frac{\lambda_k(2t+h)}{2}d\eta(\lambda)\right)^2$$

$$= \sum_{k=0}^{N-1}\int\limits_{\lambda_k}^{\lambda_{k+1}} 4\sin^2\frac{\lambda_k(2t+h)}{2}\sin^2\frac{\lambda_k h}{2}dF(\lambda)$$

$$+\sum_{k=0}^{N-1}\int\limits_{\lambda_k}^{\lambda_{k+1}} 4\sin^2\frac{\lambda_k h}{2}\cos^2\frac{\lambda_k(2t+h)}{2}dF(\lambda)$$

$$\le 8\sum_{k=0}^{N-1}\int\limits_{\lambda_k}^{\lambda_{k+1}} \sin^2\frac{\lambda_k h}{2}dF(\lambda)$$

$$\le 8\sum_{k=0}^{N-1}\int\limits_{\lambda_k}^{\lambda_{k+1}} \left(\frac{\lambda_k h}{2}\right)^{2\beta}dF(\lambda) \le 2^{3-2\beta}\Lambda^{2\beta}h^{2\beta}F(\Lambda), \quad 0<\beta\le 1,$$

that

$$\left(\mathbf{E}\left|\widetilde{Y}(t)-\widetilde{Y}(t+h)\right|^{ps_1l_1}\right)^{\frac{1}{l_1}} \le c_{ps_1l_1}^{\frac{1}{l_1}}\left(2^{3-2\beta}\Lambda^{2^\beta}h^{2\beta}B(0)\right)^{\frac{ps_1}{2}}.$$

In addition,

$$\left(\mathbf{E}e^{ps_1l_2\left|\widetilde{Y}(t)-\widetilde{Y}(t+h)\right|}\right)^{\frac{1}{l_2}} \le 2^{\frac{1}{l_2}}e^{2(ps_1)^2l_2B(0)}.$$

Thereby

$$\mathbf{E}\left|e^{\widetilde{Y}(t)-\widetilde{Y}(t+h)}-1\right|^{ps_1} \le c_{ps_1l_1}^{\frac{1}{l_1}}\left(2^{3-2\beta}\Lambda^{2^\beta}h^{2\beta}B(0)\right)^{\frac{ps_1}{2}}2^{\frac{1}{l_2}}e^{2(ps_1)^2l_2B(0)}. \quad [6.15]$$

Let us estimate $\mathbf{E}\left|e^{Y(u)-Y(t)}-e^{\widetilde{Y}(u)-\widetilde{Y}(t)}\right|^{ps_2}$:

$$\mathbf{E}\left|e^{Y(u)-Y(t)}-e^{\widetilde{Y}(u)-\widetilde{Y}(t)}\right|^{ps_2}$$

$$\le \mathbf{E}\left(\left|Y(u)-Y(t)-\widetilde{Y}(u)+\widetilde{Y}(t)\right|e^{\max\left(Y(u)-Y(t),\widetilde{Y}(u)-\widetilde{Y}(t)\right)}\right)^{ps_2}$$

$$\leq \left(\mathbf{E}\left|Y(u)-Y(t)-\widetilde{Y}(u)+\widetilde{Y}(t)\right|^{ps_2m_1}\right)^{\frac{1}{m_1}}$$

$$\left(\mathbf{E}e^{ps_2m_2\max\left(Y(u)-Y(t),\widetilde{Y}(u)-\widetilde{Y}(t)\right)}\right)^{\frac{1}{m_2}},$$

$\frac{1}{m_1}+\frac{1}{m_2}=1$. Since it was proved in lemma 6.1,

$$\mathbf{E}\left|Y(u)-Y(t)-\widetilde{Y}(u)+\widetilde{Y}(t)\right|^2 \leq A_{N,u,t},$$

$$A_{N,u,t}=2\frac{\Lambda^2}{N^2}(u+t)^2F(\Lambda)+8\left(F(\infty)-F(\Lambda)\right),$$

that

$$\left(\mathbf{E}\left|Y(u)-Y(t)-\widetilde{Y}(u)+\widetilde{Y}(t)\right|^{ps_2m_1}\right)^{\frac{1}{m_1}} \leq c_{ps_2m_1}^{\frac{1}{m_1}}A_{N,u,t}^{\frac{ps_2}{2}}.$$

$$\left(\mathbf{E}e^{ps_2m_2\max\left(Y(u)-Y(t),\widetilde{Y}(u)-\widetilde{Y}(t)\right)}\right)^{\frac{1}{m_2}} \leq 2^{\frac{1}{m_2}}e^{2(ps_2)^2m_2B(0)}.$$

It follows from the two last inequalities that

$$\mathbf{E}\left|e^{Y(u)-Y(t)}-e^{\widetilde{Y}(u)-\widetilde{Y}(t)}\right|^{ps_2} \leq c_{ps_2m_1}^{\frac{1}{m_1}}A_{N,u,t}^{\frac{ps_2}{2}}2^{\frac{1}{m_2}}e^{2(ps_2)^2m_2B(0)}$$

Using [6.15] and the preceding inequality, put $s_1=s_2=l_1=l_2=m_1=m_2=2$. It follows from [6.14] that:

$$\left(\mathbf{E}\left|\left(e^{\widetilde{Y}(t)-\widetilde{Y}(t+h)}-1\right)\int_0^T\left(e^{Y(u)-Y(t)}-e^{\widetilde{Y}(u)-\widetilde{Y}(t)}\right)du\right|^p\right)^{\frac{1}{p}}$$

$$\leq \int_0^T\left(c_{ps_1l_1}^{\frac{1}{l_1s_1}}\left(2^{3-2\beta}\Lambda^{2\beta}h^{2\beta}B(0)\right)^{\frac{p}{2}}2^{\frac{1}{l_2s_1}}e^{2p^2s_1l_2B(0)}\right.$$

$$\left.c_{ps_2m_1}^{\frac{1}{m_1s_2}}A_{N,u,t}^{\frac{p}{2}}2^{\frac{1}{m_2s_2}}e^{2p^2s_2m_2B(0)}\right)^{\frac{1}{p}}du$$

$$\leq \int_0^T\left(\left(\sqrt{2}\,(ps_1l_1)^{\frac{ps_1l_1}{2}}e^{-\frac{ps_1l_1}{2}}\right)^{\frac{1}{l_1s_1}}\left(2^{3-2\beta}\Lambda^{2\beta}h^{2\beta}B(0)\right)^{\frac{p}{2}}2^{\frac{1}{l_2s_1}}e^{2p^2s_1l_2B(0)}\right.$$

$$\left.\times\left(\sqrt{2}\,(ps_2m_1)^{\frac{ps_2m_1}{2}}e^{-\frac{ps_2m_1}{2}}\right)^{\frac{1}{m_1s_2}}A_{N,u,t}^{\frac{p}{2}}2^{\frac{1}{m_2s_2}}e^{2p^2s_2m_2B(0)}\right)^{\frac{1}{p}}du$$

$$\leq h^\beta\int_0^T\left(2^{\frac{3}{4p}}4A_{N,u,t}^{\frac{1}{2}}\left(2^{3-2\beta}\Lambda^{2\beta}B(0)\right)^{\frac{1}{2}}p\,e^{16pB(0)-1}\right)du.$$

Estimating the expression under integral and taking the integral, we obtain that:

$$\left(\mathbf{E}\left|\left(e^{\widetilde{Y}(t)-\widetilde{Y}(t+h)}-1\right)\int_0^T\left(e^{Y(u)-Y(t)}-e^{\widetilde{Y}(u)-\widetilde{Y}(t)}\right)du\right|^p\right)^{\frac{1}{p}}$$
$$\leq h^{\beta}2^{\frac{3}{4p}}4A_{N,t}\left(2^{3-2\beta}\Lambda^{2^{\beta}}B(0)\right)^{\frac{1}{2}}pe^{16pB(0)-1},$$

where $A_{N,t}=\left(\frac{T^2}{2}+Tt\right)\sqrt{2F(\Lambda)\frac{\Lambda}{N}}+2T\sqrt{2(F(\infty)-F(\Lambda))}$. Set $r_1=r_2=1$. The assertion of lemma follows from [6.8] if we take into account [6.13] and the last inequality. □

LEMMA 6.3.– Let $Y(t)$ be a homogeneous, centered, separable, continuous in mean square, Gaussian process with spectral function $F(\lambda)$. Let there exist a spectral moment $\int\limits_0^\infty\lambda^{2\beta}dF(\lambda)$, $0<\beta\leq 1$, and the partition of domain $D_\Lambda\ [0,\Lambda],\ \Lambda\in\mathbf{R}$ is such that $\lambda_{k-1}-\lambda_k=\frac{\Lambda}{N}$, $N\in\mathbf{N}$. If $S_N<\alpha\exp\left\{-\frac{64B(0)}{\beta}\right\}$, then:

$$\mathbf{P}\left\{\sup_{0\leq\zeta\leq 1}\left|G^{(-1)}(\zeta)-\widetilde{G}^{(-1)}(\zeta)\right|>\alpha\right\}$$
$$\leq\left(\left(\frac{\ln\frac{\alpha}{S_N}}{32B(0)}\right)^{\frac{\ln\frac{\alpha}{S_N}}{64B(0)}}+\left(\frac{3\beta}{2}\right)^{\frac{1}{\beta}}T\left(\frac{\ln\frac{\alpha}{S_N}}{32B(0)}\right)^{\frac{\ln\frac{\alpha}{S_N}}{32B(0)}+\frac{1}{\beta}}\right)\exp\left\{-\frac{\ln^2\frac{\alpha}{S_N}}{64B(0)}\right\},\qquad [6.16]$$

where:

$$S_N=\max\{S_{N,1},S_{N,2}\}\quad S_{N,1}=\frac{4\sqrt{2}A_{N,0}}{\sqrt{7e}},$$

$$S_{N,2}=\frac{6\left(T\sqrt{2P_N}+\left(2^{7-2\beta}\Lambda^{2\beta}B(0)\right)^{\frac{1}{2}}A_{N,T}\right)}{\sqrt{e}},$$

$$A_{N,0}=\frac{T^2}{2}\sqrt{2F(\Lambda)\frac{\Lambda}{N}}+2T\sqrt{2(F(\infty)-F(\Lambda))},$$

$$A_{N,T}=\frac{3T^2}{2}\sqrt{2F(\Lambda)\frac{\Lambda}{N}}+2T\sqrt{2(F(\infty)-F(\Lambda))},$$

$$P_N=2^{5-4\beta}\left(\left(\frac{\Lambda}{N}\right)^{\beta}+2^{\beta-1}T\frac{\Lambda^{\beta+1}}{N}\right)^2F(\Lambda)+2^{3-2\beta}\int\limits_\Lambda^\infty\lambda^{2\beta}dF(\lambda).$$

PROOF.– Note that $\left(G^{-1}(\zeta)\right)' = \frac{1}{G'(G^{-1}(\zeta))} = \int\limits_0^T e^{Y(u)}du e^{-Y\left(G^{-1}(\zeta)\right)}$. It follows from the Lagrange formula that:

$$\sup_{0\le\zeta\le1}\left|G^{(-1)}(\zeta) - \widetilde{G}^{(-1)}(\zeta)\right|$$

$$= \sup_{0\le\zeta\le1}\left|G^{-1}(0) - \widetilde{G}^{-1}(0) + \zeta\left(\left(G^{-1}\left(\hat{\zeta}\right)\right)' - \left(\widetilde{G}^{-1}\left(\hat{\zeta}\right)\right)'\right)\right|$$

$$\le \sup_{0\le\hat{\zeta}\le1}\left|\left(G^{(-1)}\left(\hat{\zeta}\right)\right)' - \left(\widetilde{G}^{(-1)}\left(\hat{\zeta}\right)\right)'\right|$$

$$= \sup_{0\le t\le T}\left|\int\limits_0^T e^{Y(u)-Y(t)}du - \int\limits_0^T e^{\widetilde{Y}(u)-\widetilde{Y}(t)}du\right|.$$

Thereby,

$$\mathbf{P}\left\{\sup_{0\le\zeta\le1}\left|G^{(-1)}(\zeta) - \widetilde{G}^{(-1)}(\zeta)\right| > \alpha\right\}$$

$$\le \mathbf{P}\left\{\sup_{0\le t\le T}\left|\int\limits_0^T e^{Y(u)-Y(t)}du - \int\limits_0^T e^{\widetilde{Y}(u)-\widetilde{Y}(t)}du\right| > \alpha\right\}.$$

If follows from lemma 6.1 that

$$\inf_{0\le t\le T}\left(\mathbf{E}\left|\int\limits_0^T e^{Y(u)-Y(t)}du - \int\limits_0^T e^{\widetilde{Y}(u)-\widetilde{Y}(t)}du\right|^p\right)^{\frac{1}{p}}$$

$$\le 2^{\frac{1}{p}}\sqrt{\upsilon_1}A_{N,0}\,p^{\frac{1}{2}}\exp\left\{2p\upsilon_2 B(0) - \frac{1}{2}\right\}, \qquad [6.17]$$

where $A_{N,0} = A_{N,t}\big|_{t=0}$. Using by 6.2, we estimate the entropy integral in corollary 1.16.

$$\int\limits_0^{\theta\varepsilon_0}\left(\frac{T}{\sigma^{(-1)}(\varepsilon)}\right)^{\frac{1}{p}}d\varepsilon \le \int\limits_0^{\theta\varepsilon_0}\left(T\left(\frac{G_{N,p}}{\varepsilon}\right)^{\frac{1}{\beta}}\right)^{\frac{1}{p}}d\varepsilon = \frac{\theta^{1-\frac{1}{p\beta}}T^{\frac{1}{p}}G_{N,p}}{1-\frac{1}{p\beta}},$$

$1-\frac{1}{p\beta}>0$, $G_{N,p}=G_{N,t,p}|_{t=T}$. Taking into account that function $f(\theta)=\frac{1}{\theta^{\frac{1}{p\beta}}(1-\theta)}$ takes a minimum value in the point $\theta_0=\frac{1}{p\beta+1}$, and also $\theta_0<\frac{\sigma\left(\frac{T}{2}\right)}{\varepsilon_0}$, after elementary manipulations we obtain:

$$\inf_{0<\theta<1}\frac{1}{\theta^{\frac{1}{p\beta}}(1-\theta)}\frac{T^{\beta}G_{N,p}}{1-\frac{1}{p\beta}}\leq T^{\frac{1}{p}}G_{N,p}\frac{(p\beta+1)^{1+\frac{1}{p\beta}}}{p\beta-1}.$$

In consideration of [6.17], we obtained estimation and inequality $(a+b)^p\leq 2^{p-1}(a^p+b^p)$; based on corollary 1.16, we have:

$$\mathbf{P}\left\{\sup_{0\leq t\leq T}\left|\int_0^T e^{Y(u)-Y(t)}du-\int_0^T e^{\widetilde{Y}(u)-\widetilde{Y}(t)}du\right|>\alpha\right\}$$
$$\leq\frac{2^{p-1}\left(2^{\frac{1}{p}}\sqrt{v_1}A_{N,0}p^{\frac{1}{2}}\exp\left\{2pv_2B(0)-\frac{1}{2}\right\}\right)^p}{\alpha^p}+\frac{2^{p-1}\left(T^{\frac{1}{p}}G_{N,p}\frac{(p\beta+1)^{1+\frac{1}{p\beta}}}{(p\beta-1)}\right)^p}{\alpha^p}.$$

By describing $G_{N,p}$, and taking into consideration that under $p\beta\geq 2$ $\left(\frac{p}{p\beta-1}\right)^p\leq\frac{2^p}{\beta^p}$ and $(p\beta+1)^{p+\frac{1}{\beta}}\leq(p\beta)^{p+\frac{1}{\beta}}\left(\frac{3}{2}\right)^{p+\frac{1}{\beta}}$ and by setting $v_2=8$, after elementary manipulations we obtain:

$$\mathbf{P}\left\{\sup_{0\leq t\leq T}\left|\int_0^T e^{Y(u)-Y(t)}du-\int_0^T e^{\widetilde{Y}(u)-\widetilde{Y}(t)}du\right|>\alpha\right\}$$
$$\leq\frac{S_{N,1}^p p^{\frac{p}{2}}\exp\left\{2p^2v_2B(0)\right\}}{\alpha^p}+\frac{TS_{N,2}^p\left(\frac{3\beta}{2}\right)^{\frac{1}{\beta}}p^{p+\frac{1}{\beta}}e^{16p^2B(0)}}{\alpha^p}$$
$$\leq\frac{S_N^p\left(p^{\frac{p}{2}}+\left(\frac{3\beta}{2}\right)^{\frac{1}{\beta}}Tp^{p+\frac{1}{\beta}}\right)\exp\left\{16p^2B(0)\right\}}{\alpha^p},$$

where $S_N=\max\left\{S_{N,1},S_{N,2}\right\}$, $S_{N,1}=\frac{4\sqrt{2}A_{N,0}}{\sqrt{7e}}$, $S_{N,2}=\frac{6\left(\left(2^{7-2\beta}\Lambda^{2\beta}B(0)\right)^{\frac{1}{2}}A_{N,T}\right)}{\sqrt{e}}+\frac{6\left(T\sqrt{2P_N}\right)}{\sqrt{e}}$, $P_N=P_{N,t}|_{t=T}$, $A_N=A_{N,t}|_{t=T}$. We calculate the value of the right-hand side of the last estimation in the point $p_0=\frac{\ln\frac{\alpha}{S_N}}{32B(0)}$. It is the point of minimum of function $\frac{S_N^p\left(p^{\frac{p}{2}}+\left(\frac{3\beta}{2}\right)^{\frac{1}{\beta}}Tp^{p+\frac{1}{\beta}}\right)\exp\left\{16p^2B(0)\right\}}{\alpha^p}$. The condition $p\beta>2$ to secure that $1-\frac{1}{p\beta}>0$ holds true. By virtue of corollary 1.16, we obtain [6.16] and the lemma is proved. □

THEOREM 6.2.– Let $Y(t)$ be a homogeneous, centered, separable, continuous in mean square, Gaussian process with spectral function $F(\lambda)$. Let there exist a spectral moment $\int\limits_0^\infty \lambda^{2\beta} dF(\lambda)$, $0 < \beta \leq 1$, a partition D_Λ of domain $[0, \Lambda]$, $\Lambda \in \mathbf{R}$ is such that $\lambda_{k-1} - \lambda_k = \frac{\Lambda}{N}$, $N \in \mathbf{N}$, then the model of Cox process $\{\widetilde{\nu}(B), B \in \mathfrak{B}\}$, directed by the log Gaussian process $\exp\left\{\widetilde{Y}(t)\right\}$, approximates them with accuracy α and reliability $1-\gamma$ if the following conditions hold true:

$$S_N < \alpha \exp\left\{-\frac{64B(0)}{\beta}\right\},$$

$$\left(\left(\frac{\ln\frac{\alpha}{S_N}}{32B(0)}\right)^{\frac{\ln\frac{\alpha}{S_N}}{64B(0)}} + \left(\frac{3\beta}{2}\right)^{\frac{1}{\beta}} T\left(\frac{\ln\frac{\alpha}{S_N}}{32B(0)}\right)^{\frac{\ln\frac{\alpha}{S_N}}{32B(0)}+\frac{1}{\beta}}\right) \exp\left\{-\frac{\ln^2\frac{\alpha}{S_N}}{64B(0)}\right\} \leq \gamma,$$

where S_N is defined in assertion of lemma 6.3.

PROOF.– The assertion of the theorem is a corollary of definition 6.3 and lemma 6.3. □

6.3. Simplified method of simulating log Gaussian Cox processes

The method of simulation considered in the previous section is complicated to realize on a computer. If a model of log Gaussian Cox process is obtained only under a large value of N, then simulation of random variables with cumulative distribution function $\widetilde{G}(x) = \frac{\int_0^x \exp\{\widetilde{Y}(t)\}dt}{\int_T \exp\{\widetilde{Y}(u)\}du}$ is a long time process. That is why we propose in this section another method of simulation. This method does not demand to simulate random variables with the above distribution function.

The model of log Gaussian Cox process is constructed in two steps. First, we simulate Gaussian process $Y(t)$. We consider some partition of domain $\mathbf{T} = [0, T]$ on k domains by length $d = \frac{T}{k} : 0 = t_0 < t_1 < \ldots < t_k = T$, $t_{i+1} - t_i = d$, $i = \overline{0, k-1}$. Let $B_i = [t_i, t_{i+1}]$, $\widetilde{\mu}(B_i) = \int\limits_{B_i} \exp\left\{\widetilde{Y}(t)\right\} dt$ and $\widetilde{Y}(t)$ is a model of $Y(t)$. Second, for each $i = \overline{0, k-1}$, we construct a model of log Gaussian Cox process $\widetilde{\nu}(B_i)$, that is a model of Poisson random variables with mean $\widetilde{\mu}(B_i)$. Since $\widetilde{\nu}(B_i)$ is a number of points of the model that belong to domain B_i, we allocate these points in B_i by all means. If $\widetilde{\nu}(B_i) = 1$, we place this point in the center of the domain.

It is evident that the model $\widetilde{\nu}(B_i)$ is admissible if the conditional probabilities $p_{kY}(B_i) = \mathbf{P}\left\{\nu(B_i) = k/Y(t), t \in \mathbf{T}\right\}$ and $\widetilde{p}_{kY}(B_i) = \mathbf{P}\left\{\widetilde{\nu}(B_i) = k/\widetilde{Y}(t),\right.$

$t \in \mathbf{T}\}$ differ little, and the probability of the event that the number of points $\nu(B_i)$ (respectively, $\widetilde{\nu}(B_i)$) is more than one is also small. Therefore, the problem of simulation of the log Gaussian Cox process consists of two problems. The first is the problem of the choice of domain $\mathbf{T}$ partitioning, and the second is a construction of the model of the field $Y(t)$.

Partitioning of the domain $\mathbf{T}$ (that is d or k) is chosen in such way that the following inequality holds true:

$$\mathbf{P}\{\nu(B_i) > 1\} < \delta, \qquad [6.18]$$

where δ is given and small (for example $\delta = 0,01$).

THEOREM 6.3.– Let $\{\nu(B), B \subset \mathfrak{B}\}$ be a Cox process driven by log Gaussian, homogeneous process $\exp\{Y(t)\}$. The inequality [6.18] holds true if we set

$$d = \frac{T}{k} \leq [2\delta \exp\{-2B(0)\}]^{\frac{1}{2}}. \qquad [6.19]$$

PROOF.– Since

$$\mathbf{P}\{\nu(B_i) > 1\} = \mathbf{E}(1 - \exp\{-\mu(B_i)\} - \mu(B_i)\exp\{-\mu(B_i)\}),$$

it is sufficient to choose such a partitioning that the following inequality holds true:

$$\mathbf{E}(1 - \exp\{-\mu(B_i)\} - \mu(B_i)\exp\{-\mu(B_i)\}) < \delta.$$

By virtue of $1 - \exp\{-x\}(1+x) \leq \frac{x^2}{2}$ as $x > 0$, the preceding inequality holds true if

$$\mathbf{E}[\mu(B_i)]^2 < 2\delta. \qquad [6.20]$$

For $\xi = N(0, \sigma^2)$ we have $\mathbf{E}\exp\{\lambda\xi\} = \exp\left\{\frac{\lambda^2\sigma^2}{2}\right\}$ and:

$$\begin{aligned}
\mathbf{E}[\mu(B_i)]^2 &= \mathbf{E}\left[\int_{B_i} \exp\{Y(t)\}\,dt\right]^2 \\
&= \mathbf{E}\int_{B_i} \exp\{Y(t)\}\,dt \int_{B_i} \exp\{Y(s)\}\,ds \\
&= \iint_{B_i \times B_i} E\exp\{Y(t) + Y(s)\}\,dtds \\
&= \iint_{B_i \times B_i} \exp\left\{\frac{\mathbf{E}(Y(t)+Y(s))^2}{2}\right\} dtds \\
&= \iint_{B_i \times B_i} \exp\left\{\frac{\mathbf{E}(Y(t))^2}{2} + \mathbf{E}Y(t)Y(s) + \frac{\mathbf{E}(Y(s))^2}{2}\right\} dtds
\end{aligned}$$

$$
\begin{aligned}
&= \iint\limits_{B_i \times B_i} \exp\{B(0) + B(t-s)\}\, dtds \\
&= \exp\{B(0)\} \iint\limits_{B_i \times B_i} \exp\{B(t-s)\}\, dtds \\
&\le d^2 \exp\{2B(0)\}.
\end{aligned}
$$

The assertion of the theorem follows from the last inequality and [6.20]. □

We want to construct such a model of the log Gaussian Cox process $Y(t)$ that the conditional probabilities p_{kY} and $\widetilde{p}_{kY}$ differ little with probability close to one for all $i = \overline{0,\, k-1}$.

DEFINITION 6.4.– *The model of Cox process* $\{\widetilde{\nu}(B),\ B \in \mathfrak{B}\}$ *driven by the log Gaussian process* $\exp\left\{\widetilde{Y}(t)\right\}$ *approximates the process with accuracy* α, $0 < \alpha < 1$ *and reliability* $1-\gamma$, $0 < \gamma < 1$, *if the following inequality holds true:*

$$
\mathbf{P}\left\{ \max_{B_i \in \mathfrak{B}\ i=\overline{0,\, k-1}} \left| p_{kY}(B_i) - \widetilde{p}_{kY}(B_i) \right| > \alpha \right\} < \gamma.
$$

LEMMA 6.4.– Let $Y(t)$ be a homogeneous, centered, continuous in mean square Gaussian process with spectral function $F(\lambda)$, the partition D_Λ of domain $[0, \Lambda]$, $\Lambda \in \mathbf{R}$ is such that $\lambda_{k-1} - \lambda_k = \frac{\Lambda}{N}$, $N \in \mathbf{N}$, then for all $p > 1$, the next estimation holds true

$$
\left(\mathbf{E}\left|\exp\{Y(t)\} - \exp\left\{\widetilde{Y}(t)\right\}\right|^p\right)^{\frac{1}{p}} \le 2^{\frac{1}{p}} \widehat{A}_{N,t}^{\frac{1}{2}} (pv_1)^{\frac{1}{2}} \exp\left\{\frac{pv_2}{2} B(0) - \frac{1}{2}\right\},
$$

where

$$
\widehat{A}_{N,t} = B(0) - F(\Lambda) + 2^{2-2b} t^{2b} \left(\frac{\Lambda}{N}\right)^{2b} F(\Lambda),
$$

$b \in [0, 1]$, v_1 and v_2 are such numbers that $\frac{1}{v_1} + \frac{1}{v_2} = 1$.

PROOF.– It follows from $|\exp\{x\} - \exp\{y\}| \le |x-y| \exp\{\max(x, y)\}$ and the *Hölder* inequality that:

$$
\begin{aligned}
&\left(\mathbf{E}\left|\exp\{Y(t)\} - \exp\left\{\widetilde{Y}(t)\right\}\right|^p\right)^{\frac{1}{p}} \\
&\le \left(\mathbf{E}\left|Y(t) - \widetilde{Y}(t)\right|^p \exp\left\{p \max\left(Y(t), \widetilde{Y}(t)\right)\right\}\right)^{\frac{1}{p}} \\
&\le \left(\mathbf{E}\left|Y(t) - \widetilde{Y}(t)\right|^{pv_1}\right)^{\frac{1}{pv_1}} \left(\mathbf{E}\exp\left\{pv_2 \max\left(Y(t), \widetilde{Y}(t)\right)\right\}\right)^{\frac{1}{pv_2}},
\end{aligned}
\qquad [6.21]
$$

$\frac{1}{v_1} + \frac{1}{v_2} = 1$. By virtue of [6.3],

$$\mathbf{E}\left|Y(t) - \widetilde{Y}(t)\right|^{pv_1} = \left(\mathbf{E}\left|Y(t) - \widetilde{Y}(t)\right|^2\right)^{\frac{pv_1}{2}} c_{pv_1},$$

where the value $c.$ is from [6.4]. Since for Gaussian, homogeneous, centered random processes, we have $\mathbf{E}(Y(t))^2 = B(0)$, $\mathbf{E}\left(\widetilde{Y}(t)\right)^2 = F(\Lambda)$, such that:

$$\mathbf{E}\left|Y(t) - \widetilde{Y}(t)\right|^2 = B(0) + F(\Lambda) - 2EY(t)\widetilde{Y}(t).$$

$$\begin{aligned}
\mathbf{E}Y(t)\widetilde{Y}(t) = \mathbf{E}&\left(\sum_{k=0}^{N-1}\int_{\lambda_k}^{\lambda_{k+1}} \cos\lambda t\, d\xi(\lambda) + \sum_{k=0}^{N-1}\int_{\lambda_k}^{\lambda_{k+1}} \sin\lambda t\, d\eta(\lambda)\right. \\
&\left. + \int_{\Lambda}^{\infty} \cos\lambda t\, d\xi(\lambda) + \int_{\Lambda}^{\infty} \sin\lambda t\, d\eta(\lambda)\right) \\
&\times\left(\sum_{k=0}^{N-1}\int_{\lambda_k}^{\lambda_{k+1}} \cos\lambda_k t\, d\xi(\lambda) + \sum_{k=0}^{N-1}\int_{\lambda_k}^{\lambda_{k+1}} \sin\lambda_k t\, d\eta(\lambda)\right) \\
&= \sum_{k=0}^{N-1}\int_{\lambda_k}^{\lambda_{k+1}} \cos\lambda t\cos\lambda_k t\, dF(\lambda) + \sum_{k=0}^{N-1}\int_{\lambda_k}^{\lambda_{k+1}} \sin\lambda t\sin\lambda_k t\, dF(\lambda) \\
&= \sum_{k=0}^{N-1}\int_{\lambda_k}^{\lambda_{k+1}} \cos\, t(\lambda - \lambda_k)\, dF(\lambda).
\end{aligned}$$

That is why

$$\begin{aligned}
&\mathbf{E}\left|Y(t) - \widetilde{Y}(t)\right|^2 \\
&= B(0) - F(\Lambda) + 2F(\Lambda) - 2\sum_{k=0}^{N-1}\int_{\lambda_k}^{\lambda_{k+1}} \cos\, t(\lambda - \lambda_k)\, dF(\lambda) \\
&= B(0) - F(\Lambda) + 2\sum_{k=0}^{N-1}\int_{\lambda_k}^{\lambda_{k+1}} 2\sin^2\left(\frac{t(\lambda-\lambda_k)}{2}\right) dF(\lambda) \\
&\le B(0) - F(\Lambda) + 4\sum_{k=0}^{N-1}\int_{\lambda_k}^{\lambda_{k+1}} \frac{t^{2b}(\lambda-\lambda_k)^{2b}}{2^{2b}}\, dF(\lambda), \quad 0 < b < 1.
\end{aligned}$$

Since $\lambda - \lambda_k \le \lambda_{k+1} - \lambda_k = \frac{\Lambda}{N}$, $k = \overline{0,\ N-1}$, then

$$\begin{aligned}
&\mathbf{E}\left|Y(t) - \widetilde{Y}(t)\right|^2 \le \widehat{A}_{N,t}, \\
&\widehat{A}_{N,t} = B(0) - F(\Lambda) + 2^{2-2b}t^{2b}\left(\tfrac{\Lambda}{N}\right)^{2b} F(\Lambda), \quad b \in [0,1].
\end{aligned} \qquad [6.22]$$

Thereby:

$$\left(\mathbf{E}\left|Y(t) - \widetilde{Y}(t)\right|^{pv_1}\right)^{\frac{1}{pv_1}} \le \widehat{A}_{N,t}^{\frac{1}{2}} c_{pv_1}^{\frac{1}{pv_1}}. \qquad [6.23]$$

Let us estimate $\exp\left\{p v_2 \max\left(Y(t), \widetilde{Y}(t)\right)\right\}$. For $\xi = N(0, \sigma^2)$, we have

$$\mathbf{E}\exp\{\lambda\xi\} = \exp\left\{\frac{\lambda^2\sigma^2}{2}\right\},$$

thus:

$$\mathbf{E}\exp\left\{p v_2 \max\left(Y(t), \widetilde{Y}(t)\right)\right\} \leq \mathbf{E}\exp\{p v_2 Y(t)\} + \mathbf{E}\exp\left\{p v_2 \widetilde{Y}(t)\right\}$$
$$= \exp\left\{\frac{(p v_2)^2}{2} B(0)\right\} + \exp\left\{\frac{(p v_2)^2}{2} F(\Lambda)\right\} \leq 2\exp\left\{\frac{(p v_2)^2}{2} B(0)\right\}.$$

Thereby by using the last inequality, [6.23] and estimation [6.4] for c_p, the assertion of the lemma follows from [6.21]. □

LEMMA 6.5.– Let $Y(t)$ be a homogeneous, centered, continuous in mean square Gaussian process with spectral function $F(\lambda)$. There exists spectral moment $\int_0^\infty \lambda^{2\beta} dF(\lambda)$, $0 < \beta \leq 1$, the partition D_Λ of domain $[0, \Lambda]$, $\Lambda \in \mathbf{R}$ is such that $\lambda_{k-1} - \lambda_k = \frac{\Lambda}{N}$, $N \in \mathbf{N}$, then for all $p > 1$, the next estimation holds true:

$$\left(\mathbf{E}\left|\exp\{Y(t+h)\} - \exp\left\{\widetilde{Y}(t+h)\right\}\right.\right.$$
$$\left.\left. - \left(\exp\{Y(t)\} - \exp\left\{\widetilde{Y}(t)\right\}\right)\right|^p\right)^{\frac{1}{p}} \leq h^\beta \widehat{G}_{N,t,p},$$

where:

$$\widehat{G}_{N,t,p} = 2^{\frac{1}{p}} p \exp\left\{\frac{p r_2}{2}\left(f_1 \widehat{A}_{N,t} + f_2 B(0)\right) - \frac{1}{2}\right\} \widehat{K}_{N,t},$$

$$\widehat{K}_{N,t} = \sqrt{r_1 \widehat{P}_{N,t}} + \sqrt{2^{3-2\beta} s_1 s_2 \Lambda^{2\beta} F(\Lambda) \widehat{A}_{N,t}},$$

$$\widehat{P}_{N,t} = 2^{5-4\beta}\left(\left(\frac{\Lambda}{N}\right)^\beta + 2^{\beta-1} t \frac{\Lambda^{\beta+1}}{N}\right)^2 F(\Lambda) + 2^{3-2\beta} \int_\Lambda^\infty \lambda^{2\beta} dF(\lambda),$$

$$\widehat{A}_{N,t} = B(0) - F(\Lambda) + 2^{2-2b} t^{2b} \left(\frac{\Lambda}{N}\right)^{2b} F(\Lambda),$$

$b \in [0, 1]$, f_1, f_2, s_1, s_2, s_3, r_1, r_2 are such numbers that $r_2 = s_3$, $\frac{1}{f_1} + \frac{1}{f_2} = 1$, $\frac{1}{s_1} + \frac{1}{s_2} + \frac{1}{s_3} = 1$, $\frac{1}{r_1} + \frac{1}{r_2} = 1$.

PROOF.– It is evident that

$$\begin{aligned}&\left(\mathbf{E}\left|e^{Y(t+h)}-e^{\widetilde{Y}(t+h)}-\left(e^{Y(t)}-e^{\widetilde{Y}(t)}\right)\right|^{p}\right)^{\frac{1}{p}}\\&=\left(\mathbf{E}\left|\left(e^{Y(t+h)-\widetilde{Y}(t+h)}-e^{Y(t)-\widetilde{Y}(t)}\right)e^{\widetilde{Y}(t+h)}\right.\right.\\&\left.\left.+\left(e^{\widetilde{Y}(t+h)}-e^{\widetilde{Y}(t)}\right)\left(e^{Y(t)-\widetilde{Y}(t)}-1\right)\right|^{p}\right)^{\frac{1}{p}}\\&\le\left(\mathbf{E}\left|\Delta_1(Y)\,V_1\right|^{p}\right)^{\frac{1}{p}}+\left(\mathbf{E}\left|\Delta_2(Y)\,\Delta_3(Y)\,V_2\right|^{p}\right)^{\frac{1}{p}},\end{aligned}$$

where:

$$\begin{aligned}\Delta_1(Y)&=\left|Y(t+h)-\widetilde{Y}(t+h)-Y(t)+\widetilde{Y}(t)\right|,\\V_1&=\exp\left\{\max\left(Y(t+h)-\widetilde{Y}(t+h),\,Y(t)-\widetilde{Y}(t)\right)\right\}\exp\left\{\widetilde{Y}(t+h)\right\},\end{aligned}$$

$$\begin{aligned}\Delta_2(Y)&=\left|\widetilde{Y}(t+h)-\widetilde{Y}(t)\right|,\\\Delta_3(Y)&=\left|Y(t)-\widetilde{Y}(t)\right|,\\V_2&=\exp\left\{\max\left(\widetilde{Y}(t+h),\,\widetilde{Y}(t)\right)\right\}\exp\left\{\left|Y(t)-\widetilde{Y}(t)\right|\right\}.\end{aligned}$$

Let $\frac{1}{r_1}+\frac{1}{r_2}=1$, $\frac{1}{s_1}+\frac{1}{s_2}+\frac{1}{s_3}=1$, by using the Holder inequality:

$$\begin{aligned}&\mathbf{E}\left|\Delta_1(Y)\,V_1\right|^{p}\le\left(\mathbf{E}\left|\Delta_1(Y)\right|^{pr_1}\right)^{\frac{1}{r_1}}\left(\mathbf{E}\left|V_1\right|^{pr_2}\right)^{\frac{1}{r_2}},\\&\mathbf{E}\left|\Delta_2(Y)\,\Delta_3(Y)\,V_2\right|^{p}\le\left(\mathbf{E}\left|\Delta_2(Y)\right|^{ps_1}\right)^{\frac{1}{s_1}}\left(\mathbf{E}\left|\Delta_3(Y)\right|^{ps_2}\right)^{\frac{1}{s_2}}\left(\mathbf{E}\left|V_2\right|^{ps_3}\right)^{\frac{1}{s_3}}.\end{aligned}$$

By virtue of [6.3], $\mathbf{E}\left|\Delta_1(Y)\right|^{pr_1}=\left(\mathbf{E}\left|\Delta_1(Y)\right|^{2}\right)^{\frac{pr_1}{2}}c_{pr_1}$. This was shown during the proof of lemma 6.2

$$\mathbf{E}\left|\Delta_1(Y)\right|^{2}=\mathbf{E}\left|Y(t+h)-\widetilde{Y}(t+h)-Y(t)+\widetilde{Y}(t)\right|^{2}\le h^{2\beta}\widehat{P}_{N,t},$$

where

$$\widehat{P}_{N,t}=2^{5-4\beta}\left(\left(\frac{\Lambda}{N}\right)^{\beta}+2^{\beta-1}t\frac{\Lambda^{\beta+1}}{N}\right)^{2}F(\Lambda)+2^{3-2\beta}\int\limits_{\Lambda}^{\infty}\lambda^{2\beta}dF(\lambda),$$

$0<\beta\le1$. Thereby

$$\mathbf{E}\left|\Delta_1(Y)\right|^{pr_1}\le h^{pr_1\beta}\,\widehat{P}_{N,t}^{\frac{pr_1}{2}}\,c_{pr_1}. \qquad [6.24]$$

Let us estimate $\mathbf{E}\left|V_1\right|^{pr_2}$. Let $\frac{1}{f_1}+\frac{1}{f_2}=1$, by using the Holder inequality

$$\mathbf{E}\left|V_1\right|^{pr_2}=\mathbf{E}\left|e^{\max\left(Y(t+h)-\widetilde{Y}(t+h),\,Y(t)-\widetilde{Y}(t)\right)}\exp\left\{\widetilde{Y}(t+h)\right\}\right|^{pr_2}$$

$$\leq \left(\mathbf{E}\exp\left\{p\,r_2 f_1 \max\left(Y(t+h)-\widetilde{Y}(t+h),\, Y(t)-\widetilde{Y}(t)\right)\right\}\right)^{\frac{1}{f_1}}$$
$$\times \left(\mathbf{E}\exp\left\{p\,r_2 f_2 \widetilde{Y}(t+h)\right\}\right)^{\frac{1}{f_2}}. \qquad [6.25]$$

$$\mathbf{E}\exp\left\{p\,r_2 f_1 \max\left(Y(t+h)-\widetilde{Y}(t+h),\, Y(t)-\widetilde{Y}(t)\right)\right\}$$
$$\leq \exp\left\{\frac{(p\,r_2 f_1)^2}{2}\mathbf{E}\left|Y(t+h)-\widetilde{Y}(t+h)\right|^2\right\}$$
$$+\exp\left\{\frac{(p\,r_2 f_1)^2}{2}\mathbf{E}\left|Y(t)-\widetilde{Y}(t)\right|^2\right\}.$$

It is evident that

$$\mathbf{E}\left|Y(t+h)-\widetilde{Y}(t+h)\right|^2 \leq \widehat{A}_{N,t},$$

where $\widehat{A}_{N,t}$ is defined in [6.22]. Then:

$$\left(\mathbf{E}\exp\left\{p\,r_2 f_1 \max\left(Y(t+h)-\widetilde{Y}(t+h),\, Y(t)-\widetilde{Y}(t)\right)\right\}\right)^{\frac{1}{f_1}}$$
$$\leq 2^{\frac{1}{f_1}}\exp\left\{\frac{(p\,r_2)^2 f_1}{2}\widehat{A}_{N,t}\right\}.$$

$$\left(\mathbf{E}\exp\left\{p\,r_2 f_2 \widetilde{Y}(t+h)\right\}\right)^{\frac{1}{f_2}} \leq \left(\exp\left\{\frac{(p\,r_2 f_2)^2}{2}F(\lambda)\right\}\right)^{\frac{1}{f_2}}$$
$$= \exp\left\{\frac{(p\,r_2)^2 f_2}{2}F(\lambda)\right\}.$$

By using two preceding inequalities, it follows from [6.25] that;

$$\mathbf{E}\left|V_1\right|^{pr_2} \leq 2^{\frac{1}{f_1}}\exp\left\{\frac{(p\,r_2)^2}{2}\left(f_1\widehat{A}_{N,t}+f_2 B(0)\right)\right\}. \qquad [6.26]$$

Let us estimate $\mathbf{E}\left|\Delta_2(Y)\right|^{ps_1}$. As a result of [6.3],

$$\left|\Delta_2(Y)\right|^{ps_1} = \left(\mathbf{E}\left|\Delta_2(Y)\right|^2\right)^{\frac{ps_1}{2}} c_{ps_1}.$$

It was shown during the proof of lemma 6.2 that

$$\mathbf{E}\left|\Delta_2(Y)\right|^2 = \mathbf{E}\left|\widetilde{Y}(t+h) - \widetilde{Y}(t)\right|^2 \le 2^{3-2\beta}\Lambda^{2\beta}h^{2\beta}F(\Lambda), \quad 0 < \beta \le 1.$$

Thereby,

$$\mathbf{E}\left|\Delta_2(Y)\right|^{ps_1} \le \left(2^{3-2\beta}\Lambda^{2\beta}h^{2\beta}F(\Lambda)\right)^{\frac{ps_1}{2}} c_{ps_1}. \tag{6.27}$$

Let us estimate $\mathbf{E}\left|\Delta_3(Y)\right|^{ps_2}$. Since $\mathbf{E}\left|\Delta_3(Y)\right|^2 = \mathbf{E}\left|Y(t) - \widetilde{Y}(t)\right|^2 \le \widehat{A}_{N,t}$, and $\mathbf{E}\left|\Delta_3(Y)\right|^{ps_2} = \left(\mathbf{E}\left|\Delta_3(Y)\right|^2\right)^{\frac{ps_2}{2}} c_{ps_2}$, then

$$\mathbf{E}\left|\Delta_3(Y)\right|^{ps_2} \le \widehat{A}_{N,t}^{\frac{ps_2}{2}} c_{ps_2}. \tag{6.28}$$

Let us estimate $\mathbf{E}\left|V_2\right|^{ps_3}$. Let $\frac{1}{e_1} + \frac{1}{e_2} = 1$, by using the Holder inequality

$$\begin{aligned}\mathbf{E}\left|V_2\right|^{ps_3} &= \mathbf{E}\left|\exp\left\{\max\left(\widetilde{Y}(t+h), \widetilde{Y}(t)\right)\right\}\exp\left\{\left|Y(t) - \widetilde{Y}(t)\right|\right\}\right|^{ps_3} \\ &\le \left(\mathbf{E}\exp\left\{ps_3e_1\max\left(\widetilde{Y}(t+h), \widetilde{Y}(t)\right)\right\}\right)^{\frac{1}{e_1}} \\ &\quad\times\left(\mathbf{E}\exp\left\{ps_3e_2\left|Y(t) - \widetilde{Y}(t)\right|\right\}\right)^{\frac{1}{e_2}}.\end{aligned}$$

$$\mathbf{E}\exp\left\{ps_3e_1\max\left(\widetilde{Y}(t+h), \widetilde{Y}(t)\right)\right\} \le 2\exp\left\{\frac{(ps_3e_1)^2}{2}F(\Lambda)\right\}.$$

For Gaussian random variable ξ with parameters 0 and σ^2, we have $\mathbf{E}\exp\{\lambda|\xi|\} \le \mathbf{E}\exp\{\lambda\xi\} + \mathbf{E}\exp\{-\lambda\xi\} = 2\exp\left\{\frac{\lambda^2\sigma^2}{2}\right\}$, which is why:

$$\mathbf{E}\exp\left\{ps_3e_2\left|Y(t) - \widetilde{Y}(t)\right|\right\} \le 2\exp\left\{\frac{(ps_3e_2)^2}{2}\widehat{A}_{N,t}\right\}.$$

That is,

$$\mathbf{E}\left|V_2\right|^{ps_3} \le 2\exp\left\{\frac{(ps_3)^2}{2}\left(e_2\widehat{A}_{N,t} + e_1B(0)\right)\right\}. \tag{6.29}$$

Thereby, by using [6.24], [6.26]–[6.29], and [6.4], we obtain:

$$\left(\mathbf{E}\left|\exp\left\{Y(t+h)\right\}-\exp\left\{\widetilde{Y}(t+h)\right\}-\left(\exp\left\{Y(t)\right\}-\exp\left\{\widetilde{Y}(t)\right\}\right)\right|^{p}\right)^{\frac{1}{p}}$$

$$\leq \left(\mathbf{E}\left|\Delta_1(Y)\right|^{pr_1}\right)^{\frac{1}{pr_1}}\left(\mathbf{E}\left|V_1\right|^{pr_2}\right)^{\frac{1}{pr_2}}$$

$$+\left(\mathbf{E}\left|\Delta_2(Y)\right|^{ps_1}\right)^{\frac{1}{ps_1}}\left(\mathbf{E}\left|\Delta_3(Y)\right|^{ps_2}\right)^{\frac{1}{ps_2}}\left(\mathbf{E}\left|V_2\right|^{ps_3}\right)^{\frac{1}{ps_3}}$$

$$\leq h^{\beta}\,\widehat{P}_{N,t}^{\frac{1}{2}}\,c_{pr_1}^{\frac{1}{pr_1}}\,2^{\frac{1}{pr_2 f_1}}\,\exp\left\{\frac{pr_2}{2}\left(f_1\widehat{A}_{N,t}+f_2B(0)\right)\right\}$$

$$+\left(2^{3-2\beta}\Lambda^{2\beta}h^{2\beta}F(\Lambda)\right)^{\frac{1}{2}}\,c_{ps_1}^{\frac{1}{ps_1}}\,\widehat{A}_{N,t}^{\frac{1}{2}}\,c_{ps_2}^{\frac{1}{ps_2}}\,2^{\frac{1}{ps_3}}\exp\left\{\frac{ps_3}{2}\left(e_2\widehat{A}_{N,t}+e_1B(0)\right)\right\}$$

$$=h^{\beta}\left[\widehat{P}_{N,t}^{\frac{1}{2}}\,2^{\frac{1}{2pr_1}}\left(pr_1\right)^{\frac{1}{2}}e^{-\frac{1}{2}}2^{\frac{1}{pr_2 f_1}}\exp\left\{\frac{pr_2}{2}\left(f_1\widehat{A}_{N,t}+f_2B(0)\right)\right\}\right.$$

$$+\left(2^{3-2\beta}\Lambda^{2\beta}F(\Lambda)\,\widehat{A}_{N,t}\right)^{\frac{1}{2}}2^{\frac{1}{2ps_1}}\left(ps_1\right)^{\frac{1}{2}}e^{-\frac{1}{2}}2^{\frac{1}{2ps_2}}\left(ps_2\right)^{\frac{1}{2}}e^{-\frac{1}{2}}\,2^{\frac{1}{ps_3}}$$

$$\left.\times\exp\left\{\frac{ps_3}{2}\left(e_2\widehat{A}_{N,t}+e_1B(0)\right)\right\}\right].$$

Put $f_1 = e_2$, $f_2 = e_1$, $r_2 = s_3$. Taking into account $2^{\frac{1}{2pr_1}}\times 2^{\frac{1}{pr_2 f_1}} \leq 2^{\frac{1}{p}}$, $2^{\frac{1}{2ps_1}}\,2^{\frac{1}{2ps_2}}\,2^{\frac{1}{ps_3 e_1}} \leq 2^{\frac{1}{p}}$, we obtain:

$$\left(\mathbf{E}\left|\exp\left\{Y(t+h)\right\}-\exp\left\{\widetilde{Y}(t+h)\right\}-\left(\exp\left\{Y(t)\right\}-\exp\left\{\widetilde{Y}(t)\right\}\right)\right|^{p}\right)^{\frac{1}{p}}$$

$$\leq h^{\beta}\left[\widehat{P}_{N,t}^{\frac{1}{2}}\,2^{\frac{1}{p}}\left(pr_1\right)^{\frac{1}{2}}\exp\left\{\frac{pr_2}{2}\left(f_1\widehat{A}_{N,t}+f_2B(0)\right)-\frac{1}{2}\right\}\right.$$

$$+\left(2^{3-2\beta}\Lambda^{2\beta}F(\Lambda)\,\widehat{A}_{N,t}\right)^{\frac{1}{2}}2^{\frac{1}{p}}\left(ps_1\right)^{\frac{1}{2}}\left(ps_2\right)^{\frac{1}{2}}$$

$$\left.\times\exp\left\{\frac{pr_2}{2}\left(f_1\widehat{A}_{N,t}+f_2B(0)\right)-1\right\}\right].$$

After elementary manipulation, the assumption of the lemma follows from the last formula. □

LEMMA 6.6.– Let $Y(t)$ be a homogeneous, centered, separable, continuous in mean square Gaussian process with spectral function $F(\lambda)$. There exists spectral moment $\int_0^\infty \lambda^{2\beta}dF(\lambda)$, $0<\beta\leq 1$, partition D_Λ of domain $[0,\Lambda]$, $\Lambda\in\mathbf{R}$ is such that $\lambda_{k-1}-\lambda_k=\frac{\Lambda}{N}$, $N\in\mathbf{N}$. If

$$\widehat{S}_N<\alpha\exp\left\{-\frac{2r_2\left(f_1\widehat{A}_N+f_2B(0)\right)}{\beta}\right\},$$

then

$$\mathbf{P}\left\{\max_{B_i\in\mathfrak{B}\ i=\overline{0,\,k-1}} |\,p_{kY}(B_i)-\widetilde{p}_{kY}(B_i)\,|>\alpha\right\}$$
$$\le\left(p^{\frac{p}{2}}+T\left(\frac{3\beta}{2}\right)^{\frac{1}{\beta}}p^{p+\frac{1}{\beta}}\right)\exp\left\{-\frac{r_2\left(f_1\widehat{A}_N+f_2B(0)\right)}{2}p^2\right\}, \quad [6.30]$$

where

$$p=\frac{\ln\frac{\alpha}{\widehat{S}_N}}{r_2\left(f_1\widehat{A}_N+f_2B(0)\right)},\qquad \widehat{S}_N=\max\left\{\widehat{S}_{N,1},\widehat{S}_{N,2}\right\},$$

$$\widehat{S}_{N,1}=\frac{2d\sqrt{v_1}\,(B(0)-F(\Lambda))}{\sqrt{e}},$$

$$\widehat{S}_{N,2}=\frac{6d}{\sqrt{e}}\left(\sqrt{r_1\widehat{P}_N}+\sqrt{2^{3-2\beta}s_1s_2\Lambda^{2\beta}F(\Lambda)\,\widehat{A}_N}\right),$$

$$\widehat{P}_N=2^{5-4\beta}\left(\left(\frac{\Lambda}{N}\right)^{\beta}+2^{\beta-1}T\frac{\Lambda^{\beta+1}}{N}\right)^2F(\Lambda)+2^{3-2\beta}\int_{\Lambda}^{\infty}\lambda^{2\beta}dF(\lambda),$$

$$\widehat{A}_N=B(0)-F(\Lambda)+2^{2-2b}T^{2b}\left(\frac{\Lambda}{N}\right)^{2b}F(\Lambda),$$

$b\in[0,1]$, f_1, f_2, s_1, s_2, s_3, r_1, r_2 are such numbers that $r_2=s_3=v_2$, $\frac{1}{f_1}+\frac{1}{f_2}=1$, $\frac{1}{s_1}+\frac{1}{s_2}+\frac{1}{s_3}=1$, $\frac{1}{r_1}+\frac{1}{r_2}=1$.

PROOF.– We now estimate the difference $|\,p_{kY}(B_i)-\widetilde{p}_{kY}(B_i)\,|$ by using the mean value theorem for derivatives:

$$|\,p_{kY}(B_i)-\widetilde{p}_{kY}(B_i)\,|=\left|\frac{\exp\{-\mu(B_i)\}(\mu(B_i))^k}{k!}-\frac{\exp\{-\widetilde{\mu}(B_i)\}(\widetilde{\mu}(B_i))^k}{k!}\right|$$
$$=|\mu(B_i)-\widetilde{\mu}(B_i)|\,\frac{1}{k!}\exp\{-\hat{\mu}(B_i)\}\,(\hat{\mu}(B_i))^{k-1}\,|k-\hat{\mu}(B_i)|$$
$$=\begin{cases}|\mu(B_i)-\widetilde{\mu}(B_i)|\dfrac{1}{(k-1)!}e^{-\hat{\mu}(B_i)}(\hat{\mu}(B_i))^{k-1}\le|\mu(B_i)-\widetilde{\mu}(B_i)|,\ k\ge\hat{\mu}(B_i);\\ |\mu(B_i)-\widetilde{\mu}(B_i)|\,\dfrac{1}{k!}e^{-\hat{\mu}(B_i)}(\hat{\mu}(B_i))^{k}\le|\mu(B_i)-\widetilde{\mu}(B_i)|,\ \ k<\hat{\mu}(B_i).\end{cases}$$

If $k = 0$, then

$$\begin{aligned}|p_{0Y}(B_i) - \widetilde{p}_{0Y}(B_i)| &= |\exp\{-\mu(B_i)\} - \exp\{-\widetilde{\mu}(B_i)\}| \\ &\le |\mu(B_i) - \widetilde{\mu}(B_i)|\,|\exp\{-\hat{\mu}(B_i)\}| \le |\mu(B_i) - \widetilde{\mu}(B_i)|.\end{aligned}$$

Therefore, the difference of probabilities $|p_{kY}(B_i) - \widetilde{p}_{kY}(B_i)|$ is estimated in terms of $|\mu(B_i) - \widetilde{\mu}(B_i)|$, and we obtain:

$$\mathbf{P}\{|p_{kY}(B_i) - \widetilde{p}_{kY}(B_i)| > \alpha\} \le \mathbf{P}\{|\mu(B_i) - \widetilde{\mu}(B_i)| > \alpha\}, \qquad [6.31]$$

$i = \overline{0, k-1}$. It is easy to check that

$$\begin{aligned}&\mathbf{P}\left\{\max_{B_i \in \mathfrak{B},\, i=\overline{0,k-1}} |\mu(B_i) - \widetilde{\mu}(B_i)| > \alpha\right\} \\ &= \mathbf{P}\left\{\max_{B_i \in \mathfrak{B},\, i=\overline{0,k-1}} \left|\int_{B_i} \exp\{Y(t)\}\,dt - \int_{B_i} \exp\left\{\widetilde{Y}(t)\right\} dt\right| > \alpha\right\} \\ &\le \mathbf{P}\left\{\max_{B_i \in \mathfrak{B},\, i=\overline{0,k-1}} \int_{B_i} \sup_{t \in \mathbf{T}} \left|\exp\{Y(t)\} - \exp\left\{\widetilde{Y}(t)\right\}\right| dt > \alpha\right\} \\ &= \mathbf{P}\left\{\max_{B_i \in \mathfrak{B},\, i=\overline{0,k-1}} \int_{B_i} dt \cdot \sup_{t \in \mathbf{T}} \left|\exp\{Y(t)\} - \exp\left\{\widetilde{Y}(t)\right\}\right| > \alpha\right\} \\ &= \mathbf{P}\left\{\sup_{t \in \mathbf{T}} \left|\exp\{Y(t)\} - \exp\left\{\widetilde{Y}(t)\right\}\right| > \frac{\alpha}{d}\right\}. \qquad [6.32]\end{aligned}$$

By virtue of lemma 6.4,

$$\begin{aligned}&\inf_{0 \le t \le T}\left(\mathbf{E}\left|\exp\{Y(t)\} - \exp\left\{\widetilde{Y}(t)\right\}\right|^p\right)^{\frac{1}{p}} \\ &\le 2^{\frac{1}{p}}(B(0) - F(\Lambda))^{\frac{1}{2}}(pv_1)^{\frac{1}{2}} \exp\left\{\frac{pv_2}{2}B(0) - \frac{1}{2}\right\}. \qquad [6.33]\end{aligned}$$

By using lemma 6.5, we estimate the integral in theorem 1.16:

$$\int_0^{\theta\gamma_0} N^{\frac{1}{p}}(\varepsilon)d\varepsilon \le \int_0^{\theta\gamma_0} \left(T\left(\frac{\widehat{G}_{N,p}}{\varepsilon}\right)^{\frac{1}{\beta}}\right)^{\frac{1}{p}} d\varepsilon = \frac{\theta^{1-\frac{1}{p\beta}} T^{\frac{1}{p}} \widehat{G}_{N,p}}{1 - \frac{1}{p\beta}},$$

$1 - \frac{1}{p\beta} > 0$, where $\widehat{G}_{N,p} = \widehat{G}_{N,t,p}\Big|_{t=T}$. It is easy to check that the function $f(\theta) = \frac{1}{\theta^{\frac{1}{p\beta}}(1-\theta)}$ has its minimum value in the point $\theta_0 = \frac{1}{p\beta+1}$ and $\theta_0 < \frac{\varphi\left(\frac{T}{2}\right)}{\varepsilon_0}$. After elementary manipulations, we obtain

$$\inf_{0<\theta<1} \frac{1}{\theta^{\frac{1}{p\beta}}(1-\theta)} \frac{T^{\beta}\widehat{G}_{N,p}}{1-\frac{1}{p\beta}} \le T^{\frac{1}{p}}\widehat{G}_{N,p}\frac{(p\beta+1)^{1+\frac{1}{p\beta}}}{p\beta-1}.$$

Taking into account [6.32] and [6.33], we obtain the estimation and inequality $(a+b)^p \le 2^{p-1}(a^p+b^p)$; based on corollary 1.16, we get

$$\mathbf{P}\left\{\max_{B_i\in\mathfrak{B},\ i=\overline{0,k-1}} |\mu(B_i) - \widetilde{\mu}(B_i)| > \alpha\right\}$$

$$\frac{2^p(B(0)-F(\Lambda))^p p^{\frac{p}{2}} v_1^{\frac{p}{2}} \exp\left\{\frac{pv_2}{2}B(0) - \frac{p}{2}\right\}}{\left(\frac{\alpha}{d}\right)^p} + \frac{2^{p-1}T\widehat{G}^p_{N,p}\frac{(p\beta+1)^{p+\frac{1}{\beta}}}{(p\beta-1)^p}}{\left(\frac{\alpha}{d}\right)^p}.$$

By using the definition of $\widehat{G}_{N,p}$ and taking into consideration that under $p\beta \ge 2$ $\left(\frac{p}{p\beta-1}\right)^p \le \frac{2^p}{\beta^p}$, $(p\beta+1)^{p+\frac{1}{\beta}} \le (p\beta)^{p+\frac{1}{\beta}}\left(\frac{3}{2}\right)^{p+\frac{1}{\beta}}$ and putting $v_2 = r_2$, after elementary manipulation we obtain the following estimation:

$$\mathbf{P}\left\{\max_{B_i\in\mathfrak{B},\ i=\overline{0,k-1}} |\mu(B_i) - \widetilde{\mu}(B_i)| > \alpha\right\}$$

$$\le \frac{\widehat{S}^p_{N,1}p^{\frac{p}{2}}\exp\left\{\frac{pv_2}{2}B(0)\right\}}{\alpha^p} + \frac{T\widehat{S}^p_{N,2}\left(\frac{3\beta}{2}\right)^{\frac{1}{\beta}}p^{p+\frac{1}{\beta}}\exp\left\{\frac{p^2r_2}{2}\left(f_1\widehat{A}_N + f_2B(0)\right)\right\}}{\alpha^p}$$

$$\le \frac{\widehat{S}^p_N\left(p^{\frac{p}{2}} + T\left(\frac{3\beta}{2}\right)^{\frac{1}{\beta}}p^{p+\frac{1}{\beta}}\right)\exp\left\{\frac{p^2r_2}{2}\left(f_1\widehat{A}_N + f_2B(0)\right)\right\}}{\alpha^p},$$

where $\widehat{S}_N = \max\left\{\widehat{S}_{N,1}, \widehat{S}_{N,2}\right\}$, $\widehat{S}_{N,1} = \frac{2d\sqrt{v_1}(B(0)-F(\Lambda))}{\sqrt{e}}$, $\widehat{S}_{N,2} = \frac{6d}{\sqrt{e}}\sqrt{r_1\widehat{P}_N} + \frac{6d}{\sqrt{e}}\sqrt{2^{3-2\beta}s_1s_2\Lambda^{2\beta}F(\Lambda)\widehat{A}_N}$. By evaluating the value of the right-hand side at the point $p_0 = \frac{\ln\frac{\alpha}{\widehat{S}_N}}{r_2\left(f_1\widehat{A}_N + f_2B(0)\right)}$ that are closed to the point of minimum of the corresponding function, and taking into consideration that condition $p\beta \ge 2$ guarantees that $1 - \frac{1}{p\beta} > 0$ holds true, the assertion of the lemma follows from corollary 1.16 and [6.30]. □

THEOREM 6.4.– Let $Y(t)$ be a homogeneous, centered, separable, continuous in mean square Gaussian process with spectral function $F(\lambda)$. There exists spectral moment $\int_0^\infty \lambda^{2\beta}dF(\lambda)$, $0 < \beta \le 1$, the partition D_Λ of domain $[0,\Lambda]$, $\Lambda \in \mathbf{R}$, such that $\lambda_{k-1} - \lambda_k = \frac{\Lambda}{N}$, $N \in \mathbf{N}$, then model of Cox process $\{\widetilde{\nu}(B), B \in \mathfrak{B}\}$, directed by log Gaussian process $\exp\left\{\widetilde{Y}(t)\right\}$ approximates process ν with accuracy α and reliability $1-\gamma$ if the following inequalities hold true:

$$\widehat{S}_N < \alpha \exp\left\{-\frac{2r_2\left(f_1\widehat{A}_N + f_2B(0)\right)}{\beta}\right\},$$

$$\left(p^{\frac{p}{2}} + T\left(\frac{3\beta}{2}\right)^{\frac{1}{\beta}} p^{p+\frac{1}{\beta}}\right)\exp\left\{-\frac{r_2\left(f_1\widehat{A}_N + f_2B(0)\right)}{2}p^2\right\} < \gamma,$$

where $p = \frac{\ln\frac{\alpha}{\widehat{S}_N}}{r_2\left(f_1\widehat{A}_N + f_2B(0)\right)}$, $\widehat{S}_N, \widehat{A}_N$ are defined in 6.6, $b \in [0,1]$, $f_1, f_2, s_1, s_2, s_3, r_1, r_2$ such number, that $r_2 = s_3$, $\frac{1}{f_1} + \frac{1}{f_2} = 1$, $\frac{1}{s_1} + \frac{1}{s_2} + \frac{1}{s_3} = 1$, $\frac{1}{r_1} + \frac{1}{r_2} = 1$.

PROOF.– The assertion of theorem follows from definition 6.4 and lemma 6.6. □

Given the construction method of the models of the log Gaussian Cox processes are easier in comparison with the previous. But this method gives worse accuracy, especially in the case where one of the domains B_i consists of more than one point.

6.4. Simulation of the Cox process when density is generated by a homogeneous log Gaussian field

In this section, the method of Cox processes simulation offered in the previous section is expanded the case where density is generated by a homogeneous random field.

Let $\{Y(\vec{t}), \vec{t} \in \mathbf{T}\}$ be a centered, homogeneous, Gaussian field, simple functions of which are measurable on $\mathbf{T}$. By analogy with previous section, first we simulate field $Y(\vec{t})$, second we consider some partition $D_{\mathbf{T}}$ of domain $\mathbf{T}$, and on each element of partition $D_{\mathbf{T}}$ we construct the model of Poisson random variable with the corresponding mean.

Let $\mathbf{T} = [0, T] \times \ldots \times [0, T]$, $T \in \mathbf{R}_+$, and the partition $D_{\mathbf{T}}$ is chosen in the following way:

$$B_{i_1,\ldots,i_n} = \Big\{ \left[t_1^{i_1}, t_1^{i_1+1}\right) \times \ldots \times \left[t_n^{i_n}, t_n^{i_n+1}\right) \,\Big|\, t_m^{i_m} < t_m^{i_m+1},$$
$$t_m^{i_m+1} - t_m^{i_m} = d = \frac{T}{k}, k \in \mathbf{N},\, m = \overline{1,n},\, i_m = \overline{0,k-1}\Big\}.$$

We denote $\widetilde{Y}(\vec{t})$ by model of field $Y(\vec{t})$, $\widetilde{\mu}(B_{i_1,\dots,i_n}) = \int\limits_{B_{i_1,\dots,i_n}} \exp\left\{\widetilde{Y}(\vec{t})\right\}d\vec{t}$ and $\widetilde{\nu}(B_{i_1,\dots,i_n})$ by model $\nu(B_{i_1,\dots,i_n})$, that is the model of Poisson random variable with mean $\widetilde{\mu}(B_{i_1,\dots,i_n})$.

Since $\widetilde{\nu}(B_{i_1,\dots,i_n})$ is the number of model points that belong to the domain $B_{i_1,\dots,i_n}$, we allocate these points at $B_{i_1,\dots,i_n}$ by any way. If $\widetilde{\nu}(B_{i_1,\dots,i_n}) = 1$, we allocate one point at the center of the domain.

The partition of the domain $\mathbf{T}$ (that is d or k) we choose in such a way that the inequality

$$\mathbf{P}\{\nu(B_{i_1,\dots,i_n}) > 1\} < \delta, \qquad [6.34]$$

holds true, where δ is defined beforehand.

THEOREM 6.5.– Let $\{\nu(B_{i_1,\dots,i_n}), B_{i_1,\dots,i_n} \subset \mathfrak{B}\}$ be a Cox process, directed by log Gaussian homogeneous field $\exp\left\{Y(\vec{t})\right\}$. The inequality [6.34] holds true if we set

$$d = \frac{T}{k} \le \left[2\delta \exp\left\{-2B(\vec{0})\right\}\right]^{\frac{1}{2n}}.$$

PROOF.– The proof repeats the proof of theorem 6.3. □

DEFINITION 6.5.– *The model of Cox process* $\{\nu(B_{i_1,\dots,i_n}), B_{i_1,\dots,i_n} \subset \mathfrak{B}\}$ *directed by log Gaussian homogeneous field* $\exp\left\{Y(\vec{t})\right\}$ *approximates with accuracy* α, $0 < \alpha < 1$ *and reliability* $1-\gamma$, $0 < \gamma < 1$, *if the following inequality holds true:*

$$\mathbf{P}\left\{\max_{B_{i_1,\dots,i_n} \in \mathfrak{B}} |p_{kY}(B_{i_1,\dots,i_n}) - \widetilde{p}_{kY}(B_{i_1,\dots,i_n})| > \alpha\right\} < \gamma.$$

LEMMA 6.7.– Let $Y(\vec{t})$ be a homogeneous, centered continuous in mean square Gaussian field, then for all $\forall p > 1$, the following inequality holds true:

$$\mathbf{P}\left\{\max_{B_{i_1,\dots,i_n} \in \mathfrak{B}} |\mu(B_{i_1,\dots,i_n}) - \widetilde{\mu}(B_{i_1,\dots,i_n})| > \alpha\right\} \le \frac{2k^n W_N^p p^{\frac{p}{2}} \exp\left\{\frac{p^2 v_2}{2} B\left(\vec{0}\right) - \frac{p}{2}\right\}}{\alpha^p},$$

where

$$W_N = \sqrt{v_1} d^n J_N^{\frac{1}{2}},$$
$$J_N = 2^{2-2a} n^{2a} \frac{d^{2a}\Lambda^{2a}}{N^{2a}} \nu(\Lambda^n) + B\left(\vec{0}\right) - \nu(\Lambda^n),$$

$v_2 = \frac{v_1}{v_1 - 1}$, v_1 is any positive number more than one, $a \in [0, 1]$.

PROOF.–

$$\mathbf{P}\left\{\max_{B_{i_1,\dots,i_n}\in\mathfrak{B}} \left|\mu\left(B_{i_1,\dots,i_n}\right)-\widetilde{\mu}\left(B_{i_1,\dots,i_n}\right)\right|>\alpha\right\}$$

$$\leq \sum_{i_1,\dots,i_n=0}^{k} \mathbf{P}\left\{\left|\mu\left(B_{i_1,\dots,i_n}\right)-\widetilde{\mu}\left(B_{i_1,\dots,i_n}\right)\right|>\alpha\right\}$$

$$\leq k^n \max_{B_{i_1,\dots,i_n}\in\mathfrak{B}} \mathbf{P}\left\{\left|\mu\left(B_{i_1,\dots,i_n}\right)-\widetilde{\mu}\left(B_{i_1,\dots,i_n}\right)\right|>\alpha\right\}.$$

It follows from the Tchebychev inequality that:

$$\mathbf{P}\left\{\left|\mu\left(B_{i_1,\dots,i_n}\right)-\widetilde{\mu}\left(B_{i_1,\dots,i_n}\right)\right|>\alpha\right\} \leq \frac{\mathbf{E}\left|\mu\left(B_{i_1,\dots,i_n}\right)-\widetilde{\mu}\left(B_{i_1,\dots,i_n}\right)\right|^p}{\alpha^p}.$$

By virtue of the generated Minkovski inequality:

$$\mathbf{E}\left|\mu\left(B_{i_1,\dots,i_n}\right)-\widetilde{\mu}\left(B_{i_1,\dots,i_n}\right)\right|^p$$

$$\leq \mathbf{E}\left(\int_{B_{i_1,\dots,i_n}} \left|\exp\left\{Y\left(\vec{t}\right)\right\}-\exp\left\{\widetilde{Y}\left(\vec{t}\right)\right\}\right| d\vec{t}\right)^p$$

$$\leq \left(\int_{B_{i_1,\dots,i_n}} \left(\mathbf{E}\left|\exp\left\{Y\left(\vec{t}\right)\right\}-\exp\left\{\widetilde{Y}\left(\vec{t}\right)\right\}\right|^p\right)^{\frac{1}{p}} d\vec{t}\right)^p$$

That is, it follows from the last three inequalities that:

$$\mathbf{P}\left\{\max_{B_{i_1,\dots,i_n}\in\mathfrak{B}} \left|\mu\left(B_{i_1,\dots,i_n}\right)-\widetilde{\mu}\left(B_{i_1,\dots,i_n}\right)\right|>\alpha\right\}$$

$$\leq \frac{k^n\left(\int_{B_{i_1,\dots,i_n}} \left(\mathbf{E}\left|\exp\left\{Y\left(\vec{t}\right)\right\}-\exp\left\{\widetilde{Y}\left(\vec{t}\right)\right\}\right|^p\right)^{\frac{1}{p}} d\vec{t}\right)^p}{\alpha^p}. \qquad [6.35]$$

Let us estimate $\mathbf{E}\left|\exp\left\{Y\left(\vec{t}\right)\right\}-\exp\left\{\widetilde{Y}\left(\vec{t}\right)\right\}\right|^p$. Let $\frac{1}{v_1}+\frac{1}{v_2}=1$. By using first inequality $\left|\exp\{x\}-\exp\{y\}\right|\leq|x-y|\exp\{\max(x,\,y)\}$, and then the Holder inequality, we obtain:

$$\mathbf{E}\left|\exp\left\{Y\left(\vec{t}\right)\right\}-\exp\left\{\widetilde{Y}\left(\vec{t}\right)\right\}\right|^p$$

$$\leq \mathbf{E}\left|Y\left(\vec{t}\right)-\widetilde{Y}\left(\vec{t}\right)\right|^p \exp\left\{p\max\left(Y\left(\vec{t}\right),\,\widetilde{Y}\left(\vec{t}\right)\right)\right\}$$

$$\leq \left(\mathbf{E}\left|Y\left(\vec{t}\right)-\widetilde{Y}\left(\vec{t}\right)\right|^{pv_1}\right)^{\frac{1}{v_1}}\left(\mathbf{E}\exp\left\{p\,v_2\max\left(Y\left(\vec{t}\right),\,\widetilde{Y}\left(\vec{t}\right)\right)\right\}\right)^{\frac{1}{v_2}}. \qquad [6.36]$$

By virtue of [6.3],

$$\mathbf{E}\left|Y\left(\vec{t}\right)-\widetilde{Y}\left(\vec{t}\right)\right|^{pv_1}=c_{pv_1}\left(\mathbf{E}\left|Y\left(\vec{t}\right)-\widetilde{Y}\left(\vec{t}\right)\right|^{2}\right)^{\frac{pv_1}{2}}.$$

Since for the Gaussian, homogeneous, centered random field $\mathbf{E}\left(Y\left(\vec{t}\right)\right)^2=B\left(\vec{0}\right)$, $\mathbf{E}\left(\widetilde{Y}\left(\vec{t}\right)\right)^2=\Phi\left(\Lambda^n\right)$, then

$$\mathbf{E}\left|Y\left(\vec{t}\right)-\widetilde{Y}\left(\vec{t}\right)\right|^{2}=B\left(\vec{0}\right)+\Phi\left(\Lambda^n\right)-2\,\mathbf{E}\,Y\left(\vec{t}\right)\widetilde{Y}\left(\vec{t}\right).$$

By using representation [4.9] of field $Y\left(\vec{t}\right)$ and model [4.10], we obtain:

$$\mathbf{E}Y\left(\vec{t}\right)\widetilde{Y}\left(\vec{t}\right)=\mathbf{E}\left(\sum_{i_1,\ldots,i_n=0}^{N-1}\int\limits_{\Delta(i_1,\ldots,i_n)}\cos\left(\vec{t},\vec{\lambda}\right)dZ_1\left(\vec{\lambda}\right)\right.$$

$$+\sum_{i_1,\ldots,i_n=0}^{N-1}\int\limits_{\Delta(i_1,\ldots,i_n)}\sin\left(\vec{t},\vec{\lambda}\right)dZ_2\left(\vec{\lambda}\right)$$

$$\left.+\int\limits_{\mathbf{R}^n\setminus\Lambda^n}\cos\left(\vec{t},\vec{\lambda}\right)dZ_1\left(\vec{\lambda}\right)+\int\limits_{\mathbf{R}^n\setminus\Lambda^n}\sin\left(\vec{t},\vec{\lambda}\right)dZ_2\left(\vec{\lambda}\right)\right)$$

$$\times\left(\sum_{i_1,\ldots,i_n=0}^{N-1}\int\limits_{\Delta(i_1,\ldots,i_n)}\cos\left(\vec{t},\vec{\lambda}\left(\lambda_1^{i_1},\ldots,\lambda_n^{i_n}\right)\right)dZ_1\left(\vec{\lambda}\right)\right.$$

$$\left.+\sum_{i_1,\ldots,i_n=0}^{N-1}\int\limits_{\Delta(i_1,\ldots,i_n)}\sin\left(\vec{t},\vec{\lambda}\left(\lambda_1^{i_1},\ldots,\lambda_n^{i_n}\right)\right)dZ_2\left(\vec{\lambda}\right)\right)$$

$$=\sum_{i_1,\ldots,i_n=0}^{N-1}\int\limits_{\Delta(i_1,\ldots,i_n)}\cos\left(\vec{t},\vec{\lambda}\right)\cos\left(\vec{t},\vec{\lambda}\left(\lambda_1^{i_1},\ldots,\lambda_n^{i_n}\right)\right)d\Phi\left(\vec{\lambda}\right)$$

$$+\sum_{i_1,\ldots,i_n=0}^{N-1}\int\limits_{\Delta(i_1,\ldots,i_n)}\sin\left(\vec{t},\vec{\lambda}\right)\sin\left(\vec{t},\vec{\lambda}\left(\lambda_1^{i_1},\ldots,\lambda_n^{i_n}\right)\right)d\Phi\left(\vec{\lambda}\right)$$

$$= \sum_{i_1,\ldots,i_n=0}^{N-1} \int_{\Delta(i_1,\ldots,i_n)} \cos\left(\vec{t}, \vec{\lambda} - \vec{\lambda}\left(\lambda_1^{i_1},\ldots,\lambda_n^{i_n}\right)\right) d\Phi\left(\vec{\lambda}\right).$$

Thus, taking into consideration the above-presented relation:

$$\mathbf{E}\left|Y\left(\vec{t}\right) - \widetilde{Y}\left(\vec{t}\right)\right|^2 = 2\,\Phi\left(\Lambda^n\right) - 2\,\mathbf{E}Y\left(\vec{t}\right)\widetilde{Y}\left(\vec{t}\right) + B\left(\vec{0}\right) - \Phi\left(\Lambda^n\right)$$

$$= 2 \sum_{i_1,\ldots,i_n=0}^{N-1} \int_{\Delta(i_1,\ldots,i_n)} \left(1 - \cos\left(\vec{t}, \vec{\lambda} - \vec{\lambda}\left(\lambda_1^{i_1},\ldots,\lambda_n^{i_n}\right)\right)\right) d\Phi\left(\vec{\lambda}\right) + B\left(\vec{0}\right) - \Phi\left(\Lambda^n\right)$$

$$= 4 \sum_{i_1,\ldots,i_n=0}^{N-1} \int_{\Delta(i_1,\ldots,i_n)} \sin^2 \frac{\left(\vec{t}, \vec{\lambda} - \vec{\lambda}\left(\lambda_1^{i_1},\ldots,\lambda_n^{i_n}\right)\right)}{2} d\Phi\left(\vec{\lambda}\right) + B\left(\vec{0}\right) - \Phi\left(\Lambda^n\right)$$

$$\leq 4 \sum_{i_1,\ldots,i_n=0}^{N-1} \int_{\Delta(i_1,\ldots,i_n)} \left(\frac{\vec{t}, \vec{\lambda} - \vec{\lambda}\left(\lambda_1^{i_1},\ldots,\lambda_n^{i_n}\right)}{2}\right)^{2a} d\Phi\left(\vec{\lambda}\right) + B\left(\vec{0}\right) - \Phi\left(\Lambda^n\right),$$

$a \in [0, 1]$. By using inequality $\left(\vec{e}, \vec{f}\right) \leq \left(\sum_{i=1}^{n} e_i^2\right)^{\frac{1}{2}} \left(\sum_{i=1}^{n} f_i^2\right)^{\frac{1}{2}}$ and taking into consideration that $\lambda_m - \lambda_m^{i_m} \leq \lambda_m^{i_m+1} - \lambda_m^{i_m} = \frac{\Lambda}{N}$,

$$\mathbf{E}\left|Y\left(\vec{t}\right) - \widetilde{Y}\left(\vec{t}\right)\right|^2$$

$$\leq 4 \sum_{i_1,\ldots,i_n=0}^{N-1} \int_{\Delta(i_1,\ldots,i_n)} \frac{\left(\sum_{m=1}^{n} t_m^2\right)^a \left(\sum_{m=1}^{n} \left(\lambda_m - \lambda_m^{i_m}\right)^2\right)^a}{2^{2a}} d\Phi\left(\vec{\lambda}\right) + B\left(\vec{0}\right) - \Phi\left(\Lambda^n\right)$$

$$= 2^{2-2a} \sum_{i_1,\ldots,i_n=0}^{N-1} \int_{\Delta(i_1,\ldots,i_n)} \left(nd^2\right)^a \left(n\frac{\Lambda^2}{N^2}\right)^a d\Phi\left(\vec{\lambda}\right) + B\left(\vec{0}\right) - \Phi\left(\Lambda^n\right)$$

$$= 2^{2-2a} n^{2a} \frac{d^{2a}\Lambda^{2a}}{N^{2a}} \Phi\left(\Lambda^n\right) + B\left(\vec{0}\right) - \Phi\left(\Lambda^n\right).$$

Hereby,

$$\mathbf{E}\left|Y\left(\vec{t}\right) - \widetilde{Y}\left(\vec{t}\right)\right|^{pv_1} \leq c_{pv_1} J_N^{\frac{pv_1}{2}},$$

$$J_N = 2^{2-2a} n^{2a} \frac{d^{2a}\Lambda^{2a}}{N^{2a}} \Phi\left(\Lambda^n\right) + B\left(\vec{0}\right) - \Phi\left(\Lambda^n\right). \quad [6.37]$$

Let us estimate $\mathbf{E}\exp\left\{p\,v_2 \max\left(Y\left(\vec{t}\right), \widetilde{Y}\left(\vec{t}\right)\right)\right\}$.

$$\begin{aligned}
&\mathbf{E}\exp\left\{p\,v_2 \max\left(Y\left(\vec{t}\right), \widetilde{Y}\left(\vec{t}\right)\right)\right\} \\
&\leq \mathbf{E}\exp\left\{p\,v_2 Y\left(\vec{t}\right)\right\} + \mathbf{E}\exp\left\{p\,v_2 \widetilde{Y}\left(\vec{t}\right)\right\} \\
&= \exp\left\{\frac{(p\,v_2)^2}{2} B\left(\vec{0}\right)\right\} + \exp\left\{\frac{(p\,v_2)^2}{2}\Phi\left(\Lambda^n\right)\right\} \\
&\leq 2\exp\left\{\frac{(p\,v_2)^2}{2} B\left(\vec{0}\right)\right\}.
\end{aligned} \qquad [6.38]$$

Taking into considerations [6.37] and [6.38], it follows from [6.36] that

$$\mathbf{E}\left|\exp\left\{Y\left(\vec{t}\right)\right\} - \exp\left\{\widetilde{Y}\left(\vec{t}\right)\right\}\right|^p \leq c_{pv_1}^{\frac{1}{v_1}} J_N^{\frac{p}{2}} 2^{\frac{1}{v_2}} \exp\left\{\frac{p^2 v_2}{2} B\left(\vec{0}\right)\right\}. \qquad [6.39]$$

The statement of the lemma follows from [6.39], [6.4] and [6.35]. □

LEMMA 6.8.– Let $Y\left(\vec{t}\right)$ be a homogeneous, centered, continuous in mean square Gaussian field. If $W_N < \alpha \exp\left\{\frac{1}{2} - v_2 B\left(\vec{0}\right)\right\}$, then the next inequality holds true:

$$\begin{aligned}
&\mathbf{P}\left\{\max_{B_{i_1,\ldots,i_n} \in \mathfrak{B}} \left| p_{kY}\left(B_{i_1,\ldots,i_n}\right) - \widetilde{p}_{kY}\left(B_{i_1,\ldots,i_n}\right)\right| > \alpha\right\} \\
&\leq 2k^n \left(\frac{\frac{1}{2} - \ln\frac{W_N}{\alpha}}{v_2 B\left(\vec{0}\right)}\right)^{\frac{\frac{1}{2} - \ln\frac{W_N}{\alpha}}{2v_2 B\left(\vec{0}\right)}} \exp\left\{-\frac{\left(\frac{1}{2} - \ln\frac{W_N}{\alpha}\right)^2}{2v_2 B\left(\vec{0}\right)}\right\},
\end{aligned}$$

where W_N is defined in lemma 6.7, $v_2 = \frac{v_1}{v_1 - 1}$, v_1 is any positive real number more than one.

PROOF.– It was shown at the proof of lemma 6.6:

$$\begin{aligned}
&\mathbf{P}\left\{\max_{B_{i_1,\ldots,i_n} \in \mathfrak{B}} \left| p_{kY}\left(B_{i_1,\ldots,i_n}\right) - \widetilde{p}_{kY}\left(B_{i_1,\ldots,i_n}\right)\right| > \alpha\right\} \\
&\leq \mathbf{P}\left\{\max_{B_{i_1,\ldots,i_n} \in \mathfrak{B}} \left| \mu\left(B_{i_1,\ldots,i_n}\right) - \widetilde{\mu}\left(B_{i_1,\ldots,i_n}\right)\right| > \alpha\right\}.
\end{aligned}$$

By virtue of lemma 6.7:

$$\mathbf{P}\left\{\max_{B_{i_1,\ldots,i_n}\in\mathfrak{B}} \left|p_{kY}\left(B_{i_1,\ldots,i_n}\right)-\widetilde{p}_{kY}\left(B_{i_1,\ldots,i_n}\right)\right|>\alpha\right\}$$
$$\leq \frac{2k^n W_N^p p^{\frac{p}{2}} \exp\left\{\frac{p^2 v_2}{2}B\left(\vec{0}\right)-\frac{p}{2}\right\}}{\alpha^p}.$$

Let $p_0 = \frac{\frac{1}{2}-\ln\frac{W_N}{\alpha}}{v_2 B(\vec{0})}$ and substitute this value in the right-hand side of the inequality above. This point is near to the point of minimum. Taking into consideration that p_0 must be more than 1, we obtain the assertion of the lemma. □

THEOREM 6.6.– Let $Y\left(\vec{t}\right)$ be a homogeneous, centered, continuous in mean square random field, then the model of Cox process $\{\widetilde{\nu}\left(B_{i_1,\ldots,i_n}\right),\ B_{i_1,\ldots,i_n}\subset\mathfrak{B}\}$, driven by log Gaussian homogeneous field $\exp\left\{\widetilde{Y}\left(\vec{t}\right)\right\}$, approximates it with accuracy α and reliability $1-\gamma$, if the following inequalities hold true:

$$W_N < \alpha\exp\left\{\frac{1}{2}-v_2 B\left(\vec{0}\right)\right\},$$

$$2k^n\left(\frac{\frac{1}{2}-\ln\frac{W_N}{\alpha}}{v_2 B(\vec{0})}\right)^{\frac{\frac{1}{2}-\ln\frac{W_N}{\alpha}}{2v_2 B(\vec{0})}}\exp\left\{-\frac{\left(\frac{1}{2}-\ln\frac{W_N}{\alpha}\right)^2}{2v_2 B(\vec{0})}\right\}<\gamma,$$

where W_N is defined in lemma 6.7, $v_2 = \frac{v_1}{v_1-1}$, v_1 is any positive real number more than 1.

PROOF.– The theorem is a corollary of definition 6.5 and lemma 6.8. □

EXAMPLE 6.1.– Let random field $\left\{Y\left(\vec{t}\right),\vec{t}\in\mathbf{T}\right\}$, $\mathbf{T} = [0,T]\times[0,T]$, $T\in\mathbf{R}$ satisfy assertions of theorem 6.6 and have spectral density $f\left(\lambda_1,\lambda_2\right) = \exp\left\{-\beta\left(\lambda_1^2+\lambda_2^2\right)\right\}$. In Table 6.1, it is shown to find the value of N for such a process under given accuracy α and reliability $1-\gamma$. All models are constructed in the domain $\mathbf{T} = [0,10]\times[0,10]$.

In Figure 6.1 sample functions of Gaussian field $Y(t)$ and generated models of log Gaussian process $\nu(B)$ are shown in the case when $\beta = 10$ and the values $\delta, \alpha, 1-\gamma$ are taken from the last four rows of Table 6.1 respectively.

6.5. Simulation of log Gaussian Cox process when the density is generated by the inhomogeneous field

In this section simplified method of simulation is considered in the case when the density isn't homogeneous field. That is as opposed to the case of homogeneous field, procedure of simulation differs only by the way of construction of models of field $Y(\vec{t})$. That is why we formulate main results without describing the simulation procedure. Let $\{Y(\vec{t}), \vec{t} \in \mathbf{T}\}$ be centered, Gaussian field, in which simple function of them are measurable on $\mathbf{T}$. The partition of domain $\mathbf{T}$ and all denotations remain as in earlier section.

δ	α	$1-\gamma$	β	d	N
0.01	0.01	0.99	1	0.253928	12,386
0.01	0.01	0.97	1	0.253928	9,736
0.01	0.03	0.97	1	0.253928	3,015
0.01	0.05	0.95	1	0.253928	1,552
0.01	0.03	0.97	10	0.361579	95
0.01	0.05	0.95	10	0.361579	54
0.02	0.03	0.97	10	0.429992	156
0.02	0.05	0.95	10	0.429992	87

Table 6.1. *The result of simulation of log Gaussian Cox process*

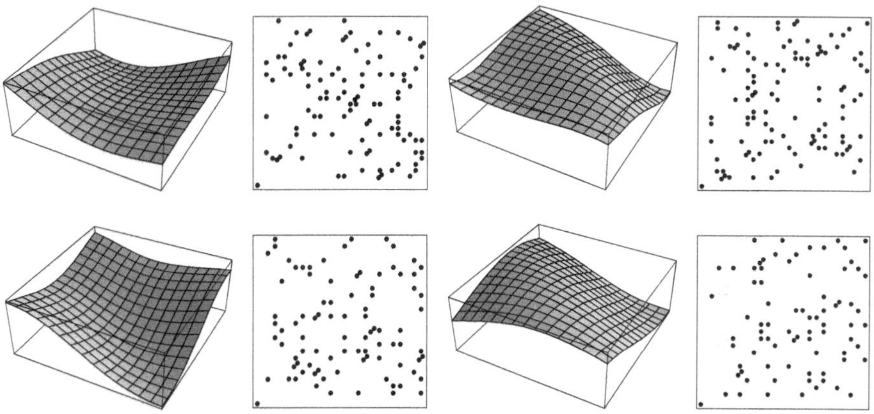

Figure 6.1. *Sample functions of Gaussian field and generated models of log Gaussian Cox process*

THEOREM 6.7.– Let $\{\nu(B_{i_1,\ldots,i_n}), B_{i_1,\ldots,i_n} \subset \mathfrak{B}\}$ be a Cox process driven by log Gaussian inhomogeneous field $\exp\{Y(\vec{t})\}$, the eigenvalue of integral equation [4.11] is bounded,

$$|\phi_k(\vec{t})| \leq L, \quad \forall \vec{t} \in \mathbf{T}, \forall k \in \mathbf{N}.$$

The inequality

$$\mathbf{P}\{\nu(B_{i_1,\ldots,i_n}) > 1\} < \delta, \qquad [6.40]$$

holds true if

$$d = \frac{T}{k} \leq \left[2\delta \exp\left\{ -2L^2 \sum_{k=1}^{\infty} \frac{1}{\lambda_k} \right\} \right]^{\frac{1}{2n}}.$$

PROOF.– Since

$$\mathbf{P}\{\nu(B_{i_1,\dots,i_n}) > 1\}$$
$$= \mathbf{E}\left(1 - \exp\{-\mu(B_{i_1,\dots,i_n})\} - \mu(B_{i_1,\dots,i_n}) \exp\{-\mu(B_{i_1,\dots,i_n})\}\right)$$

and under $x > 0$ $1 - \exp\{-x\}(1+x) \leq \frac{x^2}{2}$, then it is sufficient to choose such a partition that the following inequality holds true:

$$\mathbf{E}\left[\mu(B_{i_1,\dots,i_n})\right]^2 < 2\delta.$$

By virtue of $\xi = N(0, \sigma^2)$, the next relation holds true $\mathbf{E}\exp\{\lambda\xi\} = \exp\left\{\frac{\lambda^2\sigma^2}{2}\right\}$, we have:

$$\mathbf{E}\left[\mu(B_{i_1,\dots,i_n})\right]^2 = \mathbf{E} \int_{B_{i_1,\dots,i_n}} \exp\{Y(\vec{t})\} d\vec{t} \int_{B_{i_1,\dots,i_n}} \exp\{Y(\vec{s})\} d\vec{s}$$
$$= \iint_{B_{i_1,\dots,i_n} \times B_{i_1,\dots,i_n}} \mathbf{E}\exp\{Y(\vec{t}) + Y(\vec{s})\} d\vec{t} d\vec{s}$$
$$= \iint_{B_{i_1,\dots,i_n} \times B_{i_1,\dots,i_n}} \exp\left\{\frac{\mathbf{E}\left(Y(\vec{t}) + Y(\vec{s})\right)^2}{2}\right\} d\vec{t} d\vec{s}$$
$$= \iint_{B_{i_1,\dots,i_n} \times B_{i_1,\dots,i_n}} \exp\left\{\frac{\mathbf{E}\left(Y(\vec{t})\right)^2}{2} + \mathbf{E}Y(\vec{t})\,Y(\vec{s}) + \frac{\mathbf{E}(Y(\vec{s}))^2}{2}\right\} d\vec{t} d\vec{s}.$$

By using representation [4.12] of correlation function of field $Y(\vec{t})$,

$$\mathbf{E}\left[\mu(B_{i_1,\dots,i_n})\right]^2$$
$$\leq \iint_{B_{i_1,\dots,i_n} \times B_{i_1,\dots,i_n}} \exp\left\{\frac{1}{2}\sum_{k=1}^{\infty} \frac{1}{\lambda_k}\left(\varphi_k^2(\vec{t}) + 2\varphi_k(\vec{t})\,\varphi_k(\vec{s}) + \varphi_k^2(\vec{s})\right)\right\} d\vec{t}\vec{s}$$
$$= \iint_{B_{i_1,\dots,i_n} \times B_{i_1,\dots,i_n}} \exp\left\{\frac{1}{2}\sum_{k=1}^{\infty} \frac{1}{\lambda_k}\left(\varphi_k(\vec{t}) + \varphi_k(\vec{s})\right)^2\right\} d\vec{t}\vec{s}$$
$$\leq d^{2n} \exp\left\{2L^2 \sum_{k=1}^{\infty} \frac{1}{\lambda_k}\right\}.$$

The last inequality proof the theorem. □

LEMMA 6.9.– Let $Y(\vec{t})$ be Gaussian, centered, continuous in mean square random field with the eigenfunctions of integral equation [4.11] restricted:

$$\left|\phi_k(\vec{t})\right| \leq L \quad \forall \vec{t} \in \mathbf{T},\ k \in \mathbf{N},$$

then $\forall p > 1$, the next inequality holds true:

$$\mathbf{P}\left\{\max_{B_{i_1,\dots,i_n} \in \mathfrak{B}} \left|\mu(B_{i_1,\dots,i_n}) - \widetilde{\mu}(B_{i_1,\dots,i_n})\right| > \alpha\right\} \leq \frac{2k^n \widehat{W}_N^p p^{\frac{p}{2}} \exp\left\{-\frac{p}{2} + \frac{p^2 v_2 L^2}{2} \sum_{k=1}^{\infty} \frac{1}{\lambda_k}\right\}}{\alpha^p},$$

where

$$\widehat{W}_N = L\, d^n\, v_1^{\frac{1}{2}} \left(\sum_{k=N+1}^{\infty} \frac{1}{\lambda_k}\right)^{\frac{1}{2}},$$

$v_2 = \frac{v_1}{v_1 - 1}$, v_1 is any positive real number more than 1.

PROOF.– It was shown under proof of lemma 6.7:

$$\mathbf{P}\left\{\max_{B_{i_1,\dots,i_n} \in \mathfrak{B}} \left|\mu(B_{i_1,\dots,i_n}) - \widetilde{\mu}(B_{i_1,\dots,i_n})\right| > \alpha\right\} \leq \frac{k^n \left(\int_{B_{i_1,\dots,i_n}} \left(\mathbf{E}\left|\exp\{Y(\vec{t})\} - \exp\{\widetilde{Y}(\vec{t})\}\right|^p\right)^{\frac{1}{p}} d\vec{t}\right)^p}{\alpha^p}. \qquad [6.41]$$

$$\mathbf{E}\left|\exp\{Y(\vec{t})\} - \exp\{\widetilde{Y}(\vec{t})\}\right|^p \leq \left(\mathbf{E}\left|Y(\vec{t}) - \widetilde{Y}(\vec{t})\right|^{pv_1}\right)^{\frac{1}{v_1}} \left(\mathbf{E}\exp\left\{p\, v_2 \max\left(Y(\vec{t}), \widetilde{Y}(\vec{t})\right)\right\}\right)^{\frac{1}{v_2}}, \qquad [6.42]$$

$\frac{1}{v_1} + \frac{1}{v_1} = 1$. By virtue of [6.4]:

$$\mathbf{E}\left|Y(\vec{t}) - \widetilde{Y}(\vec{t})\right|^{pv_1} = c_{pv_1} \left(\mathbf{E}\left|Y(\vec{t}) - \widetilde{Y}(\vec{t})\right|^2\right)^{\frac{pv_1}{2}}.$$

Taking into consideration that in representation [4.13] of field $Y(\vec{t})$ $\mathbf{E}\xi_k\xi_l = \delta_{kl}$, when δ_{kl} is a symbol of Kroneker, we have:

$$\mathbf{E}\left|Y(\vec{t}) - \widetilde{Y}(\vec{t})\right|^2 = \mathbf{E}\left|\sum_{k=1}^{\infty} \frac{\xi_k}{\sqrt{\lambda_k}}\phi_k(\vec{t}) - \sum_{k=1}^{N} \frac{\xi_k}{\sqrt{\lambda_k}}\phi_k(\vec{t})\right|^2$$

$$= \mathbf{E}\left|\sum_{k=N+1}^{\infty} \frac{\xi_k}{\sqrt{\lambda_k}}\phi_k(\vec{t})\right|^2 = \sum_{k=N+1}^{\infty} \frac{\phi_k^2(\vec{t})}{\lambda_k} \le L^2 \sum_{k=N+1}^{\infty} \frac{1}{\lambda_k}.$$

That is,

$$\mathbf{E}\left|Y(\vec{t}) - \widetilde{Y}(\vec{t})\right|^{pv_1} \le c_{pv_1} L^{pv_1} \left(\sum_{k=N+1}^{\infty} \frac{1}{\lambda_k}\right)^{\frac{pv_1}{2}}$$

Let us estimate $\mathbf{E}\exp\left\{p\,v_2 \max\left(Y(\vec{t}), \widetilde{Y}(\vec{t})\right)\right\}$.

$$\mathbf{E}\exp\left\{pv_2 \max\left(Y(\vec{t}), \widetilde{Y}(\vec{t})\right)\right\} \le \exp\left\{\frac{(pv_2)^2}{2}\mathbf{E}\left(\sum_{k=1}^{\infty} \frac{\xi_k}{\sqrt{\lambda_k}}\phi_k(\vec{t})\right)^2\right\}$$

$$+ \exp\left\{\frac{(p\,v_2)^2}{2}\mathbf{E}\left(\sum_{k=1}^{N} \frac{\xi_k}{\sqrt{\lambda_k}}\phi_k(\vec{t})\right)^2\right\}$$

$$\le 2\exp\left\{\frac{(pv_2)^2}{2}\sum_{k=1}^{\infty} \frac{\phi_k^2(\vec{t})}{\lambda_k}\right\} \le 2\exp\left\{\frac{(pv_2 L)^2}{2}\sum_{k=1}^{\infty} \frac{1}{\lambda_k}\right\}.$$

Taking into consideration two last inequalities, it follows from [6.42] that:

$$\mathbf{E}\left|\exp\left\{Y(\vec{t})\right\} - \exp\left\{\widetilde{Y}(\vec{t})\right\}\right|^p$$

$$\le c_{pv_1}^{\frac{1}{v_1}} L^p \left(\sum_{k=N+1}^{\infty} \frac{1}{\lambda_k}\right)^{\frac{p}{2}} 2^{\frac{1}{v_2}} \exp\left\{\frac{p^2 v_2 L^2}{2}\sum_{k=1}^{\infty} \frac{1}{\lambda_k}\right\}.$$

Taking into consideration estimation [6.4], the assumption of the lemma follows from [6.41]. □

LEMMA 6.10.– Let $Y(\vec{t})$ be a Gaussian, centered, continuous in mean square random field, with eigenfunctions of integral equation [4.11] restricted:

$$\left|\phi_k(\vec{t})\right| \le L \quad \forall \vec{t} \in \mathbf{T},\ k \in \mathbf{N}.$$

If $\widehat{W}_N < \alpha \exp\left\{\frac{1}{2} - v_2 L^2 \sum\limits_{k=1}^{\infty} \frac{1}{\lambda_k}\right\}$, then:

$$\mathbf{P}\left\{\max_{B_{i_1,\ldots,i_n}\in\mathfrak{B}} |p_{kY}(B_{i_1,\ldots,i_n}) - \widetilde{p}_{kY}(B_{i_1,\ldots,i_n})| > \alpha\right\}$$

$$\leq 2k^n \left(\frac{1-2\ln\frac{\widehat{W}_N}{\alpha}}{2v_2L^2\sum\limits_{k=1}^{\infty}\frac{1}{\lambda_k}}\right)^{\frac{1-2\ln\frac{\widehat{W}_N}{\alpha}}{4v_2L^2\sum\limits_{k=1}^{\infty}\frac{1}{\lambda_k}}} \exp\left\{-\frac{\left(1-2\ln\frac{\widehat{W}_N}{\alpha}\right)^2}{8v_2L^2\sum\limits_{k=1}^{\infty}\frac{1}{\lambda_k}}\right\},$$

where $\widehat{W}_N$ is defined in lemma 6.9, $v_2 = \frac{v_1}{v_1-1}$, v_1 is any positive real number more than 1.

PROOF.– It was shown under proof of lemma 6.6:

$$\mathbf{P}\left\{\max_{B_{i_1,\ldots,i_n}\in\mathfrak{B}} |p_{kY}(B_{i_1,\ldots,i_n}) - \widetilde{p}_{kY}(B_{i_1,\ldots,i_n})| > \alpha\right\}$$

$$\leq \mathbf{P}\left\{\max_{B_{i_1,\ldots,i_n}\in\mathfrak{B}} |\mu(B_{i_1,\ldots,i_n}) - \widetilde{\mu}(B_{i_1,\ldots,i_n})| > \alpha\right\},$$

that is why by virtue of lemma 6.9, we get:

$$\mathbf{P}\left\{\max_{B_{i_1,\ldots,i_n}\in\mathfrak{B}} |p_{kY}(B_{i_1,\ldots,i_n}) - \widetilde{p}_{kY}(B_{i_1,\ldots,i_n})| > \alpha\right\}$$

$$\leq \frac{2k^n\widehat{W}_N^p p^{\frac{p}{2}} \exp\left\{-\frac{p}{2} + \frac{p^2 v_2 L^2}{2}\sum\limits_{k=1}^{\infty}\frac{1}{\lambda_k}\right\}}{\alpha^p}.$$

Consider the point $p_0 = \frac{1-2\ln\frac{\widehat{W}_N}{\alpha}}{2v_2L^2\sum\limits_{k=1}^{\infty}\frac{1}{\lambda_k}}$, that is close to minimum point of the right-hand side of the last estimation. Since p_0 is greater than one then we can substitute p_0 for p in the above inequality. After this manipulation the statement of the lemma is completely proved. □

THEOREM 6.8.– Let $Y(\vec{t})$ be a Gaussian, centered, continuous in mean square random field, with eigenfunctions of integral equation [4.11] restricted:

$$\left|\phi_k(\vec{t})\right| \leq L \quad \forall \vec{t} \in \mathbf{T},\ k \in \mathbf{N},$$

then the model of Cox process $\{\widetilde{\nu}(B_{i_1,\dots,i_n}), B_{i_1,\dots,i_n} \subset \mathfrak{B}\}$, generated by log Gaussian inhomogeneous field $\exp\left\{\widetilde{Y}(\vec{t})\right\}$, approximates them with accuracy α and reliability $1-\gamma$ if the following inequalities hold true:

$$\widehat{W}_N < \alpha \exp\left\{\frac{1}{2} - v_2 L^2 \sum_{k=1}^{\infty} \frac{1}{\lambda_k}\right\},$$

$$2k^n \left(\frac{1-2\ln\frac{\widehat{W}_N}{\alpha}}{2v_2 L^2 \sum_{k=1}^{\infty} \frac{1}{\lambda_k}}\right)^{\frac{1-2\ln\frac{\widehat{W}_N}{\alpha}}{4v_2 L^2 \sum_{k=1}^{\infty}\frac{1}{\lambda_k}}} \exp\left\{-\frac{\left(1-2\ln\frac{\widehat{W}_N}{\alpha}\right)^2}{8v_2 L^2 \sum_{k=1}^{\infty}\frac{1}{\lambda_k}}\right\} < \gamma,$$

where $\widehat{W}_N$ is defined in lemma 6.9, $v_2 = \frac{v_1}{v_1-1}$, v_1 is any positive real number more than 1.

PROOF.– It is evident that the assumption of the theorem is a corollary of lemma 6.10 and definition 6.5. □

6.6. Simulation of the Cox process when the density is generated by the square Gaussian random process

This section is a logical continuation of section 6.3. It uses a simplified method of simulation. The difference is that the density of the Cox process $\mu(\cdot)$ in this case is generated by the square Gaussian process, that is $\mu(B) = \int_B Y^2(t)\,dt$, where $Y(t)$ is a centered, homogeneous, Gaussian process.

Since the procedure of the simulation was already described in section 6.3, we formulate the result at once in the same way as in section 6.3.

THEOREM 6.9.– Let $\{\nu(B_i), B_i \in \mathfrak{B}\}$ be a Cox process, driven by square Gaussian process $Y^2(t)$. The inequality

$$\mathbf{P}\{\nu(B_i) > 1\} < \delta,$$

holds true if we set

$$d = \frac{T}{k} \le \left(\frac{2\delta}{3B^2(0)}\right)^{\frac{1}{2}}.$$

PROOF.– It was shown under proof of theorem 6.3,

$$\mathbf{P}\{\nu(B_i) > 1\} \le \frac{\mathbf{E}[\mu(B_i)]^2}{2}. \qquad [6.43]$$

$$\mathbf{E}\left[\mu\left(B_i\right)\right]^2=\mathbf{E}\left[\int\limits_{B_i} Y^2(t)\,dt\right]^2=\mathbf{E}\int\limits_{B_i} Y^2(t)\,dt\int\limits_{B_i} Y^2(s)\,ds$$

$$=\mathbf{E}\iint\limits_{B_i\times B_i} Y^2(t)\,Y^2(s)\,dtds=\iint\limits_{B_i\times B_i}\mathbf{E}Y^2(t)\,Y^2(s)\,dtds.$$

[6.44]

By virtue of the Isserlis formula:

$$\begin{aligned}\mathbf{E}Y^2(t)\,Y^2(s)&=\mathbf{E}Y^2(t)\,\mathbf{E}Y^2(s)+2\left(\mathbf{E}Y(t)\,Y(s)\right)^2\\&=B^2(0)+2B^2(t-s)\le 3B^2(0).\end{aligned}$$

Taking into consideration the last estimation, the assertion of the theorem follows from [6.43] and [6.44]. □

LEMMA 6.11.– Let $Y(t)$ be a homogeneous, centered, continuous in mean square Gaussian process with spectral function $F(\lambda)$, a partition D_Λ of domain $[0,\Lambda]$, $\Lambda\in\mathbf{R}$ such that $\lambda_{k-1}-\lambda_k=\frac{\Lambda}{N}$, $N\in\mathbf{N}$, then

$$\left(\operatorname{Var}\left(Y^2(t)-\widetilde{Y}^2(t)\right)\right)^{\frac{1}{2}}\le 8\sqrt[4]{2}\exp\{-1\}\,B^{\frac{1}{2}}(0)\,\widehat{A}^{\frac{1}{2}}_{N,t},$$

where

$$\widehat{A}_{N,t}=B(0)-F(\Lambda)+2^{2-2b}t^{2b}\left(\frac{\Lambda}{N}\right)^{2b}F(\Lambda),\quad b\in[0,1].$$

PROOF.– By using the Holder inequality,

$$\left(\operatorname{Var}\left(Y^2(t)-\widetilde{Y}^2(t)\right)\right)^{\frac{1}{2}}=\left(\mathbf{E}\left|Y^2(t)-\widetilde{Y}^2(t)\right|^2\right)^{\frac{1}{2}}$$

$$=\left(\mathbf{E}\left|Y(t)-\widetilde{Y}(t)\right|^2\left|Y(t)+\widetilde{Y}(t)\right|^2\right)^{\frac{1}{2}}$$

$$\le\left(\mathbf{E}\left|Y(t)-\widetilde{Y}(t)\right|^{2v_1}\right)^{\frac{1}{2v_1}}\left(\mathbf{E}\left|Y(t)+\widetilde{Y}(t)\right|^{2v_2}\right)^{\frac{1}{2v_2}},\quad [6.45]$$

$\frac{1}{v_1}+\frac{1}{v_2}=1$. By virtue of [6.3] proofed in lemma 6.1:

$$\mathbf{E}\left|Y(t)-\widetilde{Y}(t)\right|^{2v_1}=c_{2v_1}\left(\mathbf{E}\left|Y(t)-\widetilde{Y}(t)\right|^2\right)^{v_1},$$

$$\mathbf{E}\left|Y(t)+\widetilde{Y}(t)\right|^{2v_2}=c_{2v_2}\left(\mathbf{E}\left|Y(t)+\widetilde{Y}(t)\right|^2\right)^{v_2}$$

Under proof of lemma 6.4 it was shown that:

$$\mathbf{E}\left|Y(t)-\widetilde{Y}(t)\right|^2 \le \widehat{A}_{N,t},$$

where $\widehat{A}_{N,t} = B(0) - F(\Lambda) + 2^{2-2b}t^{2b}\left(\frac{\Lambda}{N}\right)^{2b} F(\Lambda)$, $b \in [0,1]$. It is evident that

$$\mathbf{E}\left|Y(t)+\widetilde{Y}(t)\right|^2 \le 4B(0).$$

Thus, by estimating c_{2v_1} and c_{2v_2} it follows from [6.45] that:

$$\left(\mathrm{Var}\left(Y^2(t)-\widetilde{Y}^2(t)\right)\right)^{\frac{1}{2}} = \left(c_{2v_1}\widehat{A}_{N,t}^{v_1}\right)^{\frac{1}{2v_1}}\left(c_{2v_2}\left(4B(0)\right)^{v_2}\right)^{\frac{1}{2v_2}}$$

$$\le \left(\sqrt{2}\left(2v_1\right)^{v_1}\exp\{-v_1\}\right)^{\frac{1}{2v_1}}\widehat{A}_{N,t}^{\frac{1}{2}}\left(\sqrt{2}\left(2v_2\right)^{v_2}\exp\{-v_2\}\right)^{\frac{1}{2v_2}} 2\sqrt{B(0)}$$

$$= 4\sqrt[4]{2}\left(v_1v_2B(0)\right)^{\frac{1}{2}}\exp\{-1\}\widehat{A}_{N,t}^{\frac{1}{2}}.$$

If we set $v_1 = v_2 = 2$, we obtain the assertion of the lemma. □

LEMMA 6.12.– Let $Y(t)$ be a homogeneous, centered, continuous in mean square Gaussian process with spectral function $F(\lambda)$, there exists spectral moment $\int_0^\infty \lambda^{2\beta}dF(\lambda)$, $0 < \beta \le 1$, a partition D_Λ of domain $[0,\Lambda]$, $\Lambda \in \mathbf{R}$ such that $\lambda_{k-1} - \lambda_k = \frac{\Lambda}{N}$, $N \in \mathbf{N}$, then the following inequality holds true:

$$\left(\mathrm{Var}\left(Y^2(t+h)-\widetilde{Y}^2(t+h)-\left(Y^2(t)-\widetilde{Y}^2(t)\right)\right)\right)^{\frac{1}{2}} \le H_{N,t}h^{\beta},$$

where

$$H_{N,t} = 2^{3\frac{1}{4}}\exp\{-1\}\sqrt{B(0)}\left(\sqrt{P_{N,t}}+\left(\frac{\Lambda}{2}\right)^{\beta}\sqrt{R_N}\right),$$

$$P_{N,t} = 2^{5-4\beta}\left(\left(\frac{\Lambda}{N}\right)^{\beta}+2^{\beta-1}t\frac{\Lambda^{\beta+1}}{N}\right)^2 F(\Lambda) + 2^{3-2\beta}\int_\Lambda^\infty \lambda^{2\beta}dF(\lambda),$$

$$R_N = 8T^2\frac{\Lambda^2}{N^2}F(\Lambda) + 8\left(F(\infty)-F(\Lambda)\right).$$

PROOF.–

$$\left(\mathrm{Var}\left(Y^2(t+h)-\widetilde{Y}^2(t+h)-\left(Y^2(t)-\widetilde{Y}^2(t)\right)\right)\right)^{\frac{1}{2}}$$

$$=\left(\mathbf{E}\Big|(Y(t+h)-Y(t))(Y(t+h)+Y(t))\right.$$

$$\left.-\left(\widetilde{Y}(t+h)-\widetilde{Y}(t)\right)\left(\widetilde{Y}(t+h)+\widetilde{Y}(t)\right)\Big|^2\right)^{\frac{1}{2}}$$

$$=\left(\mathbf{E}\Big|\Big[Y(t+h)-Y(t)-\left(\widetilde{Y}(t+h)-\widetilde{Y}(t)\right)\Big](Y(t+h)+Y(t))\right.$$

$$+\left(\widetilde{Y}(t+h)-\widetilde{Y}(t)\right)(Y(t+h)+Y(t))$$

$$\left.-\left(\widetilde{Y}(t+h)-\widetilde{Y}(t)\right)\left(\widetilde{Y}(t+h)+\widetilde{Y}(t)\right)\Big|^2\right)^{\frac{1}{2}}$$

$$\leq\left(\mathbf{E}\Big|\left(Y(t+h)-\widetilde{Y}(t+h)-\left(Y(t)-\widetilde{Y}(t)\right)\right)\Big[Y(t+h)+Y(t)\Big]\Big|^2\right)^{\frac{1}{2}}$$

$$+\left(\mathbf{E}\Big|\left(Y(t+h)-\widetilde{Y}(t+h)+\left(Y(t)-\widetilde{Y}(t)\right)\right)\Big[\widetilde{Y}(t+h)-\widetilde{Y}(t)\Big]\Big|^2\right)^{\frac{1}{2}}. \quad [6.46]$$

Let us estimate each of the two summands of the right-hand side of the last estimation. For v_1, v_2 such that $\frac{1}{v_1}+\frac{1}{v_2}=1$, we have:

$$\mathbf{E}\Big|\left(Y(t+h)-\widetilde{Y}(t+h)-\left(Y(t)-\widetilde{Y}(t)\right)\right)\Big[Y(t+h)+Y(t)\Big]\Big|^2$$

$$\leq\left(\mathbf{E}\Big|Y(t+h)-\widetilde{Y}(t+h)-\left(Y(t)-\widetilde{Y}(t)\right)\Big|^{2v_1}\right)^{\frac{1}{v_1}}$$

$$\times\left(\mathbf{E}\Big|Y(t+h)+Y(t)\Big|^{2v_2}\right)^{\frac{1}{v_2}}. \quad [6.47]$$

It was shown under proof of lemma 6.2 that:

$$\mathbf{E}\Big|Y(t+h)-\widetilde{Y}(t+h)-\left(Y(t)-\widetilde{Y}(t)\right)\Big|^2\leq h^{2\beta}P_{N,t},$$

where $P_{N,t} = 2^{5-4\beta}\left(\left(\frac{\Lambda}{N}\right)^{\beta} + 2^{\beta-1}t\frac{\Lambda^{\beta+1}}{N}\right)^2 F(\Lambda) + 2^{3-2\beta}\int\limits_{\Lambda}^{\infty}\lambda^{2\beta}dF(\lambda)$, $0 < \beta \leq 1$. This is why:

$$\left(\mathbf{E}\left|Y(t+h) - \widetilde{Y}(t+h) - \left(Y(t) - \widetilde{Y}(t)\right)\right|^{2v_2}\right)^{\frac{1}{v_1}} \leq c_{2v_1}^{\frac{1}{v_1}} h^{2\beta} P_{N,t} \quad [6.48]$$

Since $\mathbf{E}\left|Y(t+h) + Y(t)\right|^2 \leq 4B(0)$, we get:

$$\left(\mathbf{E}\left|Y(t+h) + Y(t)\right|^{2v_2}\right)^{\frac{1}{v_2}} \leq c_{2v_2}^{\frac{1}{v_2}} 4B(0). \quad [6.49]$$

If we put $v_1 = v_2 = 2$ and taking into consideration [6.48] and [6.49] after elementary manipulation, it follows from [6.47] that:

$$\mathbf{E}\left|\left(Y(t+h) - \widetilde{Y}(t+h) - \left(Y(t) - \widetilde{Y}(t)\right)\right)\left[Y(t+h) + Y(t)\right]\right|^2$$
$$\leq 2^{6\frac{1}{2}} B(0) \exp\{-2\} P_{N,t} h^{2\beta}. \quad [6.50]$$

It is evident that:

$$\mathbf{E}\left|Y(t+h) - \widetilde{Y}(t+h) + \left(Y(t) - \widetilde{Y}(t)\right)\right|^2 \leq R_N,$$
$$R_N = 8T^2\frac{\Lambda^2}{N^2}F(\Lambda) + 8\left(F(\infty) - F(\Lambda)\right).$$

By using representation of process $Y(t)$, it is evident that:

$$\mathbf{E}\left|\widetilde{Y}(t+h) - \widetilde{Y}(t)\right|^2 = 2F(\Lambda) - 2\sum_{k=0}^{N-1}\int_{\lambda_k}^{\lambda_{k+1}} \cos\lambda_k h\, dF(\lambda)$$
$$= 4\sum_{k=0}^{N-1}\int_{\lambda_k}^{\lambda_{k+1}} \sin^2\frac{\lambda_k h}{2} dF(\lambda) \leq 2^{2-2\beta}\Lambda^{2\beta}F(\Lambda)\, h^{2\beta}, \quad 0 < \beta \leq 1.$$

By virtue of the two last inequalities:

$$\mathbf{E}\left|\left(Y(t+h) - \widetilde{Y}(t+h) + \left(Y(t) - \widetilde{Y}(t)\right)\right)\left[\widetilde{Y}(t+h) - \widetilde{Y}(t)\right]\right|^2$$
$$\leq 2^{6\frac{1}{2}-2\beta} \exp\{-2\}\Lambda^{2\beta}F(\Lambda)\, R_N\, h^{2\beta}. \quad [6.51]$$

Taking into consideration [6.50] and [6.51], the assertion of the lemma follows from [6.46]. □

LEMMA 6.13.– Let $Y(t)$ be a homogeneous, centered, separable, continuous in mean square Gaussian process with spectral function $F(\lambda)$, there exists spectral moment $\int\limits_0^\infty \lambda^{2\beta} dF(\lambda)$, $\frac{1}{2} < \beta \leq 1$, a partition D_Λ of domain $[0, \Lambda]$, $\Lambda \in \mathbf{R}$ such that $\lambda_{k-1} - \lambda_k = \frac{\Lambda}{N}$, $N \in \mathbf{N}$, then:

$$\mathbf{P}\left\{\max_{B_i \in \mathfrak{B}\, i=\overline{0,\, k-1}} |\, p_{kY}(B_i) - \widetilde{p}_{kY}(B_i)\,| > \alpha\right\}$$

$$\leq \frac{2^{4+\frac{1}{2\beta}} \beta^2}{(2\beta-1)^2} \left(1 - \frac{(\sqrt{2}-1)\alpha}{U_N}\right)^{-\frac{1}{2}}$$

$$\times \exp\left\{-\frac{(\sqrt{2}-1)^2 \alpha^2}{2d \max(\delta_{0,N}, t_{0,N})\, U_N} - \frac{(\sqrt{2}-1)\alpha}{2U_N}\right\},$$

where:

$$U_N = d \max(\delta_{0,N}, t_{0,N}) + (\sqrt{2}-1)\alpha,$$

$$\delta_{0,N} = 8\sqrt[4]{2} \exp\{-1\} B^{\frac{1}{2}}(0) \widehat{A}_N^{\frac{1}{2}},$$

$$t_{0,N} = 2^{3\frac{1}{4}-\beta} T^\beta \exp\{-1\} \sqrt{B(0)} \left(\sqrt{P_{N,\frac{T}{2}}} + \left(\frac{\Lambda}{2}\right)^\beta \sqrt{R_N}\right),$$

$$\widehat{A}_N = B(0) - F(\Lambda) + 2^{2-2b} T^{2b} \left(\frac{\Lambda}{N}\right)^{2b} F(\Lambda), \quad b \in [0,1],$$

$$P_{N,\frac{T}{2}} = 2^{5-4\beta} \left(\left(\frac{\Lambda}{N}\right)^\beta + 2^{\beta-2} T \frac{\Lambda^{\beta+1}}{N}\right)^2 F(\Lambda) + 2^{3-2\beta} \int\limits_\Lambda^\infty \lambda^{2\beta} dF(\lambda),$$

$$R_N = 8T^2 \frac{\Lambda^2}{N^2} F(\Lambda) + 8(F(\infty) - F(\Lambda)).$$

PROOF.– It was shown under proof of lemma 6.6:

$$\mathbf{P}\left\{\max_{B_i \in \mathfrak{B}\, i=\overline{0,\, k-1}} |p_{kY}(B_i) - \widetilde{p}_{kY}(B_i)| > \alpha\right\}$$

$$\leq \mathbf{P}\left\{\max_{B_i \in \mathfrak{B}\, i=\overline{0,\, k-1}} |\mu(B_i) - \widetilde{\mu}(B_i)| > \alpha\right\},$$

that is for the square Gaussian Cox process, we have:

$$\mathbf{P}\left\{\max_{B_i \in \mathfrak{B}\, i=\overline{0,\,k-1}} \left|p_{kY}(B_i) - \widetilde{p}_{kY}(B_i)\right| > \alpha\right\}$$
$$\leq \mathbf{P}\left\{\sup_{t \in \mathbf{T}} \left|Y^2(t) - \widetilde{Y}^2(t)\right| > \frac{\alpha}{d}\right\}. \quad [6.52]$$

Let us estimate the entropy integral from corollary 3.5. Using lemma 6.12, we have:

$$\int_0^{t_{0}p} r\left(N\left(\sigma^{(-1)}(v)\right)\right) dv \leq \int_0^{t_{0}p} \left(T\frac{H_N^{\frac{1}{\beta}}}{v^{\frac{1}{\beta}}}\right)^{\frac{1}{2}} dv = \frac{T^{\frac{1}{2}} H_N^{\frac{1}{2\beta}} (t_0 p)^{-\frac{1}{2\beta}+1}}{-\frac{1}{2\beta}+1},$$
$$-\frac{1}{2\beta}+1 > 0, H_N = H_{N,t}\big|_{t=\frac{T}{2}}.$$

Thus, after elementary manipulation:

$$r^{(-1)}\left(\frac{1}{t_0 p}\int_0^{t_{0}p} r\left(N\left(\sigma^{(-1)}(v)\right)\right) dv\right) \leq \frac{8\beta^2}{(2\beta-1)^2 p^{\frac{1}{\beta}}}, \beta > \frac{1}{2}.$$

Taking into consideration lemmas 6.11 and 6.12, the last inequality and by set $p = \frac{1}{\sqrt{2}}$, the assertion of the lemma follows from [6.52] and corollary 3.5. □

THEOREM 6.10.– Let $Y(t)$ be a homogeneous, centered, separable, continuous in mean square, Gaussian process with spectral function $F(\lambda)$. There exists spectral moment $\int_0^\infty \lambda^{2\beta} dF(\lambda)$, $\frac{1}{2} < \beta \leq 1$, a partition D_Λ of domain $[0, \Lambda]$, $\Lambda \in \mathbf{R}$ such that $\lambda_{k-1} - \lambda_k = \frac{\Lambda}{N}$, $N \in \mathbf{N}$, then a model of Cox process $\{\widetilde{\nu}(B),\ B \in \mathfrak{B}\}$, generated by square Gaussian process $\widetilde{Y}^2(t)$, approximates them with accuracy α and reliability $1-\gamma$ if:

$$\frac{2^{4+\frac{1}{2\beta}}\beta^2}{(2\beta-1)^2}\left(1 - \frac{(\sqrt{2}-1)\alpha}{U_N}\right)^{-\frac{1}{2}}$$
$$\times \exp\left\{-\frac{\left(\sqrt{2}-1\right)^2 \alpha^2}{2d \max\left(\delta_{0,N}, t_{0,N}\right) U_N} - \frac{(\sqrt{2}-1)\alpha}{2U_N}\right\} < \gamma, \quad [6.53]$$

where U_N, $\delta_{0,N}$, $t_{0,N}$ are defined in lemma 6.13.

PROOF.– The assertion of the theorem follows from definition 6.4 and lemma 6.13. □

6.7. Simulation of the square Gaussian Cox process when density is generated by a homogeneous field

In this section, we consider the square Gaussian Cox process when the density $\mu(\cdot)$ is generated by a homogeneous random field ($\mu(B) = \int_B Y^2(\vec{t})\, d\vec{t}$, where $Y(\vec{t})$ is Gaussian, homogeneous random field). The simplified method of simulation described in section 6.3 is used. Let us formulate the results.

THEOREM 6.11.– Let $\{\nu(B_{i_1,\dots,i_n}), B_{i_1,\dots,i_n} \subset \mathfrak{B}\}$ be a Cox process, driven by square Gaussian homogeneous field $Y^2(\vec{t})$. The inequality

$$\mathbf{P}\{\nu(B_{i_1,\dots,i_n}) > 1\} < \delta,$$

holds true if we set

$$d = \frac{T}{k} \leq \left(\frac{2\delta}{3B^2\left(\vec{0}\right)}\right)^{\frac{1}{2n}}.$$

PROOF.– Proof is analogical to the proof of theorem 6.9. □

LEMMA 6.14.– Let $Y(\vec{t})$ be a homogeneous, centered, continuous in mean square Gaussian field, and for all $\forall p > 1$ the following inequality holds true:

$$\mathbf{P}\left\{\max_{B_{i_1,\dots,i_n} \in \mathfrak{B}} |\mu(B_{i_1,\dots,i_n}) - \widetilde{\mu}(B_{i_1,\dots,i_n})| > \alpha\right\} \leq \frac{\sqrt{2}k^n d^{np}\left(4B\left(\vec{0}\right)v_1v_2\right)^{\frac{p}{2}} J_N^{\frac{p}{2}} p^p \exp\{-p\}}{\alpha^p},$$

where

$$J_N = 2^{2-2a} n^{2a} \frac{d^{2a}\Lambda^{2a}}{N^{2a}} \Phi(\Lambda^n) + B\left(\vec{0}\right) - \Phi(\Lambda^n),$$

$a \in [0, 1]$, v_1, v_2 are such numbers that $\frac{1}{v_1} + \frac{1}{v_2} = 1$.

PROOF.– Analogical to [6.35], we have:

$$\mathbf{P}\left\{\max_{B_{i_1,\dots,i_n} \in \mathfrak{B}} |\mu(B_{i_1,\dots,i_n}) - \widetilde{\mu}(B_{i_1,\dots,i_n})| > \alpha\right\} \leq \frac{k^n \left(\int\limits_{B_{i_1,\dots,i_n}} \left(\mathbf{E}\left|Y^2(\vec{t}) - \widetilde{Y}^2(\vec{t})\right|^p\right)^{\frac{1}{p}} d\vec{t}\right)^p}{\alpha^p}. \quad [6.54]$$

By virtue of the Holder inequality for v_1 and v_2 such that $\frac{1}{v_1} + \frac{1}{v_2} = 1$,

$$\mathbf{E}|Y^2(\vec{t}) - \widetilde{Y}^2(\vec{t})|^p \leq (\mathbf{E}|Y(\vec{t}) - \widetilde{Y}(\vec{t})|^{pv_1})^{\frac{1}{v_1}} (\mathbf{E}|Y(\vec{t}) + \widetilde{Y}(\vec{t})|^{pv_2})^{\frac{1}{v_2}}. \quad [6.55]$$

Since $\mathbf{E}\left|Y\left(\vec{t}\right)-\widetilde{Y}\left(\vec{t}\right)\right|^{pv_1} = c_{pv_1}\left(\mathbf{E}\left|Y\left(\vec{t}\right)-\widetilde{Y}\left(\vec{t}\right)\right|^2\right)^{\frac{pv_1}{2}}$, by using the estimation from lemma 6.7 for $\mathbf{E}\left|Y\left(\vec{t}\right)-\widetilde{Y}\left(\vec{t}\right)\right|^2$, we have

$$\mathbf{E}\left|Y\left(\vec{t}\right)-\widetilde{Y}\left(\vec{t}\right)\right|^{pv_1} \leq c_{pv_1} J_N^{\frac{pv_1}{2}},$$

$$J_N = 2^{2-2a} n^{2a} \frac{d^{2a}\Lambda^{2a}}{N^{2a}} \Phi\left(\Lambda^n\right) + B\left(\vec{0}\right) - \Phi\left(\Lambda^n\right), \quad [6.56]$$

$a \in [0,1]$. It is evident that

$$\mathbf{E}\left|Y\left(\vec{t}\right)+\widetilde{Y}\left(\vec{t}\right)\right|^{pv_2} \leq c_{pv_2}\left(4B\left(\vec{0}\right)\right)^{\frac{pv_2}{2}}. \quad [6.57]$$

Taking into considerations [6.56] and [6.57] and also estimation [6.4] for c_{pv_1} and c_{pv_2} after elementary manipulation, it follows from [6.55] that:

$$\mathbf{E}\left|Y^2\left(\vec{t}\right)-\widetilde{Y}^2\left(\vec{t}\right)\right|^p \leq \sqrt{2}\left(4B\left(\vec{0}\right)v_1v_2\right)^{\frac{p}{2}} p^p \exp\left\{-p\right\} J_N^{\frac{p}{2}}.$$

Taking into consideration the last estimation, the assertion of the lemma follows from [6.54]. □

LEMMA 6.15.– Let $Y\left(\vec{t}\right)$ be homogeneous, centered, continuous in mean square of a Gaussian field, then if $\alpha > 2d^n\left(B\left(\vec{0}\right)J_N\right)^{\frac{1}{2}}$, then there is a valuation:

$$\mathbf{P}\left\{\max_{B_{i_1,\ldots,i_n}\in\mathfrak{B}} \left|p_{kY}\left(B_{i_1,\ldots,i_n}\right)-\widetilde{p}_{kY}\left(B_{i_1,\ldots,i_n}\right)\right| > \alpha\right\}$$

$$\leq \sqrt{2}k^n \exp\left\{-\frac{\alpha}{2d^n\left(B\left(\vec{0}\right)J_N\right)^{\frac{1}{2}}}\right\},$$

where J_N is defined by the condition of lemma 6.14.

PROOF.– By using the proof in lemma 6.6 inequality, we get:

$$\mathbf{P}\left\{\max_{B_{i_1,\ldots,i_n}\in\mathfrak{B}} \left|p_{kY}\left(B_{i_1,\ldots,i_n}\right)-\widetilde{p}_{kY}\left(B_{i_1,\ldots,i_n}\right)\right| > \alpha\right\}$$

$$\leq \mathbf{P}\left\{\max_{B_{i_1,\ldots,i_n}\in\mathfrak{B}} \left|\mu\left(B_{i_1,\ldots,i_n}\right)-\widetilde{\mu}\left(B_{i_1,\ldots,i_n}\right)\right| > \alpha\right\}. \quad [6.58]$$

We find the minimum of function $\frac{\sqrt{2}k^n d^{np}(4B(\vec{0})v_1v_2)^{\frac{p}{2}} J_N^{\frac{p}{2}} p^p \exp\{-p\}}{\alpha^p}$ on variable p. By set $v_1 = v_2 = 2$, it is evident that the given lemma is a corollary of lemma 6.14.

□

THEOREM 6.12.– Let $Y(\vec{t})$ be a homogeneous, centered, continuous in mean square Gaussian field, a model of random Cox process $\{\nu(B_{i_1,\ldots,i_n}), B_{i_1,\ldots,i_n} \subset \mathfrak{B}\}$, driven by square Gaussian homogeneous field $\widetilde{Y}^2(\vec{t})$, approximates them with accuracy α and reliability $1-\gamma$, if the following inequalities hold true:

$$\alpha > 2d^n \left(B\left(\vec{0}\right) J_N\right)^{\frac{1}{2}},$$

$$\sqrt{2}k^n \exp\left\{-\frac{\alpha}{2d^n\left(B(\vec{0})J_N\right)^{\frac{1}{2}}}\right\} < \gamma,$$

where J_N is defined in lemma 6.14.

PROOF.– It is evident that the theorem follows from definition 6.5 and lemma 6.15.

□

6.8. Simulation of the square Gaussian Cox process when the density is generated by an inhomogeneous field

In this case, the algorithm of simulation of the square Gaussian Cox process differs only by construction of the inhomogeneous field model $\{Y(\vec{t}), \vec{t} \in \mathbf{T}\}$.

THEOREM 6.13.– Let $\{\nu(B_{i_1,\ldots,i_n}), B_{i_1,\ldots,i_n} \subset \mathfrak{B}\}$ be a Cox process, directed by square Gaussian inhomogeneous field $Y^2(\vec{t})$, with the eigenfunction of integral equation [4.11] restricted,

$$\left|\phi_k\left(\vec{t}\right)\right| \le L, \quad \forall \vec{t} \in \mathbf{T}, \forall k \in \mathbf{N}.$$

The inequality

$$\mathbf{P}\{\nu(B_{i_1,\ldots,i_n}) > 1\} < \delta,$$

holds true if we set

$$d = \frac{T}{k} \le \left(\frac{\delta \exp\{2\}}{8\sqrt{2}\left(L^2 \sum_{k=1}^{\infty} \frac{1}{\lambda_k}\right)^2}\right)^{\frac{1}{2n}}.$$

PROOF.– Under proof of theorem 6.9 for the square Gaussian Cox process, we obtain estimation:

$$\mathbf{P}\left\{\nu\left(B_{i_1,\ldots,i_n}\right) > 1\right\} \le \iint\limits_{B_{i_1,\ldots,i_n}\times B_{i_1,\ldots,i_n}} \frac{\mathbf{E}Y^2\left(\vec{t}\right)\mathbf{E}Y^2\left(\vec{s}\right)}{2}d\vec{t}d\vec{s}.$$

By virtue of Holder inequality for u_1, u_2 such that $\frac{1}{u_1}+\frac{1}{u_2}=1$,

$$\mathbf{P}\left\{\nu\left(B_{i_1,\ldots,i_n}\right) > 1\right\} \le \iint\limits_{B_{i_1,\ldots,i_n}\times B_{i_1,\ldots,i_n}} \frac{\left(\mathbf{E}Y^{2u_1}\left(\vec{t}\right)\right)^{\frac{1}{u_1}}\left(\mathbf{E}Y^{2u_2}\left(\vec{s}\right)\right)^{\frac{1}{u_2}}}{2}d\vec{t}d\vec{s}. \qquad [6.59]$$

Since by the use of representation [4.14],

$$\mathbf{E}Y^2\left(\vec{t}\right) = \mathbf{E}\left(\sum_{k=1}^{\infty}\frac{\xi_k}{\sqrt{\lambda_k}}\phi_k\left(\vec{t}\right)\right)^2 = \sum_{k=1}^{\infty}\frac{\phi_k^2\left(\vec{t}\right)}{\lambda_k} \le L^2\sum_{k=1}^{\infty}\frac{1}{\lambda_k},$$

then by virtue of [6.3],

$$\mathbf{E}Y^{2u_1}\left(\vec{t}\right) = c_{2u_1}\left(L^2\sum_{k=1}^{\infty}\frac{1}{\lambda_k}\right)^{u_1}.$$

Analogically, $\mathbf{E}Y^{2u_2}\left(\vec{s}\right) = c_{2u_2}\left(L^2\sum_{k=1}^{\infty}\frac{1}{\lambda_k}\right)^{u_2}$. We estimate c_{2u_1} and c_{2u_2} by using [6.4]. By set $u_1=u_2=2$, the assertion of the lemma follows from [6.59]. □

LEMMA 6.16.– Let $Y\left(\vec{t}\right)$ be a centered, continuous in mean square random field, with the eigenfunction of integral equation [4.11] restricted,

$$\left|\phi_k\left(\vec{t}\right)\right| \le L, \quad \forall\vec{t}\in\mathbf{T}, \forall k\in\mathbf{N}.$$

then $\forall p>1$ the following estimation holds true:

$$\mathbf{P}\left\{\max_{B_{i_1,\ldots,i_n}\in\mathfrak{B}}\left|\mu\left(B_{i_1,\ldots,i_n}\right)-\widetilde{\mu}\left(B_{i_1,\ldots,i_n}\right)\right| > \alpha\right\}$$

$$\le \frac{\sqrt{2}k^n d^{np}(v_1v_2)^{\frac{p}{2}}\left(2L^2\sqrt{\left(\sum\limits_{k=N+1}^{\infty}\frac{1}{\lambda_k}\right)\left(\sum\limits_{k=1}^{\infty}\frac{1}{\lambda_k}\right)}\right)^p p^p\exp\{-p\}}{\alpha^p},$$

where v_1, v_2 are such numbers that $\frac{1}{v_1}+\frac{1}{v_2}=1$.

PROOF.– Under proof of lemma 6.14 for v_1, v_2 such that $\frac{1}{v_1} + \frac{1}{v_2} = 1$, the following inequality is proofed:

$$\mathbf{P}\left\{\max_{B_{i_1,\ldots,i_n} \in \mathfrak{B}} \left|\mu\left(B_{i_1,\ldots,i_n}\right) - \widetilde{\mu}\left(B_{i_1,\ldots,i_n}\right)\right| > \alpha\right\}$$

$$\leq \frac{k^n \left(\int\limits_{B_{i_1,\ldots,i_n}} \left(\mathbf{E}\left|Y\left(\vec{t}\right) - \widetilde{Y}\left(\vec{t}\right)\right|^{pv_1}\right)^{\frac{1}{pv_1}} \left(\mathbf{E}\left|Y\left(\vec{t}\right) + \widetilde{Y}\left(\vec{t}\right)\right|^{pv_2}\right)^{\frac{1}{pv_2}} d\vec{t}\right)^p}{\alpha^p}. \quad [6.60]$$

Since $\mathbf{E}\left|Y\left(\vec{t}\right) - \widetilde{Y}\left(\vec{t}\right)\right|^{pv_1} = c_{pv_1}\left(\mathbf{E}\left|Y\left(\vec{t}\right) - \widetilde{Y}\left(\vec{t}\right)\right|^2\right)^{\frac{pv_1}{2}}$, then by using estimation $\mathbf{E}\left|Y\left(\vec{t}\right) - \widetilde{Y}\left(\vec{t}\right)\right|^2 \leq L^2 \sum\limits_{k=N+1}^{\infty} \frac{1}{\lambda_k}$ obtained in lemma 6.9, we have:

$$\mathbf{E}\left|Y\left(\vec{t}\right) - \widetilde{Y}\left(\vec{t}\right)\right|^{pv_1} \leq c_{pv_1} L^{pv_1} \left(\sum_{k=N+1}^{\infty} \frac{1}{\lambda_k}\right)^{\frac{pv_1}{2}}. \quad [6.61]$$

By using representations [4.13] and [4.14] of the field and their models, we get:

$$\begin{aligned}
\mathbf{E}\left|Y\left(\vec{t}\right) + \widetilde{Y}\left(\vec{t}\right)\right|^2 &= \mathbf{E}\left|\sum_{k=1}^{\infty} \frac{\xi_k}{\sqrt{\lambda_k}} \phi_k\left(\vec{t}\right) + \sum_{k=1}^{N} \frac{\xi_k}{\sqrt{\lambda_k}} \phi_k\left(\vec{t}\right)\right|^2 \\
&= \mathbf{E}\left|2\sum_{k=1}^{N} \frac{\xi_k}{\sqrt{\lambda_k}} \phi_k\left(\vec{t}\right) + \sum_{k=N+1}^{\infty} \frac{\xi_k}{\sqrt{\lambda_k}} \phi_k\left(\vec{t}\right)\right|^2 \\
&= 4\sum_{k=1}^{N} \frac{\phi_k^2\left(\vec{t}\right)}{\lambda_k} + \sum_{k=N+1}^{\infty} \frac{\phi_k^2\left(\vec{t}\right)}{\lambda_k} \leq 4L^2 \sum_{k=1}^{\infty} \frac{1}{\lambda_k}.
\end{aligned}$$

Taking into consideration the last inequality,

$$\begin{aligned}
\mathbf{E}\left|Y\left(\vec{t}\right) + \widetilde{Y}\left(\vec{t}\right)\right|^{pv_2} &= c_{pv_2}\left(\mathbf{E}\left|Y\left(\vec{t}\right) + \widetilde{Y}\left(\vec{t}\right)\right|^2\right)^{\frac{pv_2}{2}} \\
&\leq c_{pv_2} (2L)^{pv_2} \left(\sum_{k=1}^{\infty} \frac{1}{\lambda_k}\right)^{\frac{pv_2}{2}}. \quad [6.62]
\end{aligned}$$

By using [6.61] and [6.62] and also estimation [6.4] for c_{pv_1} and c_{pv_2} after elementary manipulation, the assertion of the lemma follows from [6.60]. □

LEMMA 6.17.– Let $Y(\vec{t})$ be a centered, continuous in mean square random field, with the eigenfunctions of integral equation [4.11] restricted:

$$\left|\phi_k(\vec{t})\right| \leq L \quad \forall \vec{t} \in \mathbf{T},\, k \in \mathbf{N}.$$

If $\alpha > 4d^n L^2 \sqrt{\left(\sum_{k=N+1}^{\infty} \frac{1}{\lambda_k}\right)\left(\sum_{k=1}^{\infty} \frac{1}{\lambda_k}\right)}$, then

$$\mathbf{P}\left\{\max_{B_{i_1,\ldots,i_n} \in \mathfrak{B}} \left|p_{kY}(B_{i_1,\ldots,i_n}) - \widetilde{p}_{kY}(B_{i_1,\ldots,i_n})\right| > \alpha\right\}$$

$$\leq \sqrt{2}k^n \exp\left\{-\frac{\alpha}{4d^n L^2 \sqrt{\left(\sum_{k=N+1}^{\infty} \frac{1}{\lambda_k}\right)\left(\sum_{k=1}^{\infty} \frac{1}{\lambda_k}\right)}}\right\}.$$

PROOF.– Find the minimum of function:

$$\frac{\sqrt{2}k^n d^{np}(v_1 v_2)^{\frac{p}{2}} \left(2L^2 \sqrt{\left(\sum_{k=N+1}^{\infty} \frac{1}{\lambda_k}\right)\left(\sum_{k=1}^{\infty} \frac{1}{\lambda_k}\right)}\right)^p}{\alpha^p} \times p^p \exp\{-p\}$$

on variable p. If we put $v_1 = v_2 = 2$, it is evident that the given lemma is a corollary of [6.58] and lemma 6.16. □

THEOREM 6.14.– Let $Y(\vec{t})$ be a centered, continuous in mean square Gaussian field, with the eigenfunctions of integral equation [4.11] restricted:

$$\left|\phi_k(\vec{t})\right| \leq L \quad \forall \vec{t} \in \mathbf{T},\, k \in \mathbf{N},$$

then the model of Cox process $\{\widetilde{\nu}(B_{i_1,\ldots,i_n}), B_{i_1,\ldots,i_n} \subset \mathfrak{B}\}$, generated by square Gaussian inhomogeneous field $\widetilde{Y}^2(\vec{t})$, approximates them with accuracy α and reliability $1-\gamma$, if the next condition holds true:

$$\alpha > 4d^n L^2 \sqrt{\left(\sum_{k=N+1}^{\infty} \frac{1}{\lambda_k}\right)\left(\sum_{k=1}^{\infty} \frac{1}{\lambda_k}\right)},$$

$$\sqrt{2}k^n \exp\left\{-\frac{\alpha}{4d^n L^2 \sqrt{\left(\sum_{k=N+1}^{\infty} \frac{1}{\lambda_k}\right)\left(\sum_{k=1}^{\infty} \frac{1}{\lambda_k}\right)}}\right\} < \gamma.$$

PROOF.– The assertion of the theorem follows from lemma 6.17 and definition 6.5. □

7

On the Modeling of Gaussian Stationary Processes with Absolutely Continuous Spectrum

A model of a Gaussian stationary process with absolutely continuous spectrum is proposed that simulates the process with given reliability and accuracy in $L^2(0,T)$. Under certain restrictions on the covariance function of the process, formulas for computing the parameters of the model are described.

Let $\xi(t)$ be a Gaussian stationary random process, $E\xi(t) = 0$, with a continuous covariance function $R(\tau) = E\xi(t+\tau)\xi(t)$ and a spectral function $F(\lambda)$, i.e. $R(\tau) = \int\limits_0^\infty \cos\lambda\tau \, dF(\lambda)$. Assume that there exists an integral $\int\limits_0^\lambda f(u)\,du$, where $f(\lambda)$ is a spectral density of $\xi(t)$.

Consider the model of the process in such way

$$\xi_N(t) = \sum_{k=0}^{N-1} \tau_k \left(\eta_k^{(1)} \cos\lambda_k t + \eta_k^{(2)} \sin\lambda_k t\right), \qquad [7.1]$$

where $\eta_k^{(i)}$, $i = 1, 2$, are independent Gaussian variables such that $E\eta_k^{(i)} = 0$ and $Var\eta_k^{(i)} = 1$, and $0 = \lambda_0 < \lambda_1 < \cdots < \lambda_{N-1} < \lambda_N = \Lambda$ is a partition of the interval $[0,\Lambda]$. We shall assume that $\lambda_k - \lambda_{k-1} = \Lambda/N$, where Λ and N are chosen so as to ensure a given accuracy and reliability of the model, and

$$\tau_k^2 = F(\lambda_{k+1}) - F(\lambda_k) = \int_{\lambda_k}^{\lambda-k+1} f(u)\,du$$

$$= \frac{2}{\pi}\int_0^\infty (\sin\lambda_{k+1}\tau - \sin\lambda_k\tau) R(\tau)\,d\tau.$$

Also, we shall suppose that the covariance function $R(\tau)$ is known, whereas $f(\lambda)$ cannot be computed explicitly in general. The essential results of [KOZ 94] will be used to compute all parameters of model [7.1] for random processes for which $R(0) - R(\tau)$ increases a neighborhood of zero.

Let $F(u)$ be the distribution function of some random variable and let $\varphi(t)$ be its characteristic function.

LEMMA 7.1.– For any $k > 0$ and $a > 0$, the following equality holds:

$$\int_0^{2a} \Big(\int_{v_1-a}^{v_2+a} \Big(\int_{v_2-a}^{v_2-a} \cdots \Big(\int_{v_k-a}^{v_k+a} (\text{sign } v_{k+1} - R(v_{(k+1)})\, dv_{k+1} \Big) dv_k \Big) \cdots dv_1 \Big)$$
$$= \frac{2^{k+1}a^{k+1}}{\pi} \int_{-\infty}^{\infty} \frac{\sin^{k+2} u}{u^{k+2}} \Big(1 - \varphi\big(\frac{u}{a}\big)\Big)\, du, \qquad [7.2]$$

where $R(v) = F(v) - F(-v)$.

PROOF.– For any distribution function $F(v)$ and its characteristic function $\varphi(t)$, the relation

$$\int_0^{2a} [F(V) - F(-V)]\, dv = \frac{1}{\pi} \int_{-\infty}^{\infty} \frac{1 - \cos 2at}{t^2} \varphi(t)\, dt$$
$$= \frac{2}{\pi} \int_{-\infty}^{\infty} \frac{\sin^2 at}{t^2} \varphi(t)\, dt \qquad [7.3]$$

holds true. Now replace $\varphi(y)$ by the characteristic function $(\sin at/(at))^k \varphi(t)$. Observe that $\sin at/(at)$ is the characteristic function of the uniform distribution on $[-a, a]$. Therefore, $(\sin at/(at))^k \varphi(t)$ is the characteristic function of the random variable $\eta = \xi + \sum_{s=1}^{k} \theta_s$, where ξ and θ_s, $s = 1, 2, \ldots, k$, are independent random variables. The distribution function of ξ is $F(x)$. The random variables θ_s are uniformly distributed on the interval $[-a, a]$.

Let $F_1(x)$ be the distribution function of the sum $\xi + \theta_1$. Then

$$F_1(x) = \frac{1}{2a} \int_{-a}^{a} F(x - y)\, dy = \frac{1}{2a} \int_{x-a}^{x+a} F(u)\, du$$

and

$$F_1(-x) = \frac{1}{2a} \int_{x-a}^{x+a} F(-u)\, du.$$

Therefore, from [7.3] it follows that for $k = 1$

$$\frac{1}{2a}\int_0^{2a}\Big(\int_{v_1-a}^{v_1+a} R(v_2)\,dv_2\Big)\,dv_1 = \frac{2}{\pi}\int_{-\infty}^{\infty}\frac{\sin^3 at}{t^2 a}\varphi(t)\,dt. \qquad [7.4]$$

Now substitute the characteristic function $(\sin at/(at))\varphi(t)$ for $\varphi(t)$ in [7.4] to obtain

$$\int_0^{2a}\Big(\int_{v_1-a}^{v_2+a}\Big(\int_{v_2-a}^{v_2+a}\cdots\Big(\int_{v_k-a}^{v_k+a} R(v_{k+1})\,dv_{k+1}\Big)\,dv_k\Big)\cdots\Big)\,dv_1$$
$$= \frac{2^{k+1}}{\pi}\int_{-\infty}^{\infty}\Big(\frac{\sin at}{t}\Big)^{k+2}\varphi\,dt. \qquad [7.5]$$

If $\varphi(t) \equiv 1$, then

$$\int_0^{2a}\Big(\int_{v_1-a}^{v_2+a}\Big(\int_{v_2-a}^{v_2+a}\cdots\Big(\int_{v_k-a}^{v_k+a} \text{sign }(v_{k+1}\,dv_{k+1}\Big)\,dv_k\Big)\cdots\Big)\,dv_1$$
$$= \frac{2^{k+1}}{\pi}\int_{-\infty}^{\infty}\Big(\frac{\sin at}{t}\Big)^{k+2}\,dt. \qquad [7.6]$$

since $R(u)$ =sign u.

Next, by subtracting [7.5] from [7.6] and making a change in variables in the integral on the right-hand side of the equality obtained, we obtain [4.2]. □

COROLLARY 7.1.– Let a random variable ξ have a symmetric distribution. Then, $\varphi(t)$ is an even real function and the following equality holds:

$$\int_0^{2a}\Big(\int_{v_1-a}^{v_2+a}\Big(\int_{v_2-a}^{v_2-a}\cdots\Big(\int_{v_k-a}^{v_k+a}(\text{sign } v_{k+1} - R(v_{(k+1)})\,dv_{k+1}\Big)\,dv_k\Big)\cdots dv_1\Big)$$
$$= \frac{2^{k+1}a^{k+1}}{\pi}\int_{-\infty}^{\infty}\Big(\frac{\sin\, u}{u}\Big)^{k+2}\Big(1-\varphi\big(\frac{u}{a}\big)\Big)\,du. \qquad [7.7]$$

LEMMA 7.2.– Let $\varphi(t)$ be the characteristic function of a symmetric random variable and let $\psi(t)$ be a monotonically increasing function such that $\psi(0) = 0$ for $0 \le t < t_0$ and $\psi(uv) \le \psi_1(u)\psi_2(v)$ for $|uv| < t_0$, where $\psi_i(u)$, $i = 1, 2$, are monotonically increasing functions such that $\psi_i(0) = 0$. If for $|t| < t_0$ the inequality $1-\varphi(t) \leq \psi(t)$ holds, then for any $a > t_0$ and $k = 0, 1, 2, \ldots$

$$\Big|\int_{-\infty}^{\infty} \Big(\frac{\sin u}{u}\Big)^{k+2}\Big(1-\varphi\big(\frac{u}{a}\big)\Big)\, du\Big| \leq \delta_{k,\psi}(a),$$

where

$$\delta_{k,\psi}(a) = \psi_1\Big(\frac{1}{a}\Big)\Big(\int_0^1 \psi_2(u)\, du + \int_1^{at_0} \frac{\psi_2(u)}{u^{k+2}}\, du\Big) + \frac{2}{(k+1)(at_0)^{k+2}}.$$

PROOF.– By the properties of function $\psi(uv)$, we obtain

$$\begin{aligned}&\Big|\int_{-\infty}^{\infty} \Big(\frac{\sin u}{u}\Big)^{k+2}\Big(1-\varphi\big(\frac{u}{a}\big)\Big)\, du\Big| \\ &\leq \int_0^1 \Big|\Big(\frac{\sin u}{u}\Big)^{k+2}\Big|\psi\Big(\frac{u}{a}\Big)\, du + \int_1^{at_0} \psi\Big(\frac{u}{a}\Big)\frac{1}{u^{k+2}}\, du + \int_{at_0}^{\infty} \frac{2}{u^{k+2}}\, du \\ &\leq \psi_1\Big(\frac{1}{a}\Big)\int_0^1 \psi_2(u)\, du + \int_1^{at_0} \psi_1\Big(\frac{1}{a}\Big)\frac{\psi_2(u)}{u^{k+2}}\, du + \frac{2}{(k+1)(at_0)^{k+2}}.\end{aligned}$$

□

COROLLARY 7.2.– If for $|t| < t_0$ and some functions $\psi^s(t)$, $s = 1, 2, \ldots, M$, satisfying the assumptions of lemma 7.2, the inequality $1 - \varphi(t) \leq \sum_{s=1}^{M} \psi^s(t)$ holds, then for $a > t_0^{-1}$ the following inequality:

$$\Big|\int_{-\infty}^{\infty} \Big(\frac{\sin u}{u}\Big)^{k+2}\Big(1-\varphi\big(\frac{u}{a}\big)\Big)\, du\Big| \leq \sum_{s=1}^{M} \frac{\delta_{k,\psi}}{a}$$

is valid.

EXAMPLE 7.1.– If the assumptions of lemma 7.2 are satisfied for $\psi(t) = c_\gamma |t|^\gamma$, $0 < \gamma \leq 2$, then for $k > \gamma - 1$

$$\delta_{k,\psi}(a) = \frac{c_\gamma}{a^\gamma}\Big(\frac{1}{\gamma+1} + \frac{1}{k+1-\gamma}\Big) + \frac{1}{(at_0)^{k+1}}\Big(\frac{2}{k+1} - \frac{c_\gamma t_0^\gamma}{k+1-\gamma}\Big).$$

LEMMA 7.3.– Let $R(v) = F(v) - F(-v)$, where $F(v)$ is distribution function of some random variable. Then for each $a > 0$, the following inequalities hold:

$$\int_0^{2a} (\operatorname{sign} v_1 - R(v_1))\, dv_1 \geq (1 - R(2a))2a, \qquad [7.8]$$

$$\int_0^{2a} \Big(\int_{v_1-a}^{v_1+a} (\text{sign } v_2 - R(v_2))\, dv_2 \Big)\, dv_1 \geq (1 - R(3a))3a^2, \qquad [7.9]$$

and

$$\int_0^{2a} \Big(\int_{v_1-a}^{v_1+a} \Big(\int_{v_2-a}^{v_2+a} (\text{sign } v_3 - R(v_3))dv_3 \Big) dv_2 \Big) dv_1 \geq (1 - R(4a))\frac{16}{3}\, a^3. \quad [7.10]$$

PROOF.– Since the function $R(v)$ is monotonically for positive values of v, we have

$$\int_0^{2a} (\text{sign } v_1 - R(v_1))\, dv_1 = \int_0^{2a} (1 - R(v_1))\, dv_1 \geq (1 - R(2a))2a.$$

Inequality [7.8] is proved. To derive the second inequality, we note that $-a \leq v_2 \leq 3a$, since $0 \leq v_1 \leq 2a$. For $v_1 > a$,

$$\int_{v_1-a}^{v_1+a} (\text{sign } v_2 - R(v_2))\, dv_2 = \int_{v_1-a}^{v_1+a} (1 - R(v_2))\, dv_2 \geq (1 - R(v_1 + a))2a.$$

For $0 \leq v_1 \leq a$

$$\begin{aligned}
&\int_{v_1-a}^{v_1+a} (\text{sign } v_2 - R(v_2))\, dv_2 \\
&= \int_{v_1-a}^{0} (-1 - R(v_2))\, dv_2 + \int_0^{v_1+a} (1 - R(v_2))\, dv_2 \\
&= \int_0^{v_1-a} (-1 + R(v_2))\, dv_2 + \int_0^{v_1+a} (1 - R(v_2))\, dv_2 \\
&= \int_{a-v_1}^{a+v_1} (1 - R(v_2))\, dv_2 \geq (1 - R(a + v_1))2v_1.
\end{aligned}$$

Therefore,

$$\begin{aligned}
&\int_0^{2a} \Big(\int_{v_1-a}^{v_1+a} (\text{sign } v_2 - R(v_2))\, dv_2 \Big)\, dv_1 \\
&\geq \int_a^{2a} (1 - R(v_1 + a))\, 2a\, dv_1 \\
&+ \int_0^{a} (1 - R(v_1 + a))\, 2v_1\, dv_1 \geq (1 - R(3a))\, 3a^2.
\end{aligned}$$

Inequality [7.9] is proved. Let us establish now inequality [7.10]. Set

$$I_1(a, v_2) = \int_{v_2-a}^{v_2+a} (\text{sign } v_3 - R(v_3))\, dv_3.$$

It is not hard to see that for $v_2 > a$

$$I_1(a, v_2) = \int_{v_2-a}^{v_2+a} (1 - R(v_3))\, dv_3.$$

For $0 \le v_2 < a$

$$\begin{aligned} I_1(a, v_2) &= \int_{v_2-a}^{0} (-1 - R(v_3))\, dv_3 + \int_{0}^{v_2+a} (1 - R(v_3))\, dv_3 \\ &= -\int_{0}^{a-v_2} (1 - R(v_3))\, dv_3 + \int_{0}^{v_2+a} (1 - R(v_3))\, dv_3 \\ &= \int_{a-v_2}^{a+v_2} (1 - R(v_3))\, dv_3. \end{aligned}$$

Similarly, for $-a \le v_2 < 0$

$$I_1(a, v_2) = -\int_{a+v_2}^{a-v_2} (1 - R(v_3))\, dv_3.$$

Set

$$I_2(a, v_1) = \int_{v_1-a}^{v_1+a} I_1(a, v_2)\, dv_2.$$

For $0 \le v_1 \le a$,

$$\begin{aligned} I_2(a, v_1) = &\int\limits_{v_1-a}^{0} \Big[-\int\limits_{v_2+a}^{a-v_2} (1 - R(v_3))\, dv_3\Big]\, dv_2 \\ &+ \int\limits_{0}^{a} \Big(\int\limits_{a-v_2}^{a+v_2} (1 - R(v_3))\, dv_3\Big)\, dv_2 \\ &+ \int\limits_{a}^{a+v_1} \Big(\int\limits_{v_2-a}^{v_2+a} (1 - R(v_3))\, dv_3\Big)\, dv_2 \\ = &-\int\limits_{0}^{a-v_1} \Big(\int\limits_{a-v_2}^{a+v_2} (1 - R(v_3))\, dv_3\Big)\, dv_2 \\ &+ \int\limits_{0}^{a} \Big(\int\limits_{a-v_2} a + v_2(1 - R(v_3))\, dv_3\Big)\, dv_2 \end{aligned}$$

$$+\int_{a}^{a+v_1}\Big(\int_{v_2-a}^{v_2+a}(1-R(v_3))\,dv_3\Big)\,dv_2$$

$$=\int_{a-v_1}^{a}\Big(\int_{v_2-a}^{v_2+a}(1-R(v_3))\,dv_3\Big)\,dv_2$$

$$+\int_{a}^{a+v_1}\Big(\int_{v_2-a}^{v_2+a}(1-R(v_3))\,dv_3\Big)\,dv_2.$$

For $0 \leq v_1 \leq 2a$,

$$I_2(a,v_1)=\int_{a-v_1}^{a}\Big(\int_{v_2-a}^{v_2+a}(1-R(v_3))\,dv_3\Big)\,dv_2$$

$$+\int_{a}^{a+v_1}\Big(\int_{v_2-a}^{v_2+a}(1-R(v_3))\,dv_3\Big)\,dv_2.$$

Therefore,

$$\int_{0}^{2a}\Big(\int_{v_1-a}^{v_1+a}\Big(\int_{v_2-a}^{v_2+a}(\mathrm{sign}\ v_3-R(v_3))\,dv_3\Big)\,dv_2\Big)\,dv_1$$

$$=\int_{0}^{a}\Big(\int_{a-v_1}^{a}\Big(\int_{a-v_2}^{a+v_2}(1-R(v_3))\,dv_3\Big)\,dv_2\Big)\,dv_1$$

$$+\int_{a}^{2a}\Big(\int_{v_1-a}^{a}\Big(\int_{a-v_2}^{a+v_2}(1-R(v_3))\,dv_3\Big)\,dv_2\Big)\,dv_1$$

$$+\int_{0}^{2a}\Big(\int_{a}^{a+v_1}\Big(\int_{v_2-a}^{v_2+a}(1-R(v_3))\,dv_3\Big)\,dv_2\Big)\,dv_1$$

$$\geq\int_{0}^{a}\Big(\int_{a-v_1}^{a}(1-R(a+v_2))\,2v_2\,dv_2\Big)\,dv_1$$

$$+\int_{0}^{2a}\Big(\int_{v_1-a}^{v_1+a}(1-R(a+v_2))\,2v_2\,dv_2\Big)\,dv_1$$

$$+\int_0^{2a}\Big(\int_a^{a+v_1}(1-R(a+v_2))\,2a\,dv_2\Big)\,dv_1$$

$$\geq(1-R(2a))\Big[\int_0^{a}\Big(\int_{a-v_1}^{a}2v_2\,dv_2\Big)\,dv_1+\int_a^{2a}\Big(\int_{v_1-a}^{a}2v_2\,dv_2\Big)\,dv_1\Big]$$

$$+(1-R(4a))\int_0^{2a}2av_1\,dv_1\geq(1-R(4a))\frac{16}{3}\,a^3.$$

Inequality is proved. □

COROLLARY 7.3.– Let a random variable ξ have a symmetric distribution, $h > o$, and $R(h) = F(h) - F(-h)$, where $F(x)$ is the distribution function of ξ. Then, following inequalities hold true:

$$1-R(h)\leq\frac{2}{\pi}\int_0^{\infty}\Big(\frac{\sin\,u}{u}\Big)^2\Big(1-\varphi\Big(\frac{2u}{h}\Big)\Big)\,du, \qquad [7.11]$$

$$1-R(h)\leq\frac{8}{3\pi}\int_0^{\infty}\Big(\frac{\sin\,u}{u}\Big)^3\Big(1-\varphi\Big(\frac{3u}{h}\Big)\Big)\,du, \qquad [7.12]$$

and

$$1-R(h)\leq\frac{3}{\pi}\int_0^{\infty}\Big(\frac{\sin\,u}{u}\Big)^4\Big(1-\varphi\Big(\frac{4u}{h}\Big)\Big)\,du. \qquad [7.13]$$

The above inequalities are results of [7.7]–[7.10]. With the aid of [7.11]–[7.13] and lemma 7.2, we may obtain estimates for $1 - R(h)$.

LEMMA 7.4.– If the characteristic function of the symmetric random variable satisfies the conditions of lemma 7.2, then for $h > t_k/t_0$, $k = 1, 2, 3$,

$$1-R(h)\leq\frac{s_k}{\pi}\,\delta_{k,\psi}\Big(\frac{h}{t_k}\Big), \qquad [7.14]$$

where $s_1 = 2$, $s_2 = 8/3$, $s_3 = 3$, $t_1 = 2$, $t_2 = 3$ and $t_3 = 4$.

When analyzing example 7.1, it becomes apparent that in order to obtain a better estimate for $1 - R(h)$, given a value of γ, it is necessary to use [7.14] with appropriate value of k.

Let us now estimate the increment of the spectral function, $F(\lambda)$, of stationary random processes. Let $R(\tau)$ be the covariance function. The function $R(\tau)/R(0)$ may be interpreted as the characteristic function of the symmetric random variable with the distribution function $G(\lambda)$ such that $F(\lambda)/F(+\infty) = G(\lambda) - G(-\lambda)$ for $\lambda > 0$. To estimate $F(+\infty) - F(\lambda)$, lemma 7.4 may be applied.

THEOREM 7.1.– Let $R(\tau)$ be the covariance function of a stationary random process $\xi(t)$, and let $F(\lambda)$ be the spectral function of $\xi(t)$. If the function $R(\tau)/R(0)$ satisfies the assumptions of lemma 7.2, i.e. $1 - R(\tau)/R(0) \leq \psi(\tau)$, then for $h > t_k/t_0$ the following inequality holds:

$$F(+\infty) - F(h) \leq F(+\infty)\,\frac{s_k}{\pi}\,\delta_{k,\psi}\Big(\frac{h}{t_k}\Big), \qquad [7.15]$$

where $k = 1, 2, 3$ and s_k and t_k are same as in [7.14].

This theorem follows immediately from lemma 7.2.

REMARK.– The assertion of corollary 7.2 is also applicable to the function $R(\tau)/R(0)$.

Now we shall employ the estimate for the increment $F(+\infty) - F(\lambda)$ to construct models of random processes. Let $\xi(t)$ be a stationary random process whose covariance function $R(\tau)/R(0)$ satisfies the assumptions of lemma 7.2 or corollary 7.2. Following the method proposed in [KOZ 94], we construct the model in the form [7.1], where $\lambda_k = k\Lambda/N$, $k = 0, 1, 2, \ldots, N-1$. In order that, given ε and p, the inequality

$$P\Big\{\int_0^T (\xi(t) - \xi_N(t))^2\,dt > \varepsilon\Big\} \leq p$$

hold true, Λ and N may be chosen in the following manner suggested by formulas from [KOZ 94]. Let $z_{p/2}$ be a root of the equation $\exp\{-z/2\}(z+1)^{1/2} = p/2$. Then, Λ is the minimal number satisfying

$$z_{p/2}\Big(\int_0^T\int_0^T\Big(\int_\Lambda^\infty \cos\lambda(t-s)\,dF(\lambda)\Big)^2 dt\,ds\Big)^{1/2} + T(F(+\infty) - F(\Lambda)) \leq \varepsilon/4.$$

Since

$$\Big(\int_0^T\int_0^T\Big(\int_\Lambda^\infty \cos\lambda(t-s)\,dF(\lambda)\Big)^2 dt\,ds\Big)^{1/2} \leq T(F(+\infty) - F(\Lambda)),$$

Λ is the minimal number for which

$$T(F(+\infty) - F(\Lambda))(z_{p/2} + 1) \leq \varepsilon/4, \qquad [7.16]$$

or by theorem, the minimal number for which

$$\delta_{k,\psi}\left(\frac{\Lambda}{t_k}\right) \leq \varepsilon(4T(z_{p/2}+1))^{-1}.$$

Hence, N is the least number satisfying the inequality

$$N \geq ((z_{p/2}+1)\,T^3\,\Lambda^2\,F(\Lambda)\,4\,(3\varepsilon)^{-1})^{1/2}+1.$$

According to [7.16], N may be chosen in the following way:

$$N = \left[(z_{p/2}+1)T^3\Lambda^2\left(F(+\infty)-\frac{\varepsilon}{4T(z_{p/2}+1)}\right)4(3\varepsilon)^{-1}\right]+1.$$

Thus, we have found all parameters of the model.

8

Simulation of Gaussian Isotropic Random Fields on a Sphere

The models of Gaussian isotropic random fields on an n-measurable sphere are constructed that approximate these fields with given accuracy and reliability in the space L_p, $p \geq 2$.

DEFINITION 8.1.– *A random field $\xi(x)$ on sphere S_n in n-measurable space is called isotropic in wide sense if $\mathbf{E}\,\xi(x) = const$ (further we will suppose that $\mathbf{E}\,\xi(x) = 0$) and $\mathbf{E}\,\xi(x_1)\xi(x_2) = B(\cos\theta)$, where $\cos\theta$ is the angular distance between x_1 and x_2 [YAD 93].*

We suggest that the field $\xi = \{\xi(x), x \in S_n\}$ is Gaussian and continuous in mean square. Random field ξ has a representation [5, p. 61].

$$\xi(x) = \sum_{m=0}^{\infty} \sum_{l=1}^{h(m,n)} \xi_m^l s_m^l(x), \tag{8.1}$$

where ξ_m^l is a sequence of independent Gaussian random variables such that $\mathbf{E}\,\xi_m^l = 0$, $\mathbf{E}\,\xi_m^l \xi_r^s = b_m \delta_m^r \delta_l^s$, $(m = 0, 1, 2, \ldots; l = 1, \ldots, h(m, n))$, δ_m^r is a Kronecker symbol, $b_m > 0$, $s_m^l(x) = s_m^l(\theta_1, \ldots, \theta_{n-2}, \varphi)$ are orthonormal spherical harmonics with degree m, $h(m, n) = (2m + n - 2)\frac{(m+n-3)!}{(n-2)!m!}$ is the number of such harmonics.

In addition, $\sum\limits_{m=0}^{\infty} b_m h(m, n) < \infty$. Note that $s_m^l(x)$ are trigonometric polynomials of $n - 1$ variables with degree l. The properties of $s_m^l(x)$ and the formulas for $s_m^l(x)$ can be found in [BAT 53].

Note that $P_m(x) = \sum_{l=1}^{h(m,n)} \xi_m^l s_m^l(x)$. Hence,

$$\xi(x) = \sum_{m=0}^{\infty} P_m(x). \qquad [8.2]$$

A simulation problem of the field ξ consists of a construction of some Gaussian field $\hat{\xi} = \{\hat{\xi}(x), x \in S_n\}$, that approximates in some sense random field ξ with given accuracy and reliability [ZEL 88, KOZ 94, KOZ 94, KOZ 92]. Field $\hat{\xi}$ has to accept the possibility of computer simulation.

In this section, the model is constructed that approximates random field ξ with given accuracy and reliability in space $L_p(S_n)$, $p \geq 2$, it means that the filed $\hat{\xi}$ is found that by known ε and δ inequality

$$\mathbf{P}\left\{\left(\int_{S_n} |\hat{\xi}(x) - \xi(x)|^p \, dx\right)^{\frac{1}{p}} > \varepsilon\right\} < \delta \qquad [8.3]$$

holds true.

As a model of random filed is proposed to choose such field

$$\hat{\xi}(x) = \xi_N(x) = \sum_{m=0}^{N} \sum_{l=1}^{h(m,n)} \xi_m^l s_m^l(x). \qquad [8.4]$$

Computer simulation of this random field does not present any difficulties.

The main problem is to find N such that inequality [8.3] is satisfied.

LEMMA 8.1.– Let $\xi_1, \xi_2, \ldots, \xi_n$ be independent Gaussian random variables $\mathbf{E}\,\xi_i = 0$, $D\xi_i = \mathbf{E}\xi_i^2 = \sigma_i^2$, $i = 1, \ldots, n$. Then, for arbitrary $0 \leq u < 1$ inequality

$$\mathbf{E}\exp\left\{\frac{u}{2(\sum_{i=1}^{n} \sigma_i^4)^{\frac{1}{2}}}\left(\sum_{i=1}^{n} \xi_i^2 - \sum_{i=1}^{n} \sigma_i^2\right)\right\} \leq \exp\left\{-\frac{u}{2}\right\}(1-u)^{-\frac{1}{2}} \qquad [8.5]$$

holds true.

Lemma 8.1 is a particular case of (4) in paper [KOZ 94].

REMARK 8.1.– It is easy to show that

$$\sum_{i=1}^{n} \sigma_i^2 = \mathbf{E}\Big(\sum_{i=1}^{n} \xi_i^2\Big) \quad \text{and} \quad 2\Big(\sum_{i=1}^{n} \sigma_i^4\Big)^{\frac{1}{2}} = \Big(2D\Big(\sum_{i=1}^{n} \xi_i^2\Big)\Big)^{\frac{1}{2}}.$$

COROLLARY 8.1.– Let $\xi_1, \xi_2, \ldots$ be sequence of independent Gaussian random variables such that $\mathbf{E}\,\xi_i = 0$, $\mathbf{E}\,\xi_i^2 = \sigma_i^2$. If $\sum_{i=1}^{\infty} \sigma_i^2 < \infty$, then for any $0 \le u \le 1$ inequality

$$\mathbf{E}\,\exp\Big\{\frac{u}{2(\sum_{i=1}^{\infty} \sigma_i^4)^{\frac{1}{2}}}\Big(\sum_{i=1}^{\infty} \xi_i^2 - \sum_{i=1}^{\infty} \sigma_i^2\Big)\Big\} \le \exp\Big\{-\frac{u}{2}\Big\}(1-u)^{-\frac{1}{2}} \qquad [8.6]$$

holds.

PROOF.– It is clear that the convergence of series $\sum_{i=1}^{\infty} \sigma_i^2$ provides convergence of series $\sum_{i=1}^{\infty} \xi_i^2$ with probability one [LOE 60]. Moreover,

$$\Big(\sum_{i=1}^{\infty} \sigma_i^4\Big)^{\frac{1}{2}} \le \sum_{i=1}^{\infty} \sigma_i^2 < \infty.$$

Hence, [8.6] follows from [8.5], taking a limit as $n \to \infty$ and using the Fatou lemma. □

LEMMA 8.2.– (The Nikolski Inequality) [NIK 77]. Let $T(\vec{u})$, $\vec{u}' = (u_1, \ldots, u_d)$, $0 \le u_i \le 2\pi$, $i = 1, \ldots, d$, be trigonometric polynomials from d variables of the power of $\vec{\nu}$, $\vec{\nu}' = (\nu_1, \nu_2, \ldots, \nu_d)$.

$$\|T\|_p = \Big(\int_0^{\pi} \cdots \int_0^{\pi} |T(\vec{u})|^p \, d\vec{u}\Big)^{\frac{1}{p}},$$

then for $p > r > 1$ inequality

$$\|T\|_p \le 3^d \Big(\prod_{i=1}^{d} \nu_i\Big)^{(\frac{1}{r} - \frac{1}{p})} \|T\|_r \qquad [8.7]$$

holds.

The following theorems give the possibility to find the number of items N in [8.4] to construct the model of random field [8.1] with given accuracy and reliability in $L_p(S_n)$.

THEOREM 8.1.– For any $z > 0$ inequality

$$\mathbf{P}\left\{\|\xi(x) - \xi_N(x)\|_2 > (A_N z + B_N)^{\frac{1}{2}}\right\} \le R(z) \qquad [8.8]$$

holds, where

$$B_N = \sum_{m=N+1}^{\infty} h(m,n) b_m, \qquad A_N = \left(\sum_{m=N+1}^{\infty} h(m,n) b_m^2\right)^{\frac{1}{2}},$$

$$R(z) = \exp\left\{-\frac{z}{2}\right\}(z+1)^{\frac{1}{2}}.$$

PROOF.– The functions $s_m^l(x)$, $x \in S_n$, are orthonormal, that is why

$$\|\xi(x) - \xi_N(x)\|_2^2 = \sum_{m=N+1}^{\infty} \sum_{l=1}^{h(m,n)} \left(\xi_m^l\right)^2.$$

Since $\mathbf{E}\,(\xi_m^l)^2 = b_m$, $l = 1, \dots, h(m,n)$, and ξ_m^l are independent centered Gaussian random variables, then from corollary 8.1 for any $0 \le u \le 1$, we obtain

$$\mathbf{E}\,\exp\left\{\frac{u(\|\xi(x) - \xi_N(x)\|_2^2 - B_N)}{2A_N}\right\} \le \exp\left\{-\frac{u}{2}\right\}(1-u)^{\frac{1}{2}}. \qquad [8.9]$$

It follows from the Chebyshev inequality that

$$\begin{aligned} &\mathbf{P}\left\{\|\xi(x) - \xi_N(x)\|_2 > (A_N z + B_N)^{\frac{1}{2}}\right\} \\ &\qquad \le \mathbf{P}\,\exp\left\{\frac{\|\xi(x) - \xi_N(x)\|_2^2 - B_N}{A_N} > z\right\} \\ &\qquad \le \mathbf{E}\,\exp\left\{\frac{u(\|\xi(x) - \xi_N(x)\|_2^2 - B_N)}{2A_N}\right\}\exp\left\{-\frac{uz}{2}\right\} \\ &\qquad \le \exp\left\{-\frac{u}{2}\right\}\exp\left\{-\frac{uz}{2}\right\}(1-u)^{-\frac{1}{2}}. \end{aligned}$$

Minimizing right-hand side of above inequality with respect to $(u = \frac{z}{z+1})$, we obtain [8.8]. □

Define $R_m^r(x) = \sum_{s=m}^{r} P_s(x)$, if $m < r$. Let $\{\psi(k), k = 1, \dots, \infty\}$, $\psi(k) > 0$, be some sequence. Define

$$R_m^r(x, \psi) = \sum_{s=m}^{r} \psi(s) P_s(x), \qquad B_m^s = \sum_{t=m}^{s} \psi^2(t) b_t h(t,n),$$

$$A_m^s = 2\Big(\sum_{t=m}^{s} \psi^4(t) b_t h(t,n)\Big)^{\frac{1}{2}}.$$

THEOREM 8.2.– Let there exist monotonically non-decreasing sequence $\{\psi(k), k = 1, \ldots, \infty\}$, $\psi(k) > 0$, $\psi(k) \to \infty$,as $k \to \infty$, and for any $m > 0$

$$\sum_{s=m}^{\infty} B_m^s (A_m^s)^{-\frac{1}{2}} c(s) \left(\frac{1}{\psi(s)} - \frac{1}{\psi(s+1)} \right) < \infty, \qquad [8.10]$$

where $c(s) = 3^{n-1} s^{(n-1)(1/2-1/p)}$. Moreover, as $s \to \infty$

$$B_m^s (A_M^s)^{-\frac{1}{2}} c(s) (\psi(s))^{-1} \to 0, \qquad [8.11]$$

then for any $z > 0$ and $N > 0$ inequality

$$\mathbf{P}\left\{ \|\xi(x) - \xi_N(x)\|_p > \left(z (V_{N+1}^{\infty})^2 + V_{N+1}^{\infty} W_{N+1}^{\infty} \right)^{\frac{1}{2}} \right\} \leq \exp\{-z\} (2z+1)^{\frac{1}{2}} \qquad [8.12]$$

holds, where

$$W_{N+1}^{\infty} = \sum_{s=N+1}^{\infty} B_m^s (A_m^s)^{-\frac{1}{2}} c(s) \left(\frac{1}{\psi(s)} - \frac{1}{\psi(s+1)} \right),$$

$$V_{N+1}^{\infty} = \sum_{s=N+1}^{\infty} (A_m^s)^{\frac{1}{2}} c(s) \left(\frac{1}{\psi(s)} - \frac{1}{\psi(s+1)} \right).$$

PROOF.– First note that from [8.10] follows that for any $m > 0$

$$\sum_{s=m}^{\infty} (A_m^s)^{\frac{1}{2}} c(s) \left(\frac{1}{\psi(s)} - \frac{1}{\psi(s+1)} \right) < \infty. \qquad [8.13]$$

Really, it is easy to see that $A_m^s \leq 2B_m^s$, then $(A_m^s)^{\frac{1}{2}} \leq 2B_m^s (A_m^s)^{-\frac{1}{2}}$. The following relationship is fulfilled (Abelian transform):

$$R_m^r(x) = \sum_{s=m}^{r} R_m^s(x,\psi) \left(\frac{1}{\psi(s)} - \frac{1}{\psi(s+1)} \right) + R_m^r(x,\psi) \frac{1}{\psi(r+1)}.$$

Then, we have

$$\|R_m^r(x)\|_p = \sum_{s=m}^{r} \|R_m^s(x,\psi)\| \left(\frac{1}{\psi(s)} - \frac{1}{\psi(s+1)} \right)$$

$$+ \|R_m^r(x,\psi)\|_p \frac{1}{\psi(r+1)}. \qquad [8.14]$$

Note that $R_m^s(x,\psi)$ is trigonometric polynomial from $(n-1)$ variables of the power of $\vec{\nu}$, $\vec{\nu}' = (s,s,\ldots,s)$. Hence, lemma 8.2 implies that for $p > 2$

$$\|R_m^r(x,\psi)\|_p \leq \|R_m^r(x,\psi)\|_2. \qquad [8.15]$$

We denote

$$\varphi(s) = c(s)\left[\frac{1}{\psi(s)} - \frac{1}{\psi(s+1)}\right], \qquad m \leq s < r,$$
$$\varphi(r) = c(r)\frac{1}{\psi(r)}. \qquad [8.16]$$

then from [8.14]–[8.16] follows inequality

$$\|R_m^r(x)\|_p \leq \sum_{s=m}^{r} \varphi(s)\|R_m^s(x,\psi)\|_2. \qquad [8.17]$$

Let $\delta_s > 0$ now be such numbers that $\sum\limits_{s=m}^{r} \delta_s = 1$, and $L > 0$ is a number that will be defined later; [8.17] and the *Hölder* inequality yield the following inequality:

$$\begin{aligned}\mathbf{E}\exp\{(L^{-1}\|R_m^r(x)\|_p)^2\} &\leq \mathbf{E}\exp\left\{\left(L^{-1}\sum_{s=m}^{r}\varphi(s)\|R_m^s(x,\psi)\|_2\right)^2\right\}\\ &\leq \mathbf{E}\exp\left\{\sum_{s=m}^{r}\delta_s(\varphi(s)L^{-1}\delta_s^{-1}\|R_m^s(x,\psi)\|_2)^2\right\}\\ &\leq \prod_{s=m}^{r}\left(\mathbf{E}\exp\{(\varphi(s)L^{-1}\delta_s^{-1}\|R_m^s(x,\psi)\|_2)^2\}\right)^{\delta_s}.\end{aligned} \qquad [8.18]$$

Since

$$\|R_m^s(x,\psi)\|_2^2 = \sum_{t=m}^{s}\psi^2(t)\sum_{l=1}^{h(t,n)}(\xi_t^l)^2, \qquad \mathbf{E}\,\|R_m^s(x,\psi)\|_2^2 = B_m^s$$

and $\mathbf{D}\,\|R_m^s(x,\psi)\|_2^2 = 2^{-1}(A_m^s)^2$, then by the condition

$$u_s = \varphi^2(s)A_m^sL^{-2}\delta_s^{-2} < 1 \qquad [8.19]$$

from lemma 8.1, we obtain

$$\mathbf{E}\exp\{(\varphi(s)L^{-1}\delta_s^{-1}\|R_m^s(x,\psi)\|_2)^2\} = \mathbf{E}\exp\{u_s)\|R_m^s(x,\psi)\|_2^2(A_M^s)^{-1}\}$$
$$\leq \exp\left\{\frac{B_m^s\varphi^2(s)}{L^2\delta_s^2}\right\}\exp\left\{-\frac{u_s}{2}\right\}(1-u_s)^{-\frac{1}{2}}. \qquad [8.20]$$

Hence, from [8.20], [8.18] and for such L and δ_s, [8.19] is satisfied, follows inequality

$$\mathbf{E}\exp\{\left(L^{-1}\|R_m^r(x)\|_p\right)^2\}$$
$$\leq \prod_{s=m}^{r}\left(\exp\left\{\frac{B_m^s\varphi^2(s)}{L^2\delta_s^2}\right\}\exp\left\{-\frac{u_s}{2}\right\}(1-u_s)^{-\frac{1}{2}}\right)^{\delta_s}. \qquad [8.21]$$

Denote now

$$\delta_s = (A_m^s)^{\frac{1}{2}}\frac{\varphi(s)}{L\nu^{\frac{1}{2}}}, \quad 0\leq\nu<1, \quad L=\nu^{-\frac{1}{2}}\sum_{s=m}^{r}(A_m^s)^{\frac{1}{2}}\varphi(s),$$

then $u_s = \nu < 1$. That is why [8.21] is transformed into inequality

$$\mathbf{E}\exp\left\{\frac{\nu\|R_m^2(x)\|_p^2}{\left(\sum_{s=m}^{r}(A_m^s)^{\frac{1}{2}}\varphi(s)\right)^2}\right\}$$
$$\leq \exp\left\{\frac{\nu\sum_{s=m}^{r}B_m^s(A_m^s)^{-\frac{1}{2}}\varphi(s)}{\sum_{s=m}^{r}(A_m^s)^{\frac{1}{2}}\varphi(s)}\right\}\exp\left\{-\frac{\nu}{2}\right\}(1-\nu)^{-\frac{1}{2}}. \qquad [8.22]$$

Define

$$V_m^r = \sum_{s=m}^{r}(A_m^s)^{\frac{1}{2}}\varphi(s), \qquad W_m^r = \sum_{s=m}^{r}B_m^s(A_m^s)^{-\frac{1}{2}}\varphi(s).$$

Then, inequality [8.22] can be rewritten as follows

$$\mathbf{E}\exp\left\{\frac{\nu\|R_m^r(x)\|_p^2 - V_m^rW_m^r}{(V_m^r)^2}\right\} \leq \exp\left\{-\frac{\nu}{2}\right\}(1-\nu)^{-\frac{1}{2}}. \qquad [8.23]$$

[8.23] yields inequality $(z>0)$

$$\mathbf{P}\left\{\frac{\|R_m^r(x)\|_p^2 - V_m^rW_m^r}{(V_m^r)^2} > z\right\} \leq \mathbf{E}\exp\left\{\frac{\nu\|R_m^r(x)\|_p^2 - V_m^rW_m^r}{(V_m^r)^2}\right\}\exp\{-\nu z\}$$
$$\leq \exp\left\{-\frac{\nu}{2}\right\}(1-\nu)^{-\frac{1}{2}}.\exp\{-\nu z\}.$$

If we minimize the right-hand side of the above inequality with respect to ν, then we obtain

$$\mathbf{P}\left\{\|R^r_m(x)\|_p^2 - V^r_m W^r_m > z(V^r_m)^2\right\} \leq \exp\{-z\}(2z+1)^{\frac{1}{2}}. \qquad [8.24]$$

From [8.24], [8.13] and the conditions of theorem follows that $\|R^r_m(x)\|_p \to 0$ as $m, r \to \infty$ in probability. This allows us to take in [8.24] a limit as $r \to \infty$ and to obtain the assertion of theorem if $m = N + 1$. □

The next theorem keeps all notation of previous one.

THEOREM 8.3.– Let there exist a monotonically non-decreasing sequence $\{\psi(k), k = 1, \ldots, \infty\}$, $\psi(k) > 0$, $\psi(k) \to \infty$, as $k \to \infty$, such that for any $m > 0$ the series converges

$$\sum_{s=m}^{\infty} (B^s_m)^{\frac{1}{2}} c(s)\left(\frac{1}{\psi(s)} - \frac{1}{\psi(s+1)}\right) < \infty. \qquad [8.25]$$

Then, for arbitrary $z > 0$, $N > 0$ inequality

$$\mathbf{P}\left\{\|\xi(x) - \xi_N(x)\|_p > z \sum_{s=N+1}^{\infty} (B^s_m)^{\frac{1}{2}} c(s)\left(\frac{1}{\psi(s)} - \frac{1}{\psi(s+1)}\right)\right\}$$
$$\leq \exp\left\{-\frac{z^2}{2}\right\} e^{\frac{1}{2}} z \qquad [8.26]$$

holds true.

PROOF.– First note that from [8.25] follows that

$$(B^s_m)^{\frac{1}{2}} c(s)\frac{1}{\psi(s)} \to 0, \quad \text{when} \quad s \to \infty. \qquad [8.27]$$

Really,

$$(B^s_m)^{\frac{1}{2}} c(s)\frac{1}{\psi(s)} = (B^s_m)^{\frac{1}{2}} c(s)\sum_{k=s}^{\infty}\left(\frac{1}{\psi(k)} - \frac{1}{\psi(k+1)}\right)$$
$$\leq \sum_{k=s}^{\infty}(B^k_m)^{\frac{1}{2}} c(k)\left(\frac{1}{\psi(k)} - \frac{1}{\psi(k+1)}\right) \to 0,$$

as $s \to \infty$. Further, in expression [8.21], we denote

$$\delta_s = (2B^s_m)^{\frac{1}{2}}\varphi(s)L^{-1}\nu^{-\frac{1}{2}}, \quad \text{where} \quad 0 < v < 1,$$

$$L = 2^{\frac{1}{2}}\nu^{-\frac{1}{2}} \sum_{s=m}^{r} (B_m^s)^{\frac{1}{2}}\varphi(s),$$

Then, $u_s = A_m^s(2B_m^s)^{-1}\nu \leq \nu < 1$, because of $A_m^s \leq 2B_m^s$. Therefore, inequality [8.21] is transformed into

$$\begin{aligned}
&\mathbf{E} \exp\left\{\frac{\nu}{2}\|R_m^r(x)\|_p^2\left(\sum_{s=m}^{r}(B_m^s)^{\frac{1}{2}}\varphi(s)\right)^{-2}\right\} \\
&\leq \exp\left\{\left(\frac{\nu}{2}\right)^{\frac{1}{2}}\frac{1}{L}\sum_{s=m}^{r}(B_m^s)^{\frac{1}{2}}\varphi(s)\right\}\exp\left\{-\frac{\nu}{2}\right\}(1-\nu)^{-\frac{1}{2}} \\
&= \exp\left\{\frac{\nu}{2}\right\}\exp\left\{-\frac{\nu}{2}\right\}(1-\nu)^{-\frac{1}{2}} \leq (1-\nu)^{-\frac{1}{2}}.
\end{aligned} \quad [8.28]$$

By the Chebyshev inequality and [8.28], we obtain

$$\mathbf{P}\left\{\|R_m^r(x)\|_p^2 > z^2\left(\sum_{s=m}^{r}(B_m^s)^{\frac{1}{2}}\varphi(s)\right)^2\right\} \leq \exp\left\{-\frac{\nu z^2}{2}\right\}(1-\nu)^{-\frac{1}{2}}.$$

If we minimize right-hand side of above inequality with respect to ν, then we obtain

$$\mathbf{P}\left\{\|R_m^r(x)\|_p > z\sum_{s=m}^{r}(B_m^s)^{\frac{1}{2}}\varphi(s)\right\} \leq \exp\left\{-\frac{z^2}{2}\right\} z\, e^{\frac{1}{2}}. \quad [8.29]$$

From [8.29], the same as in previous theorem, taking into account [8.27], we obtain [8.26]. □

REMARK 8.2.– Theorem 8.2 gives a more precise estimation than theorem 8.3 but under more restricted conditions. Moreover, the estimation of theorem 8.3 is more convenient in the computation of N, which defines the accuracy of approximation. In general, broad-brush estimations can be obtained by using the results of [ZEL 88].

8.1. Simulation of random field with given accuracy and reliability in $L_2(S_n)$

Consider accuracy of simulation $\varepsilon > 0$ and reliability $1 - \delta$, $1 > \delta > 0$. In inequality [8.8], set $z = z_\delta$, where δ is a root of equation $R(z) = \delta$. N we find as minimal number for which the inequality $(A_N z_\delta + B_N)^{\frac{1}{2}} < \varepsilon$ is fulfilled. Since $A_N < B_N$, then N can be found as a minimal number that

$$\sqrt{B_N} \leq \varepsilon(z_\delta + 1)^{-\frac{1}{2}}.$$

8.2. Simulation of random field with given accuracy and reliability in $L_p(S_n)$, $p \geq 2$

Let ε be accuracy of simulation, and $1-\delta$ be reliability of simulation. If we use inequality [8.12], then z_δ is found as a root of equation $\exp\{-z\}(2z+1)^{\frac{1}{2}} = \delta$ and N is as minimal number that inequality

$$\left(z_\delta (V_{N+1}^{\infty})^2 + V_{N+1}^{\infty} W_{N+1}^{\infty}\right)^{\frac{1}{2}} < \varepsilon \qquad [8.30]$$

holds.

Since the left-hand side of inequality [8.30] depends on the sequence $\psi(k)$, then choose $\psi(s) = s^{\beta}$, where $\beta > (n-1)\left(\frac{1}{2} - \frac{1}{p}\right)$. Then,

$$\begin{aligned} W_{N+1}^{\infty} &= \sum_{s=N+1}^{\infty} B_{N+1}(A_{N+1}^{s})^{-\frac{1}{2}} c(s)\left(\frac{1}{\psi(s)} - \frac{1}{\psi(s+1)}\right) \\ &\leq \sup_{s>N+1}\left(B_{N+1}^{s}(A_{N+1}^{s})^{-\frac{1}{2}}\right) \sum_{s=N+1}^{\infty} 3^{n-1} s^{(n-1)(\frac{1}{2}-\frac{1}{p})}\left(\frac{1}{s^{\beta}} - \frac{1}{(s+1)^{\beta}}\right) \\ &\leq \sup_{s>N+1}\left(B_{N+1}^{s}(A_{N+1}^{s})^{-\frac{1}{2}}\right)(N+1)^{\alpha-\beta} 3^{n-1}\frac{\beta}{\beta-\alpha}. \end{aligned}$$

where $\alpha = (n-1)\left(\frac{1}{2} - \frac{1}{p}\right)$. Similarly, $V_{N+1}^{\infty} \leq (A_{N+1}^{\infty})^{\frac{1}{2}}(N+1)^{\alpha-\beta}\beta(\beta-\alpha)^{-1}$, where

$$A_{N+1}^{\infty} = 2\left(\sum_{t=N+1}^{\infty} t^{4\beta} b_t^2 h(t,n)\right)^{\frac{1}{2}}.$$

In the same way, we can use inequality [8.26].

Bibliography

[ANT 02a] ANTONINI R., KOZACHENKO Y., TEGZA A., "Accuracy of simulation in L_p of Gaussian random processes", *Bulletin of the University of Kiev, Series: Physics & Mathem.*, vol. 5, pp. 7–14, 2002.

[ANT 02b] ANTONINI R., KOZACHENKO Y., TEGZA A., "Inequalities for norms subGaussian vectors and accuracy modeling of random processes", *Theory Probability and Mathematical Statistics*, vol. 66, pp. 58–66, 2002.

[ANT 03] ANTONINI R., KOZACHENKO Y., SOROKULOV V., "On accuracy and reliability of simulation of some random processes from the space $Sub_\varphi(\Omega)$", *Theory of Stochastic Processes*, vol. 9, nos. 3–4, pp. 50–57, 2003.

[ARM 94] ARMSTRONG M., DOWD P., *Geostatistical Simulations*, Kluwer, Dordrecht, 1994.

[BAN 83] BANDUT G., PIRSOL A., *Application of Correlation and Spectral Analysis*, Mir, Moscov, 1983.

[BAT 53] BATEMAN H., ERDELYI A., *Higher Transcendental Functions*, McGraw Hill, 1953.

[BRI 75] BRILLINDGER D., *Time Series: Data Analysis and Theory*, Holt, Rinehart and Winston, New York, 1975.

[BRO 95] BROOKER P., STEWART M., "A comparative study of simulation techniques for two dimensional data honoring specified exponential semivariograms", *J. Austr. Math. SOc. B.*, vol. 36, no. 3, pp. 249–260, 1995.

[BUL 74] BULDYGIN V., KOZACHENKO Y., "On local properties of realizations of some random processes and fields", *Theory Probability and Mathematical Statistics*, vol. 10, pp. 37–45, 1974.

[BUL 77] BULDYGIN V., "Sub-Gaussian processes and convergence random series in functional spaces", *Ukrainian Mathem. Journal*, vol. 4, pp. 337–345, 1977.

[BUL 80a] BULDYGIN V., *Random Elements Convergence in Topological Spaces*, Naukova dumka, Kiev, 1980.

[BUL 80b] BULDYGIN V., KOZACHENKO Y., "On sub-Gaussian random variables", *Ukrainian Mathem. Journal*, vol. 32, no. 6, pp. 483–489, 1980.

[BUL 87] BULDYGIN V., KOZACHENKO Y., "Sub-Gaussian random vectors and processes", *Theory Probability and Mathematical Statistics*, vol. 36, pp. 9–20, 1987.

[BUL 94] BULDYGIN V., KOZACHENKO Y., "Bershtein funtionals and exponential inequalities for the sum of ramdom variables", *Mathematica Today*, pp. 55–79, 1994.

[BUL 00] BULDYGIN V., KOZACHENKO Y., "Metric characterization of random variables and random processes," *Amer. Math. Soc.*, vol. 188, 2000.

[BYK 78] BYKOV V., *Numeric Simulation in Statistical Radio Engineering*, Soviet Radio, Moscow, 1978.

[CHI 99] CHILES J., DELFINER P., *Geostatistics: Modelling Spatial Uncertainty*, John Wiley & Sons, New York, 1999.

[CRA 67] CRAMER H., LIDBETTER M., *Stationary and Related Stochastic Process*, John Willey & Sons, New York, 1967.

[CRE 93] CRESSIE N., *Statistics for Spatial Data*, John Wiley & Sons, New York, 1993.

[DJE 68] DJENKINS G., VATTS D., *Spectral Analysis and its Applications*, Hoden-Day, San Francisco, 1968.

[DOV 10] DOVGAJ B., KOZACHENKO Y., ROZORA I., *Simulation of Random Processes in Physical Systems*, Zadruga, Kiev, 2010.

[DUD 73] DUDLEY R., "Sample functions of the Gaussian processes", *Ann. of Probab.*, vol. 1, no. 1, pp. 3–68, 1973.

[ERM 82] ERMAKOV S., MIKHAJLOV G., *Statistical Simulation*, Nauka, Moscow, 1982.

[FEL 70] FELLER W., *An Introduction to Probabibility Theory and its Application*, John Wiley & Sons, 1970.

[FER 57] DE FERIET. K., Fonctions de la physique mathématique, Centre National de la Recherche Scientifique, Paris, 1957.

[FER 75] FERNIQUE X., "Régularité des trajectoires des fonctions aléatoires gaussiennes", *Lecture Notes in Mathematics*, vol. 489, pp. 2–187, 1975.

[GIK 88] GIKHMAN I., SKOROHOD A., YADRENKO M., *Probability Theory and Mathematical Statistics*, Vyscha shkola, Kiev, 1988.

[GIK 04] GIKHMAN I., SKOROHOD A., *The Theory of Stochastic Processes*, Springer-Verlag, Berlin Heidelberg, 2004.

[HAR 66] HARDI G., ROGOSINSKI W., *Fourier Series*, Dover, 1966.

[HUN 51] HUNT G., "Random Fourier transforms", *Trans. Amer. Math. Soc.*, vol. 71, pp. 38–69, 1951.

[IBR 70] IBRAGIMOV I., ROZANOV Y., *Random Gaussian Processes*, Nauka, Moscow, 1970.

[IBR 73a] IBRAGIMOV I., "About condotions of smoothness of trajectory of random funktion", *Theory Probability and its Applicatons.*, vol. 28, no. 2, pp. 229–250, 1973.

[IBR 73b] IBRAGIMOV I., LINNIK Y.V., *Independance and Inhomogeneous Related Variables*, Nauka, Moscow, 1973.

[JAI 75] JAIN N., MARCUS M., "Integrability of infinite sums of independent vector-valued random variables", *Trans. Amer. Math. Soc.*, vol. 212, nos. 1, pp. 1–36, 1975.

[KÔN 80] KÔNO N., "Sample path properties of stochastic processes.", *J. Math. Kyoto Univ.*, vol. 20, pp. 295–313, 1980.

[KAH 60] KAHANE J., "Propriétés locales des fonctions à series de Fouries aléatories", *Studia Math.*, vol. 19, no. 1, pp. 1–25, 1960.

[KOZ 76] KOZACHENKO Y., YADRENKO M., "Local properties of sample functions of random fields 1", *Teor. Veroyat. i Mat. Stat.*, vol. 15, pp. 82–98, 1976.

[KOZ 85a] KOZACHENKO Y., OSTROVSKIJ E., "Banach spaces of random variables of type subGaussian.", *Theor. Probab. and Math. Statist.*, vol. 32, pp. 45–56, 1985.

[KOZ 85b] KOZACHENKO Y., "Random processes in Orlicz spaces. I.", *Theory Probab. Math. Stat.*, vol. 30, pp. 103–117, 1985.

[KOZ 85c] KOZACHENKO Y., "Random processes in Orlicz spaces. II.", *Theory Probab. Math. Stat.*, vol. 31, pp. 51–58, 1985.

[KOZ 88] KOZACHENKO Y., PASHKO A., "The accuracy of the simulation of random processes in norms Orlicz spaces I. (in English)", *Theor. Probab. and Math. Statist. (Kiev)*, vol. 58, pp. 45–60, 1988.

[KOZ 92] KOZACHENKO Y., KOZACHENKO L., "On the accuracy of modelling of stationary processes with absolutely continuous spectrum. (in English)", *Theor. Probab. and Math. Statist. (Kiev)*, vol. 47, pp. 47–54, 1992.

[KOZ 94] KOZACHENKO Y., KOZACHENKO L., "Simulation accuracy of Gaussian stochastic processes in $L_2(0,T)$.", *Journal of Mathematical Sciences*, vol. 72, no. 2, pp. 3040–3048, 1994.

[KOZ 98] KOZACHENKO Y., STUS O.V., "Square-Gaussian random processes and estimators of covariance functions", *Math.Communications*, vol. 3, no. 1, pp. 83–94, 1998.

[KOZ 99a] KOZACHENKO Y., MOKLYACHUK O.M., "Large deviation probabilities for square-Gaussian stochastic processes", *Extremes*, vol. 2, no. 3, pp. 269–293, 1999.

[KOZ 99b] KOZACHENKO Y., PASHKO A., *Simulation of stochastic Processes, (in Ukrainian)*, "Kyiv University", Kiev, 1999.

[KOZ 00] KOZACHENKO Y., MOKLYACHUK O.M., "Square-Gaussian stochastic processes", *Theory of Stoch. Processes*, vol. 6, no. 22, nos.3–4, pp. 98–121, 2000.

[KOZ 01] KOZACHENKO Y., KOZACHENKO L., "Modeling Gaussian isotropic random fields on a sphere", *Journal of Mathematical Sciences*, vol. 107, no. 2, pp. 3751–3757, 2001.

[KOZ 02] KOZACHENKO Y., TEGZA A.M., "Application of the theory $Sub_\varphi(\Omega)$ of spaces of random variables to the accuracy of the simulation of stationary Gaussian processes. (in English)", *Theor. Probab. and Math. Statist.*, vol. 67, pp. 71–87, 2002.

[KOZ 03] KOZACHENKO Y., ROZORA I., "Simulation of Gaussian stochastic processes", *Random Operators Stoch. Equations, 11*, vol. 3, pp. 275–296, 2003.

[KOZ 04a] KOZACHENKO Y., ROZORA I., "Simulation of Gaussian Stochastic Fields", *Theory of Stoch. Processes*, vol.10, nos.1–2, pp. 48–60, 2004.

[KOZ 04b] KOZACHENKO Y., ROZORA I., "On simulation of stochastic processes from the space $\text{Sub}_\varphi(\Omega)$", *Applied Statist. and Financial Math*, vol. 1, pp. 72–79, 2004.

[KOZ 05] KOZACHENKO Y., SOTTINEN T., VASYLYK O., "Simulation of weakly self-similar stationary increment $\text{Sub}_\varphi(\Omega)$-processes: a serias expansion approach", *Methodology and Computing in Applied Probability*, vol. 7, no. 3, pp. 379–400, 2005.

[KOZ 06a] KOZACHENKO Y., POGORILIAK O., "Simulation of Cox processes driven by random field. (in Ukrainian)", *Reports of national sience academia*, vol. 10, pp. 20–23, 2006.

[KOZ 06b] KOZACHENKO Y., ROZORA I., "Application of the theory of Square-Gaussian Processes to simulation of Stochastic Processes", *Theory of Stoch. Processes*, vol. 12(28), no. 3–4, pp. 43–54, 2006.

[KOZ 07a] KOZACHENKO Y., PASHKO A., ROZORA I., *Simulation of stochastic Processes and fields, (in Ukrainian)*, Zadruga, Kiev, 2007.

[KOZ 07b] KOZACHENKO Y., POGORILIAK O., "A method of modelling log Gaussian Cox processes.", *Theor. Probab. and Math. Statist.*, vol. 77, pp. 91–165, 2007.

[KOZ 07c] KOZACHENKO Y., POGORILIAK O., "Simulation of log Gaussian Cox processes with given reliability and accuracy. (in English)", *Theor. Probab. and Math. Statist.*, vol. 76, pp. 58–71, 2007.

[KOZ 07d] KOZACHENKO Y., ROZORA I., TURCHYN Y., "On an expansion of random processes in series", *Random operators Stoch. Equations*, vol. 15, pp. 15–33, 2007.

[KOZ 09] KOZACHENKO Y., KAMENSCHIKOVA O., "Approximation of $SSub_\varphi(\Omega)$ stochastic processes in the space $L_p(T)$", *Theory Probab. And Math. Statist.*, vol. 79, pp. 83–88, 2009.

[KOZ 11] KOZACHENKO Y., POGORILIAK O., "Simulation of Cox processes driven by random Gaussian field", *Methodology and Computing in Applied Probability*, vol. 13, no. 3, pp. 511–521, 2011.

[KOZ 12] KOZACHENKO Y., POGORILIAK O., TEGZA A., *Simulation of Gaussian random processes and Cox processes, (in Ukrainian)*, Karpaty, Uzhgorod, 2012.

[KRU 95] BARRASA DE LA KRUS E., KOZACHENKO Y., "Boundary-value problems for equations of mathematical physics with strictly Orlicz random initial conditions", *Random Oper. and Stoch. Eq.*, vol. 3, pp. 201–220, 1995.

[LAN 70] LANDAU H., SHEPP L., "On the supremum of Gaussian processes", *Sankhya, Ser. A*. vol. 32, no. 4, pp. 369–378, 1970.

[LEO 60] LEONOV V., SHIRYAEV A.N., "Sur le calcul des semi-invariants", *Theor. Probab. Appl*, pp. 319–329, 1960.

[LIN 71] LINDGREN G., "Extreme values of stationary normal processes", *Z. Wahrscheinlichkeitstheorie Verw. Geb.*, vol. 17, pp. 39–47, 1971.

[LOE 60] LOEV M., *Probability Theory*, van Nostrand, Princeton, 1960.

[LUC 70] LUCACZ E., *Characteristic Functions*, 2nd ed., Griffin, London, 1970.

[MAC 88] MACAK I., PLYCHKO A., "Some inequalities for sums of independent random variables in Banach spaces", *Theor. Probab. and Math. Statist.*, vol. 38, pp. 81–86, 1988, in Russian.

[MIH 06] MIHAYLOV G., VOITISHEK A.V., *Numerical Statistical Modelling. Monte-Carlo Methods, Textbook for University Students*, Academiya, Moscow, 2006.

[MOL 98] MOLLER J., SYVERSVEEN A., WAAGEPETERSEN R., "Log gaussian Cox Processes", *Scandinavian Journal of Statistics*, vol. 25, no. 4, pp. 451–482, 1998.

[NAN 76] NANOPOULOS C., NOBELIS P., "Regularité et propriétés limites des fonctions aléatoires", *Lecture Notes in Math.*, vol.49, pp. 567–690, 1976.

[NIK 77] NIKOLSKIY S., *Approximation of Multivariable Functions*, Nauka, Moscow, 1977.

[OGO 96] OGORODNIKOV V., PRIGARIN S., *Numerical Modeling of Random Processes and Fields: Algorithms and Applications*, VSP, Utrecht, 1996.

[OST 90] OSTROVSKYY E., "Exponential estimates of maximum distribution of non-Gaussian random field", *Probab. Theory and its Appl.*, vol 35, no. 3, pp. 482–493, 1990.

[PIS 80] PISIER G., "Conditions d'entropie assurant la continuité de certains processus et applications à l'analyse harmonique", *Semin. Anal. Fonct*, pp. 13–24, 1979–1980.

[PIT 96] PITERBARG V., *Asymptotic methods in the theory of Gaussian processes and fields.*, Translations of Mathematical Monographs. 148. Providence, AMS, 1996.

[POG 07] POGORILIAK O., "Simulation of square Gaussian Cox processes when intensity are generated of inhomogeneous field. (in Ukrainian)", *Scientific Bulletin of the Uzhgorod University. Series mathematica and informatica*, vol. 14, pp. 95–102, 2007.

[POG 09] POGORILIAK O., "Simulation of random Cox processes driven by square gaussian random field. (in Ukrainian)", *Scientific Bulletin of the Uzhgorod University. Series mathematica and informatica*, vol. 18, pp. 113–120, 2009.

[POG 11] POGORILIAK O., TEGZA A., "On the simulation method of square gaussian Cox processes. (in Ukrainian)", *Scientific Bulletin of the Uzhgorod University. Series mathematica and informatica*, vol. 22, pp. 131–137, 2011.

[PRI 92] PRIGARIN S., "Convergence of numerical models of random fields in Monte Carlo methods", *Russian Journal of Numerical Analysis and Math. Modelling*, vol 7, no. 5, pp. 441–456, 1992.

[PRI 94] PRIGARIN S., "Numerical models of random processes and fields and some applications in Monte Carlo methods", *Russian Journal of Numerical Analysis and Math. Modelling*, vol 9, no. 2, pp. 147–177, 1994.

[PRI 95] PRIGARIN S., "Spectral models of random fields and their applications", *Advances in modelling and analysis*, vol. A 29, no. 2, pp. 39–51, 1995.

[RIP 87] RIPLEY B., *Stochastic Simulation*, Wiley, New York, 1987.

[ROS 06] ROSS S.M., *Simulation*, Elsevier, 2006.

[ROZ 07] ROZORA I., "Gaussian Process simulation with application of the theory of Square-Gaussian processes", *Proceedings of 15th European Young Statisticians Meeting*, Castro Urdiales, Spain, pp. 1–7, September 10–14, 2007.

[ROZ 08] ROZORA I., "Simulation of Gaussian stochastic process with respect to its derivative", *Appl. Statistics, actuarial and financial Mathem.*, nos.1–2, pp. 139–147, 2008.

[ROZ 09a] ROZORA I., "The conditions on finding of the model of stochastic process with given accuracy and reliability in the space $C([0,T])$", *Bulletin of the University of Kyiv, Series: Physics & Mathem.*, vol. 1, pp. 139–144, 2009.

[ROZ 09b] ROZORA I., "Modelling accuracy of φ-subGaussian stochastic process in the space $L_2([0,T])$", *Bulletin of the University of Kyiv, Series: Physics & Mathem.*, vol. 2, no. 98, pp. 68–76, 2009.

[RYA 90] RYAZANTSEVA V., ZELEPUGINA I., "On the rate of convergence of certain sub-Gaussian random series in norm of Orlicz space", *Theor. Probab. and math. Statist.*, vol. 40, pp. 33–42, 1990.

[RYA 91] RYAZANTSEVA V., "Convergence of trigonometric series in norm of Orlicz space", *Ukrainian Mathematical Journal.*, vol. 41, no. 8, pp. 936–941, 1991.

[SHI 71] SHINUZUKY M., "Simulation of Multivariate Multidimensional Random Processes", *J.Acoust.-Soc.Am.*, pp. 357–367, 1971.

[TEG 01] TEGZA A., "On the accuracy and reliability of some models of Gaussian processes with a limited range. (in Ukrainian)", *Scientific Bulletin of the Uzhgorod University*, vol. 6, pp. 125–131, 2001.

[TEG 02] TEGZA A., "Finding the accuracy and reliability of models Gaussian processes with continuous spectrum. (in Ukrainian)", *Bulletin of the University of Kiev, Series: Physics & Mathem.*, vol. 4, pp. 38–43, 2002.

[TEG 04a] TEGZA A., "Models Gaussian stationary random process in space $L_p([0,T]), p > 1$. (in Ukrainian)", *Bulletin of the University of Kiev, Series: Physics & Mathem.*, vol. 4, pp. 17–21, 2004.

[TEG 04b] TEGZA A., "Simulating of Gaussian stationary process with given accuracy in C([0,T]).", *Theory of Stochastic Processes.*, vol. 10, no. 26, pp. 172–177, 2004.

[TEG 07] TEGZA A., "Construction of the model of homogeneous isotropic Gaussian random field with specified accuracy and reliability in L_p, $1 \leq p \leq 2$. (in Ukrainian)", *Bulletin of the University of Kiev, Series: Physics & Mathem.*, vol. 2, pp. 33–35, 2007.

[TEG 08] TEGZA A., "The accuracy and reliability of the model homogeneous isotropic Gaussian random field in space L_p, $p \geq 1$. (in Ukrainian)", *Scientific Bulletin of the Uzhgorod University. Series math. and inform.*, vol. 16, pp. 184–187, 2008.

[TEG 10] TEGZA A., "Some estimates for the construction of a generalized model of Gaussian stationary processes. (in Ukrainian)", *Scientific Bulletin of the Uzhgorod University. Series math. and inform.*, vol. 21, pp. 137–145, 2010.

[TEG 11] TEGZA A., FEDORYANICH N., "The accuracy and reliability of the model homogeneous isotropic Gaussian random field with a limited spectrum in the space C (T). (in Ukrainian)", *Scientific Bulletin of the Uzhgorod University. Series math. and inform.*, vol. 22, pp. 142–147, 2011.

[TEU 04] TEUGELS J.L., *Poisson Processes. Encyclopedia of Atuarial Science*, Wiley, 2004.

[TRI 60] TRIKOMI H., *Integral Equations*, Moscow, 1960.

[TRI 91] TRIGUB S., "On the rate of convergence of certain random series in terms of Orlicz space", *Ukrainian Mathematical Journal.*, vol. 43, no. 6, pp. 789–796, June 1991.

[VAS 08] VASYLYK O., KOZACHENKO Y., VASYLYK O. *et al*, *φ-sub-Gaussian Random Processes*, Kyivskyi Universytet, Kiev, 2008.

[VLA 67] VLADIMIROV V., *Mathematical Physics Equations*, Nauka, Moscow, 1967.

[VOJ 82] VOJTYSHEK A., "Weak convergence investigation of models of gaussian random field with given correlation function spectral decomposition", *Simulation on the calculated systems*, pp. 119–129, 1982.

[VOJ 83] VOJTYSHEK A., "Randomization numerical spectral model of homogeneous random function", *Mathematical and imitation models of systems*, pp. 17–25, 1983.

[VYZ 03] VYZHVA Z., "About Approximation of 3-D Random Fields and Statistical Simulation", *Random Operators Stoch. Equations*, vol. 4, no. 3, pp. 255–266, 2003.

[YAD 83] YADRENKO M., "Spectral theory of random fields", *Optimization Software*, Publication Devision, New York, 1983

[YAD 93] YADRENKO M., YADRENKO O., GRIKH Z., "About Approximation and Statistical Simulation of Izotropic Fields", *Random Operators Stoch. Equations*, vol. 1, no. 1, pp. 37–45, 1993.

[ZEL 88] ZELEPUGINA I., KOZACHENKO Y., "About estimates of accuracy of simulation of random fields in the sapce L_p, (in Russian)", *Operations research and ACS*, vol. 32, 1988.

Index

MIX

Papier | Fördert gute Waldnutzung

FSC® C083411

Kolibri 360 GmbH
Ferdinand-Jühlke-Straße 7
99095 Erfurt, Deutschland
produktsicherheit@kolibri360.de